# Finite Element Method for Solids and Structures

This innovative approach to teaching the finite element method blends theoretical, textbook-based learning with practical application using online and video resources. This hybrid teaching package features computational software such as MATLAB®, and tutorials presenting software applications such as PTC Creo Parametric, ANSYS APDL, ANSYS Workbench, and SolidWorks, complete with detailed annotations and instructions so students can confidently develop hands-on experience. Suitable for senior undergraduate- and graduate-level teaching, students will transition seamlessly between mathematical models and practical commercial software problems, empowering them to advance from basic differential equations to industry-standard modeling and analysis. Complete with over 120 end-of-chapter problems and over 200 illustrations, this accessible reference will equip students with the tools they need to succeed in the workplace.

**Sung W. Lee** received his Ph.D. from the Massachusetts Institute of Technology. He is currently Professor of Aerospace Engineering at the University of Maryland, where he has taught courses on the finite element method and carried out numerous sponsored researches on the finite element modeling of solids and structures. He has served on the editorial advisory board of such international journals as *Computational Mechanics*, *International Journal for Numerical Methods in Engineering*, and *Computer Modeling in Engineering and Sciences*. He received a lifetime achievement award for his work on computational analysis of shell structures at the 2015 International Conference on Computational and Experimental Engineering and Sciences.

**Peter W. Chung's** research has been concerned mainly with the computability of interdisciplinary and multiscale problems, with applications motivated primarily by his early career experiences in the US Army Research Laboratory. He is currently Professor of Mechanical Engineering at the University of Maryland. Through his time in both government and academic sectors, Professor Chung has had the privilege of conducting research as well as participating in significant policy-shaping activities, including service to the Materials Genome Initiative (MGI) and the Army Materiel Command Logistics Transformation Task Force. He is a recipient of the Superior Civilian Service Award, the second highest civilian award in the Department of the Army.

# Finite Element Method for Solids and Structures

## A Concise Approach

**Sung W. Lee**
University of Maryland, College Park

**Peter W. Chung**
University of Maryland, College Park

CAMBRIDGE
UNIVERSITY PRESS

# CAMBRIDGE
## UNIVERSITY PRESS

University Printing House, Cambridge CB2 8BS, United Kingdom

One Liberty Plaza, 20th Floor, New York, NY 10006, USA

477 Williamstown Road, Port Melbourne, VIC 3207, Australia

314–321, 3rd Floor, Plot 3, Splendor Forum, Jasola District Centre, New Delhi – 110025, India

79 Anson Road, #06–04/06, Singapore 079906

Cambridge University Press is part of the University of Cambridge.

It furthers the University's mission by disseminating knowledge in the pursuit of education, learning, and research at the highest international levels of excellence.

www.cambridge.org
Information on this title: www.cambridge.org/9781108497091
DOI: 10.1017/9781108683982

© Sung W. Lee and Peter W. Chung 2021

First published 2021

Printed in the United Kingdom by TJ Books Limited, Padstow Cornwall, 2021

*A catalogue record for this publication is available from the British Library.*

ISBN 978-1-108-49709-1 Hardback

Additional resources for this publication at www.cambridge.org/FEM-Resources

# Contents

# Preface

The finite element (FE) method is a powerful tool which improves the ability to engineer products or perform scientific studies through virtual prototyping and analysis. Inestimable costs associated with arduous synthesis/fabrication and testing have been saved by implementing FE analysis in engineering workflows. Today, most of the scientific and engineering workforce can access sophisticated FE software on their computers. These software tools offer capabilities backed by decades of research, enabling students and practicing engineers to easily develop and solve problems of high-level complexity. Software know-how is also not difficult to come by, as step-by-step tutorials on performing complex, expert-level modeling tasks can readily be found online.

With widespread adoption of commercial FE software, the needs of instructors and students of FE methods have also changed. The current trend is to increase student exposure to FE applications through software-based analyses. While this is both a welcome and a necessary approach, it comes at the cost of rigor and understanding the fundamentals of the FE method. At best, an early career can start with a working familiarity with the same tools used by more senior and experienced engineers. More common, however, is under-informed usage.

For most students of mechanical, aerospace, or civil engineering, a course on the FE method is often the first, if not the only, foray into computer-based mathematical modeling for engineering analysis. A solid grounding in FE theory can serve as a key launching point into other emerging computer-based modeling areas, including (at the time of writing) data science, machine learning, and artificial intelligence. The very same ideas of discrete data, linear algebra, vector computing, and computational algorithms in FE analysis are also foundational in these adjacent fields.

Accordingly, our textbook on the FE method can better prepare students for engineering practice without foregoing the essential mathematical foundations that could unlock broader career opportunities in such areas as database engineering writ large, quantitative analysis in the finance sector, and even computer graphics techniques used in movies and video games. This book can also be used by practicing engineers who want to learn more about the fundamentals of FE formulation. We feel that the best approach is to offer *a more refined and distilled perspective* of FE methods and yet be balanced with software tools, applications, and modern educational delivery platforms. We therefore planned this book with the following content:

– A text that offers a refined and distilled presentation of the FE method without compromising the theoretical foundations, but still remaining accessible to students with a basic college-level background in mathematics.

- A student's online guide comprised of (a) answers to some of the problems at the end of each chapter and (b) solutions to some of the example problems that benefit from the use of a programming language. The complete M-files will illustrate the calculations through programmable steps. This online material will be accessible via web icons in the margin of the text.
- An online instructor's supplement with additional MATLAB-based solutions and hand solutions to some of the problems at the end of each chapter, and higher-level MATLAB-based code examples. We emphasize that these materials are to provide additional resources to instructors. For instance, they may be used to (a) provide detailed solution steps that may be used by the instructor for exposition, (b) illustrate the basic implementation of FE methods in actual computer codes for advanced or graduate-level students, or (c) provide problems that can form the basis of special projects.
- A set of high-resolution, easy-to-follow walk-through videos presenting annotated examples of software applications, based on the ideas developed in the text. These use multiple modern software packages including PTC Creo Parametric, ANSYS APDL, ANSYS Workbench, and SolidWorks.

For practical reasons, we have tried to keep the online materials separate from the classical background and fundamentals in the text. History has shown that software and hardware can quickly grow obsolete, whereas the fundamentals can transcend the technologies with which we teach. By adopting this multi-pronged approach, we believe it is far easier to keep pace with evolving pedagogies while ensuring a consistent level of excellence from students. Web icons have been placed in the margins to identify where supplementary material is available online for students who would like more practice.

The authors have had the privilege of teaching courses on the FE method over many years at both undergraduate and graduate levels, while conducting research on the development of improved FE models. The present book is based on the course materials developed by the authors to help students understand the essentials of the FE method, primarily within the context of linear elasticity. Prerequisites for mathematics have been kept to the sophomore level, which includes calculus, elementary linear algebra, and ordinary differential equations. For some of the code-writing aspects, users of this book are expected to have a working knowledge of an interpreted or compiled programming language; MATLAB will be used primarily herein, but we employ a programming style that should make coding in other languages self-evident.

This book is organized into 11 chapters. Among these, the authors recommend that the first five chapters be used as materials for a one-semester course on the basics of the FE method for undergraduate students in mechanical, aerospace, and civil engineering. The instructor may also add the sections on one-dimensional (1D) heat transfer in Chapter 11 to their undergraduate course contents. After completing Chapter 1, the instructor may begin to use the video tutorials to introduce the FE analysis software and assign simple problems for practice throughout the duration of the course. The instructor may omit some of the topics to adjust to the pace and expectations of the students.

In Chapter 1, the FE formulation is introduced using a system of linear springs and a slender body undergoing uniaxial deformation. The body is divided into many segments or elements in which a linearly assumed displacement field is introduced. The concept of incremental work and strain energy is used to show how element stiffness matrices and load vectors are constructed and assembled into a global stiffness matrix and a global load vector to construct an FE equation. An example problem is used to demonstrate how well the solution obtained by the FE method compares with the exact solution.

In Chapter 2, truss structures are introduced within the context of an FE formulation in which a truss member is naturally an element. It shows how a global stiffness matrix is assembled from individual elements to determine displacements at the hinge joints and axial force in each member under both applied loads and temperature change. We then observe that torsional deformation of a slender body is mathematically equivalent to uniaxial or longitudinal deformation of the slender body.

In Chapter 3, we consider the FE formulation of slender bodies undergoing bending deformation. The Bernoulli–Euler beam bending theory is presented and exact solutions are obtained for example problems to be used as reference solutions. This is followed by the FE formulation of the two-node element for beam bending analysis. Finally, the two-node frame element is constructed combining the uniaxial element, torsional element, and bending elements.

Chapter 4 describes how the element mass matrix is constructed and assembled into a global mass matrix to set up the equation of motion, which can then be used for investigation of free vibration via eigenvalue analysis and forced vibration using numerical integration.

In Chapter 5, we consider bending of slender bodies under axial force. It is shown that the effect of axial force manifests as an effective stiffness matrix which can be used for static buckling analysis and free vibration analysis. It is then shown how the FE formulation can be used to investigate the dynamic stability of a slender body subjected to a compressive follower force.

Chapter 6 introduces the concept of virtual displacement and virtual work to express equilibrium in three-dimensional (3D) space in a scalar form to which the FE formulation can be applied. The equations described in Appendix 1 are used for this purpose.

Chapter 7 introduces mapping functions and shape functions used for the FE formulation. Mapping functions are used to map individual elements in the physical domain to the mapped domain for two-dimensional (2D) and 3D bodies. Shape functions identical to mapping functions are used for the assumed displacement in the isoparametric formulation. Integrations in the mapped domain are discussed, along with numerical integration.

Chapter 8 shows how the FE formulation is used to generate the element stiffness matrix and load vector for elements in the 2D domain. This is followed by the FE formulation in 3D space.

Chapter 9 describes FE modeling of thin plates and shells, beginning with the assumptions on kinematics of beam bending and plate bending and their effect on the strain–displacement and strain–stress relations. A salient feature of this chapter is the discussion on the solid element specifically tailored for modeling of plates and shells.

Chapter 10 describes element locking phenomena, in which elements lose the ability to deform when they are used to represent bending of slender and thin structures. It also shows that elements can exhibit locking when they are used to model incompressible solids. The effects of reduced-order integration are also discussed in conjunction with the spurious kinematic modes.

Chapter 11 describes the FE formulation of heat transfer in solids and structures, starting with steady-state heat conduction in the 1D domain. This is followed by the FE formulation of time-dependent heat transfer problems. The FE formulation is then extended to heat transfer problems in the 2D and 3D domains.

Appendix 1 provides the fundamental equations, such as the strain–displacement relation, force and moment equilibria, and constitutive equations to describe deformation of 3D solids and structures under applied loads. Appendix 2 describes Gaussian elimination with triple factorization for the solution of a system of linear equations and the skyline method for efficient storage of sparse matrices. There is a Bibliography supplied for further reading and supplementary information.

# Acknowledgments

The authors would like to express deep appreciation of their former mentors, who greatly influenced them in their formative years as young students. They would also like to thank their former students, with whom they had the fortune and privilege of sharing great lifetime experiences while pursuing common goals of learning. In particular, the authors thank Dr. Jack Draper, who read the entire manuscript and made valuable comments as well as finding numerous mistakes.

# 1     Introduction to the Finite Element Method

The finite element (FE) method is a powerful technique that can be used to transform any continuous body into a set of governing equations with a finite number of unknowns, called degrees of freedom (DOF). For solids and structures, DOF are displacements or rotational angles at discrete points in the body called nodes or nodal points. The simplest model for any structural system is a single degree of freedom (SDOF) system with a spring and a concentrated mass. For a SDOF system, the spring constant and the mass can be determined via experiments or analytical means. However, for complicated structures such as aircraft, power plants, launch vehicles, buildings, bridges, machinery, or electronic devices, we need systematic and rational methods based on the fundamental principles of solid and structural mechanics in order to construct the governing equations.

In the FE formulation, the domain of interest is first divided into a mesh of subdomains called elements. An assumed displacement is then introduced into each element to construct the element stiffness matrix and the load vector in terms of the element nodal DOF to form the equilibrium equations at the element level. Subsequently, they are assembled into the global stiffness matrix and the global load vector to construct the equilibrium equation at the global level, which can be solved for the unknown global nodal DOF.

In this chapter, we will introduce the fundamentals of the FE method using a system of linear springs and a slender linear elastic body undergoing axial deformation as examples. These simple problems are chosen to describe the essential features of the FE method which are common to analysis of more complicated structural systems such as three-dimensional (3D) bodies.

In the example problems given in this chapter, we will try to keep the solutions in non-dimensional form. This will render the solutions valid for any geometric properties such as body length and cross-sectional area, material properties, and the magnitude of applied loads. Similar practices will be followed in subsequent chapters to the extent possible.

## 1.1    Overview of the Finite Element Method

The FE method is a powerful technique that can be used to analyze mathematical models on computers. Models are highly specialized for a particular type of analysis. Throughout this book, we will focus largely on the types of mechanics problems of interest to mechanical, aerospace, and civil engineers. More broadly, however, the types of models encountered, and the ways in which they are studied, are not unlike those seen in other disciplines. We may follow a workflow like the one shown in Figure 1.1.

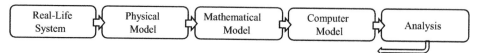

**Figure 1.1** Workflow for model creation and analysis

The workflow starts with a *real-life system.* In real life, the system of interest may be a structural member or part that serves an important mechanical purpose. It may sit within a larger system and function in a well-defined role, and it must then be designed to perform in a manner sufficient for the needs of the larger system.

The *physical model* is then identified through, for instance, a phenomenological concept such as "linear elasticity," "thermoviscoplasticity," "inviscid flow," or "thermal diffusion." This model must carefully account for the context of the later analysis. For instance, if the model is intended to study the response and failure of a structure, the physical model must possess features that could be used for that purpose.

The *mathematical model* is a complete mathematical statement of the problem in question. The model contains information about assumptions or simplifications that are very important to know as they can impose limitations on how subsequent analyses may be used. The mathematical model contains such information as the governing differential equations, boundary conditions, constitutive models, and kinematic theories. To understand how mathematical and physical models are very distinct, note that physical models are studied for physical behaviors, such as stress–strain response, whereas mathematical models must have a mathematical representation and be assured to have existing and unique solutions.

The *computer model* is enabled by the application of mathematical methods and computational algorithms. The FE method produces a type of computer model. The goal of the method is to take any mathematical model of a continuous body governed by a set of differential equations, and convert it into a problem of finding a finite number of unknowns. Much of the emphasis in this book is on how mathematical models are formulated from physical models and how they can be developed into FE computer models.

Finally, the *analysis* takes the computer model and attempts to answer questions to better understand behaviors that may occur in real life. The main purpose of performing FE analyses is to analyze and answer questions about real-life behaviors of engineered systems. We should note, however, that, owing to the many assumptions and approximations that separate the "real-life system" from the "computer model," an FE solution should be checked before being accepted as "real." The return path shown in Figure 1.1 indicates going back to any of the previous steps whenever needed. It often takes skill and effort for an engineer to define a complete set of well-posed models (i.e., physical, mathematical, computational) to ensure that the final solution informs the real-life problem.

With that perspective, let us focus on the primary purpose of the FE method, which is to express the governing equations in matrix form with a finite number of unknowns. In the FE formulation, the domain of interest is first divided into a mesh of subdomains called elements. For solids and structures, an assumed displacement is introduced into each element to construct the element stiffness matrix and the load vector for the equilibrium equation at the element level in terms of element nodal DOF. The element stiffness matrices and load vectors are then assembled into the global stiffness matrix and the global load vector for the equilibrium equation at the global level, which can be solved for the unknown global nodal DOF. For solids and structures, the DOF may be displacements or rotational angles. In general, for linear static problems, the equation constructed via the FE method can be symbolically expressed as

$$\mathbf{Kq} = \mathbf{F}. \qquad (1.1.1)$$

For a model with $N$ total DOF, $\mathbf{K}$ is an $N \times N$ global stiffness matrix, $\mathbf{q}$ is an $N \times 1$ global DOF vector, and $\mathbf{F}$ is an $N \times 1$ global load vector.

For complicated structures such as aircraft, power plants, launch vehicles, buildings, bridges, machinery, or electronic devices, we need systematic and rational methods based on the fundamental principles of solid and structural mechanics in order to construct the mathematical models and eventually convert them into forms that can be solved on computers. Procedures for performing this conversion have been an area of active research. There are various steps to achieving this, each step distinct and nuanced in ways involving nomenclatures that can at times be overwhelming for beginning students – virtual displacements and work, weighted residuals, variational methods, Galerkin method, strain energy increments, among others. The goal of such procedures is, however, largely the same – to ultimately permit the solution of differential equations using a sequence of matrix and vector arithmetic that can be performed on a computer.

**Procedures in Finite Element Analysis**

Finite element analysis entails three main stages: *preprocessing*, *solving*, and *postprocessing*. We will explore each of these in greater detail later. However, software packages available today have interfaces where users generally do not have to be aware of all of these stages. But whether the user is aware or not, the underlying structure of the codes in most software is organized with them in mind. Furthermore, new and emerging computational engineering approaches often follow similar structures. This makes the knowledge and awareness of the structure of FE analyses highly relevant as new codes and modeling approaches – including but also extending beyond FE methods – continue to push from research into commercial and industrial applications.

The *preprocessing* stage is primarily where the problem is specified and the final FE matrix equation is prepared. An understanding of the original mathematical model is important for preparing the preprocessing stage. In fact, in later chapters, the needed information of the preprocessor mirrors the information required for a complete mathematical statement of the problem. Often the existence of a solution or the stability of the algorithm will be influenced by choices at this stage. It is therefore typically the stage where most attention is paid in an introductory class. Critical information that must be specified includes:

- The **FE mesh**, which includes the coordinates of nodes and the connectivity of the elements.
- The **unique conditions of the analysis**, including boundary conditions, initial conditions, and/or duration of the simulated events (for time-dependent problems).
- The **constitutive model**, the type of phenomenology in consideration, and the values of relevant parameters.
- Other analysis parameters that are required in subsequent stages in the *solver* or *post-processor*. This is particularly relevant for advanced algorithmic or numerical features.

Techniques for generating FE meshes are beyond the scope of this book. It is a highly challenging field of research that has witnessed numerous groundbreaking developments over many decades. Most software tools available today have robust built-in mesh generators that can meet the needs of most students and early-career engineers.

All FE software packages contain a library of elements. In other words, the library of elements is the pre-programmed set of information for calculating *element stiffness matrix and load vectors*, based on the formulations developed or derived long before the point of writing code. The preparation of these formulations is where most study in formal engineering curricula is dedicated, including this book. The details of formulations can vary greatly among different problem and element types, so knowing and understanding how stiffness matrices and load vectors are developed will enhance the number of ways an engineer can effectively use the FE method to solve their specific problems. The number and types of elements available can vary widely among software products. Usually an element is specialized for a constitutive model or a unique feature of the mathematical model from which it was developed. Thus, considerable effort and attention has been paid to deriving element matrix equations, and developers of FE technologies are constantly innovating to allow models for new physics or to be more computationally efficient.

The *preprocessing* stage also entails the processes for assembly and application of boundary conditions. To early students, these can be challenging sequences of operations involving data arrays and memory management. Special attention must be paid to these book-keeping aspects, however. The genius of the FE method is that it allows a problem to be studied by assembling a domain from smaller elements. Each element has its own associated element matrix equation and the identity of each element must be preserved and recoverable after assembly. Keeping track of how the elements fit together is therefore vital and helps appreciate the need for implementing computational steps to automate the process.

The *solving* stage is where the computational task of number crunching is carried out to result in a vector of values of the DOF. At this point in an analysis, what was previously "unknown" is now "known." In structural and solid mechanics, the DOF are typically the nodal displacements or rotational angles.

Finally, the analysis is not complete until the solution vector is *postprocessed*. The *solve* step only gives us knowledge about the nodal DOF. However, this is not the only information an analyst will want to study. Other quantities – such as stress and strain – may be important, but additional operations must be performed to calculate these quantities from the solution vector. The book-keeping process that was used in the assembly procedure is very important in this stage as it can be used to deconstruct the global equations and yield information about what is happening inside each element. Postprocessing can also entail the visualization of results which, depending on the software, is based on highly sophisticated graphics algorithms for shading, occlusion, rendering, and speed to enable production-quality or highly interactive capabilities.

In this chapter, we will introduce the fundamentals of the FE method using such illustrative examples as a system of linear elastic springs or a slender elastic body undergoing axial deformation. These simple problems are used to describe the essential features of the FE method which are common to analysis of more complicated structural systems such as the 3D bodies we will study later in the book.

## 1.2 Virtual Displacement and Incremental Displacement

Before we get into the fundamentals of FE formulation, it will be useful to go through some simple exercises as follows.

(1) Given

$$ab = 0, \tag{1.2.1}$$

if $a$ is arbitrary in the sense that we can assign any value to it, then $b = 0$. To check, choose $a = 1$, then $1 \times b = 0$. Accordingly, $b = 0$.

(2) Given

$$a_1 b_1 + a_2 b_2 = 0, \tag{1.2.2}$$

if $a_1$ and $a_2$ are arbitrary, then $b_1 = 0$ and $b_2 = 0$. To check, choose first $a_1 = 1$ and $a_2 = 0$, then $b_1 = 0$ from the above equation. Choose now $a_1 = 0$ and $a_2 = 1$, then $b_2 = 0$.

In matrix form, Eq. (1.2.2) can be expressed as

$$\lfloor a_1 \ a_2 \rfloor \left\{ \begin{matrix} b_1 \\ b_2 \end{matrix} \right\} = 0 \tag{1.2.3}$$

or

$$\mathbf{a}^{\mathrm{T}} \mathbf{b} = \mathbf{0}, \tag{1.2.4}$$

where

$$\mathbf{a} = \left\{ \begin{matrix} a_1 \\ a_2 \end{matrix} \right\}, \quad \mathbf{b} = \left\{ \begin{matrix} b_1 \\ b_2 \end{matrix} \right\}. \tag{1.2.5}$$

Accordingly, $\mathbf{b} = \mathbf{0}$ for arbitrary $\mathbf{a}$. Note that Eq. (1.2.4) can be extended to column vectors $\mathbf{a}$ and $\mathbf{b}$ of any size.

(3) For the linear spring shown in Figure 1.2, the equation for static equilibrium is

$$Kq = F \tag{1.2.6}$$

or

$$Kq - F = 0. \tag{1.2.7}$$

**Figure 1.2** Linear spring subjected to tip force, and force–displacement relation

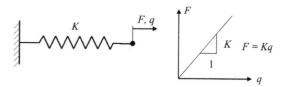

Now let's introduce an arbitrary weight $\delta q$, called a virtual displacement, and multiply Eq. (1.2.7) by $\delta q$ such that

$$(Kq - F)\delta q = 0. \tag{1.2.8}$$

Suppose now we are given Eq. (1.2.8), and we may then ask ourselves what it means. Since $\delta q$ is arbitrary in Eq. (1.2.8), we conclude that Eq. (1.2.7) holds. So, Eq. (1.2.8) is equivalent to Eq. (1.2.7). Note that the virtual displacement $\delta q$ is arbitrary and independent of the actual displacement $q$ under the applied force $F$.

Equation (1.2.8) can be written as

$$Kq\delta q - F\delta q = 0 \tag{1.2.9}$$

or

$$\delta U - \delta W = 0, \tag{1.2.10}$$

where

$$\delta U = Kq\delta q \tag{1.2.11}$$

is called "internal virtual work" and

$$\delta W = F\delta q \tag{1.2.12}$$

is called "external virtual work." The statement of equivalence between Eq. (1.2.7) and Eq. (1.2.10) is called the "principle of virtual work." We note that Eq. (1.2.7) is a statement of equilibrium in vector form involving forces, while Eq. (1.2.10) is a statement of equilibrium in scalar form.

Suppose we now choose the virtual displacement to be an infinitesimal displacement of the spring due to an external force. Then $\delta W = F\delta q$ is the increment in work done by the external force and $\delta U = Kq\delta q$ is the increment in internal energy, called the strain energy, stored in the spring. Equation (1.2.10) is then a statement of energy conservation in incremental form.

As displacement $q$ increases from $q = 0$ to $\hat{q}$, the strain energy may be determined as

$$U(\hat{q}) = \int_{q=0}^{\hat{q}} \delta U = \int_{q=0}^{\hat{q}} Kq\delta q = \frac{1}{2}Kq^2 \Big|_{q=0}^{\hat{q}} = \frac{1}{2}K\hat{q}^2. \tag{1.2.13}$$

Dropping the "hat" symbol for generality, the strain energy is expressed as

$$U(q) = \frac{1}{2}Kq^2.$$ (1.2.14)

## 1.3 A System of Linear Elastic Springs

As a way of illustrating how an FE model is constructed, let's consider a system of three linear elastic springs connected as shown in Figure 1.3. In the figure, the following labeling convention is used:

1, 2, 3, 4: global node numbers for the nodes placed at the end points of individual springs
$q_1, q_2, q_3, q_4$: global nodal DOF
$F_1, F_2, F_3, F_4$: global nodal loads
$k^{(e)}$: spring constant of spring element #e.

The leftmost end of the spring system is fixed ($q_1 = 0$). However, for convenience, we will replace this with the corresponding reaction force $R_1$ at the same location, which is included in the nodal force $F_1$.

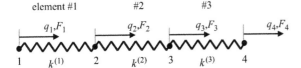

**Figure 1.3** A system of three springs

First we will consider individual springs, each of which constitutes the simplest finite element. They will then be assembled to construct the equilibrium equation in terms of the global displacements or DOF.

For the three-element model, the total strain energy $U$ is the sum of the strain energy of each element such that

$$U = \sum_{e=1}^{3} U_e,$$ (1.3.1)

where subscript "e" indicates the element number. The increment in the strain energy is then

$$\delta U = \sum_{e=1}^{3} \delta U_e.$$ (1.3.2)

Now look at an individual element #e as shown in Figure 1.4. For convenience, we introduce element node and DOF numberings in which

$u_1, u_2$: element nodal displacements or DOF.

According to Eq. (1.2.14), the strain energy of the element can be expressed as

$$U_e = \frac{1}{2}k^{(e)}(u_2 - u_1)^2,$$ (1.3.3)

where $(u_2 - u_1)$ is the net change in the spring length. The incremental strain energy is then

$$\delta U_e = k^{(e)}(u_2 - u_1)(\delta u_2 - \delta u_1). \tag{1.3.4}$$

Figure 1.4 Two-node spring element

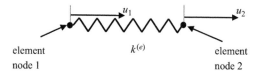

element          $k^{(e)}$          element
node 1                              node 2

Equation (1.3.4) can be rewritten in matrix form. For this, consider that in matrix form we may write

$$u_2 - u_1 = \lfloor -1 \quad 1 \rfloor \left\{ \begin{matrix} u_1 \\ u_2 \end{matrix} \right\}, \tag{1.3.5}$$

where

$$\left\{ \begin{matrix} u_1 \\ u_2 \end{matrix} \right\} : 2 \times 1 \text{ element DOF vector.}$$

Alternatively, we can write

$$u_2 - u_1 = \lfloor u_1 \quad u_2 \rfloor \left\{ \begin{matrix} -1 \\ 1 \end{matrix} \right\} \tag{1.3.6}$$

and thus

$$\delta u_2 - \delta u_1 = \lfloor \delta u_1 \quad \delta u_2 \rfloor \left\{ \begin{matrix} -1 \\ 1 \end{matrix} \right\}. \tag{1.3.7}$$

Substituting Eqs (1.3.5) and (1.3.7) into Eq. (1.3.4):

$$\delta U_e = k^{(e)} \lfloor \delta u_1 \quad \delta u_2 \rfloor \left\{ \begin{matrix} -1 \\ 1 \end{matrix} \right\} \lfloor -1 \quad 1 \rfloor \left\{ \begin{matrix} u_1 \\ u_2 \end{matrix} \right\} = k^{(e)} \lfloor \delta u_1 \quad \delta u_2 \rfloor \begin{bmatrix} 1 & -1 \\ -1 & 1 \end{bmatrix} \left\{ \begin{matrix} u_1 \\ u_2 \end{matrix} \right\} \tag{1.3.8}$$

or

$$\delta U_e = \lfloor \delta u_1 \quad \delta u_2 \rfloor \begin{bmatrix} k_{11}^e & k_{12}^e \\ k_{21}^e & k_{22}^e \end{bmatrix} \left\{ \begin{matrix} u_1 \\ u_2 \end{matrix} \right\} = \lfloor \delta u_1 \quad \delta u_2 \rfloor \mathbf{k}^e \left\{ \begin{matrix} u_1 \\ u_2 \end{matrix} \right\}, \tag{1.3.9}$$

where

$$\mathbf{k}^e = \begin{bmatrix} k_{11}^e & k_{12}^e \\ k_{21}^e & k_{22}^e \end{bmatrix} = k^{(e)} \begin{bmatrix} 1 & -1 \\ -1 & 1 \end{bmatrix} : 2 \times 2 \text{ element stiffness matrix.} \tag{1.3.10}$$

Note that the element stiffness matrix is symmetric and singular with a zero determinant.

## Equilibrium Equation at the Element Level

Consider now an element #$e$ in static equilibrium as shown in Figure 1.5. In the figure:

$A_1^e, A_2^e$ : applied forces at nodes 1 and 2
$R_1^e, R_2^e$ : reaction forces at nodes 1 and 2.

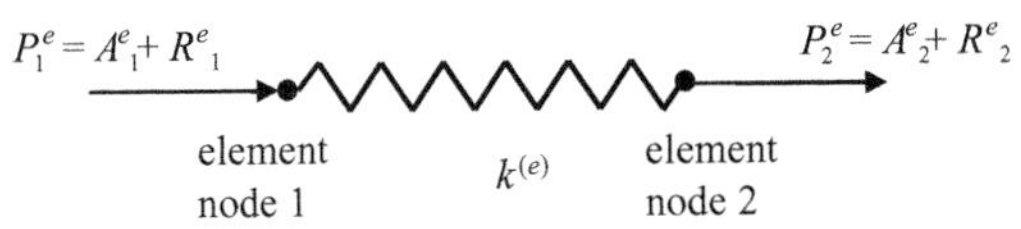

$P_1^e = A_1^e + R_1^e$

$P_2^e = A_2^e + R_2^e$

**Figure 1.5** Forces acting on the element nodes

element node 1

$k^{(e)}$

element node 2

Note that for element #1, $R_1^1$ is the reaction force $R_1$ at the left end. Reaction forces at the element boundaries cancel out when elements are joined. Treating $R_1^e$ and $R_2^e$ as the "external applied forces," the incremental work done by the applied nodal forces can be expressed as

$$\delta W_e = \left(A_1^e + R_1^e\right)\delta u_1 + \left(A_2^e + R_2^e\right)\delta u_2 = P_1^e \delta u_1 + P_2^e \delta u_2 = \lfloor \delta u_1 \quad \delta u_2 \rfloor \begin{Bmatrix} P_1^e \\ P_2^e \end{Bmatrix}, \quad (1.3.11)$$

where

$$\begin{Bmatrix} P_1^e \\ P_2^e \end{Bmatrix} = \begin{Bmatrix} A_1^e + R_1^e \\ A_2^e + R_2^e \end{Bmatrix} : \text{element load vector.} \quad (1.3.12)$$

For the element in equilibrium:

$$\delta U_e - \delta W_e = \lfloor \delta u_1 \quad \delta u_2 \rfloor \left( \begin{bmatrix} k_{11}^e & k_{12}^e \\ k_{21}^e & k_{22}^e \end{bmatrix} \begin{Bmatrix} u_1 \\ u_2 \end{Bmatrix} - \begin{Bmatrix} P_1^e \\ P_2^e \end{Bmatrix} \right) = 0. \quad (1.3.13)$$

Accordingly:

$$\begin{bmatrix} k_{11}^e & k_{12}^e \\ k_{21}^e & k_{22}^e \end{bmatrix} \begin{Bmatrix} u_1 \\ u_2 \end{Bmatrix} - \begin{Bmatrix} P_1^e \\ P_2^e \end{Bmatrix} = \begin{Bmatrix} 0 \\ 0 \end{Bmatrix} \rightarrow \begin{bmatrix} k_{11}^e & k_{12}^e \\ k_{21}^e & k_{22}^e \end{bmatrix} \begin{Bmatrix} u_1 \\ u_2 \end{Bmatrix} = \begin{Bmatrix} P_1^e \\ P_2^e \end{Bmatrix}. \quad (1.3.14)$$

Equation (1.3.14) is a statement of equilibrium; the elastic force is in balance with the "external" force at each node. Note that, for this simple spring element, we can derive Eq. (1.3.14) directly from the force equilibrium at each node.

## Equilibrium Equation for the Entire System

For the entire system with three spring elements we need to express the strain energy in terms of a global DOF numbering system as shown in Figure 1.3. The FE equilibrium equation for the entire system can be constructed from individual elements, utilizing the relationship between the element DOF and the global DOF.

For the three-element model, the relationship between element DOF and global DOF is shown below.

| Element number | Element DOF | Global DOF |
|:---:|:---:|:---:|
| 1 | $(1,2)$ | $(1,2)$ |
| 2 | $(1,2)$ | $(2,3)$ |
| 3 | $(1,2)$ | $(3,4)$ |

We may then construct a matrix called the "connectivity matrix," in which each row contains global DOF corresponding to element DOF starting from element #1 to the last element. For the present three-element model, the entries stored in the connectivity matrix are as follows:

$$\begin{bmatrix} 1 & 2 \\ 2 & 3 \\ 3 & 4 \end{bmatrix}.$$

Alternatively, we may store them in a vector array of $\lfloor 1 \quad 2 \quad 2 \quad 3 \quad 3 \quad 4 \rfloor$ called the "connectivity vector." Using the connectivity between element DOF and global DOF, the element load vectors and element stiffness matrices are assembled into the global load vector and the global stiffness matrix to construct the equilibrium equation in matrix form.

For the three-element model, the incremental work done by the applied forces can be expressed as

$$\delta W = \delta W_1 + \delta W_2 + \delta W_3 = \sum_{e=1}^{3} \delta W_e. \tag{1.3.15}$$

Since the element DOF are related to the global DOF, $\delta W$ can be expressed in an expanded form including global $\delta q_1$, $\delta q_2$, $\delta q_3$, and $\delta q_4$ such that

$$\delta W = [\delta q_1 \quad \delta q_2 \quad \delta q_3 \quad \delta q_4] \begin{Bmatrix} F_1 \\ F_2 \\ F_3 \\ F_4 \end{Bmatrix} = \delta \mathbf{q}^T \mathbf{F}, \tag{1.3.16}$$

where

$$\delta \mathbf{q} = \begin{Bmatrix} \delta q_1 \\ \delta q_2 \\ \delta q_3 \\ \delta q_4 \end{Bmatrix} : 4 \times 1 \text{ global incremental DOF vector,} \tag{1.3.17}$$

$$\mathbf{F} = \begin{Bmatrix} F_1 \\ F_2 \\ F_3 \\ F_4 \end{Bmatrix} : 4 \times 1 \text{ global load vector.} \tag{1.3.18}$$

For example, for spring element #2, $u_1 = q_2, u_2 = q_3$ and thus

$$\delta W_2 = \lfloor \delta u_1 \quad \delta u_2 \rfloor \begin{Bmatrix} P_1^2 \\ P_2^2 \end{Bmatrix} = \lfloor \delta q_2 \quad \delta q_3 \rfloor \begin{Bmatrix} P_1^2 \\ P_2^2 \end{Bmatrix}. \tag{1.3.19}$$

Accordingly, $P_1^2$ adds to $F_2$ and $P_2^2$ adds to $F_3$.

For the three-element model, the incremental strain energy is

$$\delta U = \delta U_1 + \delta U_2 + \delta U_3 = \sum_{e=1}^{3} \delta U_e. \tag{1.3.20}$$

According to Eq. (1.3.9):

$$\delta U_e = \lfloor \delta u_1 \quad \delta u_2 \rfloor \begin{bmatrix} k_{11}^e & k_{12}^e \\ k_{21}^e & k_{22}^e \end{bmatrix} \begin{Bmatrix} u_1 \\ u_2 \end{Bmatrix}. \tag{1.3.21}$$

Since the element DOF are related to the global DOF, we observe that $\delta U$ can be expressed in an expanded form including all DOF as follows:

$$\delta U = \lfloor \delta q_1 \quad \delta q_2 \quad \delta q_3 \quad \delta q_4 \rfloor \begin{bmatrix} K_{11} & K_{12} & K_{13} & K_{14} \\ K_{21} & K_{22} & K_{23} & K_{24} \\ K_{31} & K_{32} & K_{33} & K_{34} \\ K_{41} & K_{42} & K_{43} & K_{44} \end{bmatrix} \begin{Bmatrix} q_1 \\ q_2 \\ q_3 \\ q_4 \end{Bmatrix} = \delta \mathbf{q}^{\mathrm{T}} \mathbf{K} \mathbf{q}, \tag{1.3.22}$$

where

$$\mathbf{K} = \begin{bmatrix} K_{11} & K_{12} & K_{13} & K_{14} \\ K_{21} & K_{22} & K_{23} & K_{24} \\ K_{31} & K_{32} & K_{33} & K_{34} \\ K_{41} & K_{42} & K_{43} & K_{44} \end{bmatrix} : 4 \times 4 \text{ global stiffness matrix,} \tag{1.3.23}$$

$$\mathbf{q} = \begin{Bmatrix} q_1 \\ q_2 \\ q_3 \\ q_4 \end{Bmatrix} : 4 \times 1 \text{ global DOF vector.} \tag{1.3.24}$$

Then, introducing Eqs (1.3.16) and (1.3.22):

$$\delta U - \delta W = \delta \mathbf{q}^{\mathrm{T}} \mathbf{K} \mathbf{q} - \delta \mathbf{q}^{\mathrm{T}} \mathbf{F} = \delta \mathbf{q}^{\mathrm{T}} (\mathbf{K} \mathbf{q} - \mathbf{F}) = 0 \tag{1.3.25}$$

for the system in equilibrium. We can then conclude that

$$\mathbf{K} \mathbf{q} - \mathbf{F} = \mathbf{0} \rightarrow \mathbf{F} = \mathbf{K} \mathbf{q} \tag{1.3.26}$$

or

$$\mathbf{K} \mathbf{q} = \mathbf{F}. \tag{1.3.27}$$

## Assembly of Global Stiffness Matrix and Load Vector

The global load vector and global stiffness matrix can be assembled element by element as follows. First, initialize them as

$$\mathbf{F} = \begin{Bmatrix} 0 \\ 0 \\ 0 \\ 0 \end{Bmatrix}, \quad \mathbf{K} = \begin{bmatrix} 0 & 0 & 0 & 0 \\ 0 & 0 & 0 & 0 \\ 0 & 0 & 0 & 0 \\ 0 & 0 & 0 & 0 \end{bmatrix}. \tag{1.3.28}$$

(1) For element #1, $u_1 = q_1$, $u_2 = q_2$, $P_1^1$ sums to $F_1$, and $P_2^1$ sums to $F_2$. Accordingly, the equilibrium equation of element #1 is

$$\begin{Bmatrix} P_1^1 \\ P_2^1 \end{Bmatrix} = \begin{bmatrix} k_{11}^1 & k_{12}^1 \\ k_{21}^1 & k_{22}^1 \end{bmatrix} \begin{Bmatrix} u_1 \\ u_2 \end{Bmatrix} \rightarrow \begin{matrix} (1) \\ (2) \end{matrix} \begin{Bmatrix} P_1^1 \\ P_2^1 \end{Bmatrix} = \begin{matrix} (1) \\ (2) \end{matrix} \overset{(1)\ \ (2)}{\begin{bmatrix} k_{11}^1 & k_{12}^1 \\ k_{21}^1 & k_{22}^1 \end{bmatrix}} \begin{Bmatrix} q_1 \\ q_2 \end{Bmatrix}. \tag{1.3.29}$$

The numbers in parentheses are added to indicate the row number and column number of each entry in the global system. Then, after assembling element #1:

$$\begin{Bmatrix} F_1 \\ F_2 \\ F_3 \\ F_4 \end{Bmatrix} = \begin{Bmatrix} P_1^1 \\ P_2^1 \\ 0 \\ 0 \end{Bmatrix} = \begin{bmatrix} k_{11}^1 & k_{12}^1 & 0 & 0 \\ k_{21}^1 & k_{22}^1 & 0 & 0 \\ 0 & 0 & 0 & 0 \\ 0 & 0 & 0 & 0 \end{bmatrix} \begin{Bmatrix} q_1 \\ q_2 \\ q_3 \\ q_4 \end{Bmatrix}. \tag{1.3.30}$$

(2) For element #2, $u_1 = q_2$, $u_2 = q_3$, $P_1^2$ sums to $F_2$, and $P_2^2$ sums to $F_3$. Accordingly, the equilibrium equation of element #2 is

$$\begin{Bmatrix} P_1^2 \\ P_2^2 \end{Bmatrix} = \begin{bmatrix} k_{11}^2 & k_{12}^2 \\ k_{21}^2 & k_{22}^2 \end{bmatrix} \begin{Bmatrix} u_1 \\ u_2 \end{Bmatrix} \rightarrow \begin{matrix} (2) \\ (3) \end{matrix} \begin{Bmatrix} P_1^2 \\ P_2^2 \end{Bmatrix} = \begin{matrix} (2) \\ (3) \end{matrix} \overset{(2)\ \ (3)}{\begin{bmatrix} k_{11}^2 & k_{12}^2 \\ k_{21}^2 & k_{22}^2 \end{bmatrix}} \begin{Bmatrix} q_2 \\ q_3 \end{Bmatrix}. \tag{1.3.31}$$

Then, after assembling elements #1 and #2:

$$\begin{Bmatrix} F_1 \\ F_2 \\ F_3 \\ F_4 \end{Bmatrix} = \begin{Bmatrix} P_1^1 \\ P_2^1 + P_1^2 \\ P_2^2 \\ 0 \end{Bmatrix} = \begin{bmatrix} k_{11}^1 & k_{12}^1 & 0 & 0 \\ k_{21}^1 & k_{22}^1 + k_{11}^2 & k_{12}^2 & 0 \\ 0 & k_{21}^2 & k_{22}^2 & 0 \\ 0 & 0 & 0 & 0 \end{bmatrix} \begin{Bmatrix} q_1 \\ q_2 \\ q_3 \\ q_4 \end{Bmatrix}. \tag{1.3.32}$$

(3) For element #3, $u_1 = q_3$, $u_2 = q_4$, $P_1^3$ sums to $F_3$, and $P_2^3$ sums to $F_4$. Accordingly, the equilibrium equation of element #3 is

$$\begin{Bmatrix} P_1^3 \\ P_2^3 \end{Bmatrix} = \begin{bmatrix} k_{11}^3 & k_{12}^3 \\ k_{21}^3 & k_{22}^3 \end{bmatrix} \begin{Bmatrix} u_1 \\ u_2 \end{Bmatrix} \rightarrow \begin{matrix} (3) \\ (4) \end{matrix} \begin{Bmatrix} P_1^3 \\ P_2^3 \end{Bmatrix} = \begin{matrix} (3) \\ (4) \end{matrix} \overset{(3)\ \ (4)}{\begin{bmatrix} k_{11}^3 & k_{12}^3 \\ k_{21}^3 & k_{22}^3 \end{bmatrix}} \begin{Bmatrix} q_3 \\ q_4 \end{Bmatrix}. \tag{1.3.33}$$

Then, after assembling elements #1, #2, and #3:

$$
\begin{Bmatrix} F_1 \\ F_2 \\ F_3 \\ F_4 \end{Bmatrix} = \begin{Bmatrix} P_1^1 \\ P_2^1 + P_1^2 \\ P_2^2 + P_1^3 \\ P_2^3 \end{Bmatrix} = \begin{bmatrix} k_{11}^1 & k_{12}^1 & 0 & 0 \\ k_{21}^1 & k_{22}^1 + k_{11}^2 & k_{12}^2 & 0 \\ 0 & k_{21}^2 & k_{22}^2 + k_{11}^3 & k_{12}^3 \\ 0 & 0 & k_{21}^3 & k_{22}^3 \end{bmatrix} \begin{Bmatrix} q_1 \\ q_2 \\ q_3 \\ q_4 \end{Bmatrix}.
\tag{1.3.34}
$$

The above equation can be written symbolically as

$$
\begin{bmatrix} K_{11} & K_{12} & K_{13} & K_{14} \\ K_{21} & K_{22} & K_{23} & K_{24} \\ K_{31} & K_{32} & K_{33} & K_{34} \\ K_{41} & K_{42} & K_{43} & K_{44} \end{bmatrix} \begin{Bmatrix} q_1 \\ q_2 \\ q_3 \\ q_4 \end{Bmatrix} = \begin{Bmatrix} F_1 \\ F_2 \\ F_3 \\ F_4 \end{Bmatrix}.
\tag{1.3.35}
$$

To complete the problem definition, we apply the geometric boundary condition at $x = 0$ and set $q_1 = 0$ in the above equation. Since $q_1$ multiplies the entries in the first column of the $4 \times 4$ global stiffness matrix:

$$
\begin{bmatrix} K_{12} & K_{13} & K_{14} \\ K_{22} & K_{23} & K_{24} \\ K_{32} & K_{33} & K_{34} \\ K_{42} & K_{43} & K_{44} \end{bmatrix} \begin{Bmatrix} q_2 \\ q_3 \\ q_4 \end{Bmatrix} = \begin{Bmatrix} F_1 \\ F_2 \\ F_3 \\ F_4 \end{Bmatrix}.
\tag{1.3.36}
$$

The above equation represents a system of four equations for three unknown nodal displacements. We then delete the first equation with the reaction term in the load vector to match the three unknowns with the three equations as follows:

$$
\begin{bmatrix} K_{22} & K_{23} & K_{24} \\ K_{32} & K_{33} & K_{34} \\ K_{42} & K_{43} & K_{44} \end{bmatrix} \begin{Bmatrix} q_2 \\ q_3 \\ q_4 \end{Bmatrix} = \begin{Bmatrix} F_2 \\ F_3 \\ F_4 \end{Bmatrix}.
\tag{1.3.37}
$$

The above equation can then be solved for $q_2, q_3$, and $q_4$.

## Example 1.1

For the system of three spring elements as shown in Figure 1.3, the spring constants are identical such that

$$ k^{(1)} = k^{(2)} = k^{(3)} = k, $$

where $k$ is a given constant. Suppose that the load applied at each node of an element is given as

$$ \begin{Bmatrix} A_1^e \\ A_2^e \end{Bmatrix} = \begin{Bmatrix} B \\ B \end{Bmatrix}, $$

where $B$ is a given constant. An example of this type of load will be described in Section 1.6.3.

(a) Assemble the global stiffness matrix and the global load vector to set up the equilibrium equation in matrix form.

(b) Apply the geometric boundary condition ($q_1 = 0$) and solve for the nodal displacements.

## Solution:

(a) For element $i$:

$$\mathbf{k}^e = k \begin{bmatrix} 1 & -1 \\ -1 & 1 \end{bmatrix}.$$

After assembly, the global stiffness matrix is

$$\mathbf{K} = k \begin{bmatrix} 1 & -1 & 0 & 0 \\ -1 & 1+1 & -1 & 0 \\ 0 & -1 & 1+1 & -1 \\ 0 & 0 & -1 & 1 \end{bmatrix} = k \begin{bmatrix} 1 & -1 & 0 & 0 \\ -1 & 2 & -1 & 0 \\ 0 & -1 & 2 & -1 \\ 0 & 0 & -1 & 1 \end{bmatrix}.$$

After assembly of load vectors and reaction force, the global load vector is

$$\mathbf{F} = \begin{Bmatrix} B + R_1 \\ B + B \\ B + B \\ B \end{Bmatrix} = B \begin{Bmatrix} 1 + R_1/B \\ 2 \\ 2 \\ 1 \end{Bmatrix},$$

where $R_1$ is the reaction at global node 1. With the global stiffness matrix and the global load vector assembled, the equilibrium equation for the system is

$$k \begin{bmatrix} 1 & -1 & 0 & 0 \\ -1 & 2 & -1 & 0 \\ 0 & -1 & 2 & -1 \\ 0 & 0 & -1 & 1 \end{bmatrix} \begin{Bmatrix} q_1 \\ q_2 \\ q_3 \\ q_4 \end{Bmatrix} = B \begin{Bmatrix} 1 + R_1/B \\ 2 \\ 2 \\ 1 \end{Bmatrix}.$$

(b) Setting $q_1 = 0$ to apply the geometric boundary condition and deleting the first equation with the reaction term in the load vector, the equilibrium equation in reduced form is

$$k \begin{bmatrix} 2 & -1 & 0 \\ -1 & 2 & -1 \\ 0 & -1 & 1 \end{bmatrix} \begin{Bmatrix} q_2 \\ q_3 \\ q_4 \end{Bmatrix} = B \begin{Bmatrix} 2 \\ 2 \\ 1 \end{Bmatrix} \rightarrow \begin{bmatrix} 2 & -1 & 0 \\ -1 & 2 & -1 \\ 0 & -1 & 1 \end{bmatrix} \begin{Bmatrix} q_2 \\ q_3 \\ q_4 \end{Bmatrix} = \frac{B}{k} \begin{Bmatrix} 2 \\ 2 \\ 1 \end{Bmatrix}.$$

Solving the above equation:

$$\begin{Bmatrix} q_2 \\ q_3 \\ q_4 \end{Bmatrix} = \frac{B}{k} \begin{Bmatrix} 5 \\ 8 \\ 9 \end{Bmatrix}.$$

## 1.4    Slender Body under Axial Force

In order to introduce the fundamental concept of the FE formulation for solids and structures, we will consider a slender body subjected to axial forces as shown in Figure 1.6. In the figure:

$f(x)$: applied force per unit length
$P$: axial force applied at $x = L$
$A(x)$: cross-sectional area
$u(x)$: axial displacement of the cross-section located at $x$.

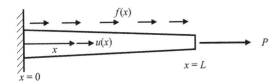

**Figure 1.6** Slender body subjected to axial forces

For the slender body under axial forces, one may assume that axial stress $\sigma_{xx}$ acting normal to a cross-section is uniform over the cross-section while all other stress components are equal to zero, and all points on the cross-section located at position $x$ translate uniformly by $u(x)$. For a mathematical description of the problem, we first consider the equilibrium of the body. For this, a free body is isolated by introducing imaginary cuts at $x$ and $x + dx$, as shown in Figure 1.7, in which $F(x)$ is the axial force acting on the surface located at $x$.

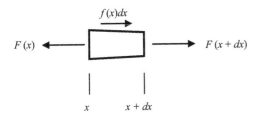

**Figure 1.7** Free body diagram of segment $dx$

Summing all forces acting on the free body:

$$F(x + dx) - F(x) + f(x)dx = 0$$

$$\rightarrow F(x) + \frac{\partial F}{\partial x}dx - F(x) + f(x)dx = 0 \tag{1.4.1}$$

$$\rightarrow \left(\frac{\partial F}{\partial x} + f\right)dx = 0.$$

For $dx \rightarrow 0$, we can conclude that

$$\frac{\partial F}{\partial x} + f = 0. \tag{1.4.2}$$

Axial force can be expressed as

$$F = \sigma_{xx} A. \tag{1.4.3}$$

Axial stress $\sigma_{xx}$ is related to axial strain $\varepsilon_{xx}$ through Young's modulus $E$ such that

$$\sigma_{xx} = E\varepsilon_{xx} \tag{1.4.4}$$

and axial strain is related to displacement such that

$$\varepsilon_{xx} = \frac{\partial u}{\partial x}. \tag{1.4.5}$$

One may combine Eqs (1.4.3), (1.4.4), and (1.4.5) such that

$$F = EA\frac{\partial u}{\partial x}. \tag{1.4.6}$$

The force equilibrium in Eq. (1.4.2), stress–strain relation in Eq. (1.4.4), and strain–displacement relation in Eq. (1.4.5) provide three equations for three unknowns – displacement, strain, and stress – for this one-dimensional (1D) problem.

To complete a problem statement, boundary conditions must be specified. The boundary conditions for the slender body in Figure 1.6 are as follows:

(1) Geometric boundary condition: at $x = 0, u = 0$.
(2) Force boundary condition: at $x = L, F(= \sigma_{xx}A) = P$.

Note that, for a body of the same geometry and material, boundary conditions make one problem different from others. In the following, we will consider a simple example problem, which permits us to find the exact solution. The same problem will be solved later using the FE method to help us appreciate the effectiveness of the method.

## Example 1.2

Consider a slender body rotating around the $z$-axis with constant angular velocity $\Omega$ (rad/s), as shown in Figure 1.8. For simplicity, assume that the body span begins at $x = 0$, where the axis of rotation is located and ends at $x = L$. Also, assume constant $EA$ and $m$, the body mass per unit length. Find displacement $u$ and stress $\sigma_{xx}$.

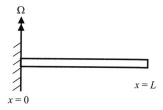

**Figure 1.8** Slender body rotating with a constant speed

Note that the centrifugal force acting on the infinitesimal segment of length $dx$ is

$$f dx = (m dx)\Omega^2 x$$

and thus

$$f = m\Omega^2 x = \rho A\Omega^2 x : \text{centrifugal force per unit length}$$

where $m = \rho A$ and $\rho$ is the mass per volume.

## Solution:

From Eq. (1.4.2):

$$\frac{\partial F}{\partial x} + f = 0 \rightarrow \frac{\partial F}{\partial x} = -f = -\rho A\Omega^2 x.$$

Integrating:

$$F = \sigma_{xx} A = -\rho A\Omega^2 \left( \frac{1}{2}x^2 + C_1 \right).$$

Applying the force boundary condition, $F = 0$ at $x = L$, to the above equation:

$$\frac{1}{2}L^2 + C_1 = 0 \rightarrow C_1 = -\frac{1}{2}L^2.$$

Accordingly:

$$\sigma_{xx} = \frac{1}{2}\rho\Omega^2 \left( L^2 - x^2 \right).$$

Also

$$\varepsilon_{xx} = \frac{\partial u}{\partial x} = \frac{1}{E}\sigma_{xx} = \frac{\rho\Omega^2}{2E}\left( L^2 - x^2 \right).$$

Integrating again:

$$u = \frac{\rho\Omega^2}{2E}\left( L^2 x - x^3/3 + C_2 \right).$$

Applying the geometric boundary conditions, $u = 0$ at $x = 0$, to the above equation, $C_2 = 0$. Accordingly:

$$u = \frac{\rho\Omega^2}{2E}\left( L^2 x - \frac{x^3}{3} \right).$$

## 1.5    Virtual Work, Incremental Work Done, and Strain Energy

As a prelude to the FE formulation, we will now construct the general expressions for the virtual work, the incremental work done by the applied loads, and the strain energy for the slender body described in Figure 1.6.

## 1.5.1  Virtual Work

We may now introduce an arbitrary weight function $\delta u(x)$, called a virtual displacement, to construct a scalar integral which is equivalent to Eq. (1.4.2). Note that $\delta u(x)$ is arbitrary and may not be related to the actual displacement $u(x)$.

By multiplying Eq. (1.4.2) by the virtual displacement $\delta u(x)$ or weight function, and integrating over the length, we obtain the following scalar integral:

$$\int_{x=0}^{x=L} \left( \frac{\partial F}{\partial x} + f \right) \delta u \, dx = 0. \tag{1.5.1}$$

It can be shown that Eq. (1.5.1) is equivalent to Eq. (1.4.2). The above equation holds for any $\delta u$. Accordingly, it holds for the following particular choice:

$$\delta u = \varepsilon \left( \frac{\partial F}{\partial x} + f \right), \tag{1.5.2}$$

where $\varepsilon$ is a constant. Placing Eq. (1.5.2) into Eq. (1.5.1):

$$\varepsilon \int_{x=0}^{x=L} \left( \frac{\partial F}{\partial x} + f \right)^2 dx = 0. \tag{1.5.3}$$

From the above equation, we conclude that

$$\frac{\partial F}{\partial x} + f = 0$$

everywhere in the body, which is identical to the equilibrium statement in Eq. (1.4.2).

Equation (1.5.1) can be transformed into a more useful form by applying integration by parts. Recall that given $G(x)$ and $H(x)$ :

$$\int_a^b d(GH) = (GH)_a^b \rightarrow \int_a^b GdH + \int_a^b HdG = (GH)_a^b$$

$$\rightarrow \int_a^b GdH = (GH)_a^b - \int_a^b HdG. \tag{1.5.4}$$

The last of the above equations is the formula for integration by parts. Applying it to the first term in Eq. (1.5.1):

$$\int_{x=0}^{x=L} \frac{\partial F}{\partial x} \delta u \, dx = \int_{x=0}^{x=L} \delta u \left( \frac{\partial F}{\partial x} dx \right) = \int_{x=0}^{x=L} \delta u \, dF = (\delta u F)_{x=0}^{x=L} - \int_{x=0}^{x=L} F d(\delta u)$$

$$= (\delta u F)_{x=0}^{x=L} - \int_{x=0}^{x=L} F \frac{\partial \delta u}{\partial x} dx = (F \delta u)_{x=L} - (F \delta u)_{x=0} - \int_{x=0}^{x=L} F \frac{\partial \delta u}{\partial x} dx. \tag{1.5.5}$$

Placing the last of Eq. (1.5.5) into Eq. (1.5.1) and changing the sign:

$$\int_{x=0}^{x=L} F\frac{\partial \delta u}{\partial x}dx - \int_{x=0}^{x=L} f\delta u dx - (F\delta u)_{x=L} + (F\delta u)_{x=0} = 0. \tag{1.5.6}$$

We now apply the force boundary condition ($F = P$ at $x = L$) to the above equation. Then

$$\int_{x=0}^{x=L} F\frac{\partial \delta u}{\partial x}dx - \int_{x=0}^{x=L} \delta u f dx - (P\delta u)_{x=L} + (F\delta u)_{x=0} = 0. \tag{1.5.7}$$

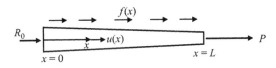

**Figure 1.9** Reaction force at the fixed end and applied forces

We may substitute the geometric boundary constraint ($u = 0$ at $x = 0$) with a reaction force $R_0$ as shown in Figure 1.9.

Placing $F = -R_0$ at $x = 0$ into Eq. (1.5.7):

$$\int_{x=0}^{x=L} F\frac{\partial \delta u}{\partial x}dx - \int_{x=0}^{x=L} \delta u f dx - (P\delta u)_{x=L} - (R_0\delta u)_{x=0} = 0. \tag{1.5.8}$$

We may express the above equation as

$$\delta U - \delta W = 0, \tag{1.5.9}$$

where

$$\delta U = \int_{x=0}^{x=L} F\frac{\partial \delta u}{\partial x}dx : \text{internal virtual work}, \tag{1.5.10}$$

$$\delta W = \int_{x=0}^{x=L} \delta u f dx + (P\delta u)_{x=L} + (R_0\delta u)_{x=0} : \text{external virtual work}. \tag{1.5.11}$$

In the above equation, reaction force $R_0$ is treated as an "external" force. Introducing Eq. (1.4.3), Eq. (1.5.10) can be expressed as

$$\delta U = \int_{x=0}^{x=L} F\frac{\partial \delta u}{\partial x}dx = \int_{x=0}^{x=L} \sigma_{xx}A\frac{\partial \delta u}{\partial x}dx = \int_{x=0}^{x=L} \sigma_{xx}\delta\varepsilon_{xx}Adx, \tag{1.5.12}$$

where

$$\delta\varepsilon_{xx} = \frac{\partial \delta u}{\partial x} : \text{virtual strain}. \tag{1.5.13}$$

Substituting the stress–strain relation in Eq. (1.4.4) into Eq. (1.5.12):

$$\delta U = \int_{x=0}^{x=L} E\varepsilon_{xx}\delta\varepsilon_{xx}Adx. \tag{1.5.14}$$

## 1.5.2  Incremental Work Done and Strain Energy

If we choose an infinitesimal increment in displacement as the virtual displacement, then $\delta W$ in Eq. (1.5.11) is the expression for infinitesimal increment in work done by applied loads and $\delta U$ in Eq. (1.5.10) or Eq. (1.5.12) is the expression for infinitesimal increment in strain energy. In this case, we could have constructed the right-hand side of Eq. (1.5.11) by inspection.

### Strain Energy Increment and Strain Energy

Equation (1.5.12) can also be expressed as

$$\delta U = \int_{x=0}^{x=L} \delta U^* Adx, \tag{1.5.15}$$

where

$$\delta U^* = \sigma_{xx}\delta\varepsilon_{xx} \tag{1.5.16}$$

is the incremental strain energy per unit volume, and

$$\delta\varepsilon_{xx} = \frac{\partial\delta u}{\partial x} = \delta\left(\frac{\partial u}{\partial x}\right) : \text{incremental strain.} \tag{1.5.17}$$

The equality between the second term and the third term of the above equation holds because the derivative accounts for a change over the space (i.e., $x$) for a given state, while "$\delta$" accounts for a change in state for a fixed position. Substituting the stress–strain relation in Eq. (1.4.4) into Eq. (1.5.16):

$$\delta U^* = E\varepsilon_{xx}\delta\varepsilon_{xx}. \tag{1.5.18}$$

Integrating from zero strain state to a finite strain state corresponding to a deformed state:

$$U^*(\hat{\varepsilon}_{xx}) = \int_{\varepsilon_{xx}=0}^{\varepsilon_{xx}=\hat{\varepsilon}_{xx}} \delta U^* = \int_{\varepsilon_{xx}=0}^{\varepsilon_{xx}=\hat{\varepsilon}_{xx}} E\varepsilon_{xx}\delta\varepsilon_{xx} = \int_{\varepsilon_{xx}=0}^{\varepsilon_{xx}=\hat{\varepsilon}_{xx}} \frac{1}{2}E\delta\left(\varepsilon_{xx}^2\right) = \frac{1}{2}E\hat{\varepsilon}_{xx}^2 = \frac{1}{2}\hat{\sigma}_{xx}\hat{\varepsilon}_{xx} \tag{1.5.19}$$

is the strain energy per unit volume. Dropping the "hat" sign for generality,

$$U^*(\varepsilon_{xx}) = \frac{1}{2}E\varepsilon_{xx}^2 = \frac{1}{2}\sigma_{xx}\varepsilon_{xx}. \tag{1.5.20}$$

The strain energy of the body is then expressed as

$$U(\varepsilon_{xx}) = \int\limits_{x=0}^{x=L} U^*(\varepsilon_{xx})A\,dx = \frac{1}{2}\int\limits_{x=0}^{x=L} EA\varepsilon_{xx}^2\,dx. \tag{1.5.21}$$

Substituting the strain–displacement relation in Eq. (1.4.5) into the above equation:

$$U = \frac{1}{2}\int\limits_{x=0}^{x=L} EA\left(\frac{\partial u}{\partial x}\right)^2 dx. \tag{1.5.22}$$

## 1.6    Element Stiffness Matrix and Load Vector

With Eqs (1.5.11) and (1.5.14), we can now introduce the FE formulation to construct the element stiffness matrix and load vector for each element. They will then be assembled into the global stiffness matrix and the global load vector.

For the slender body problem, let us first divide the domain into many segments called "elements" of finite length. One may then expect the displacement field within each element to be of very simple distribution. As an illustration, consider a three-element model as shown in Figure 1.10.

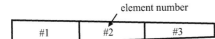

**Figure 1.10** Three-element model

A scalar integral over the body length can then be expressed as the sum of the integrals over each element. Accordingly, from Eqs (1.5.14), (1.5.21), and (1.5.11) of the previous section:

$$\delta U = \sum_{e=1}^{3} \delta U_e = \sum_{e=1}^{3}\int\limits_{e} EA\delta\varepsilon_{xx}\varepsilon_{xx}\,dx, \tag{1.6.1}$$

$$U = \sum_{e=1}^{3} U_e = \sum_{e=1}^{3}\frac{1}{2}\int\limits_{e} EA\varepsilon_{xx}^2\,dx, \tag{1.6.2}$$

$$\int\limits_{x=0}^{x=L} \delta u f\,dx = \sum_{e=1}^{3}\int\limits_{e} \delta u f\,dx, \tag{1.6.3}$$

where subscript "$e$" indicates that integration is over element number $e$.

For the FE formulation, we may select the two end points of each element and assign displacements. The selected points are called the element nodes and the displacements at the nodes are called the element nodal displacements or the element nodal DOF. For convenience, we will now introduce element node and DOF numberings as shown in Figure 1.11, in which

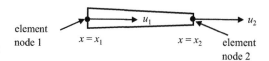

**Figure 1.11** Two-node uniaxial element

$u_1$, $u_2$: element nodal displacements or DOF
$x_1$, $x_2$: element nodal coordinates.

## 1.6.1 Mapping and Assumed Displacement

### Mapping

For convenience, let's introduce a non-dimensional coordinate $s$ defined as follows (see Figure 1.12):

$$s = \frac{x - x_1}{x_2 - x_1}. \tag{1.6.4}$$

**Figure 1.12** Definition of the non-dimensional $s$ coordinate

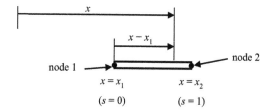

Note that $s = 0$ at node 1 and $s = 1$ at node 2. From Eq. (1.6.4):

$$x = x_1 + (x_2 - x_1)s = x_1 + ls \rightarrow x = (1 - s)x_1 + sx_2, \tag{1.6.5}$$

where

$$l = x_2 - x_1 : \text{element length}. \tag{1.6.6}$$

### Assumed Displacement

Under axial loads, the slender body deforms in the axial direction and thus displacement is the primary unknown variable as shown in Eq. (1.5.22). Accordingly, an assumed displacement is introduced in the element in terms of nodal displacements or DOF such that the assumed displacement field is continuous along the slender body.

The element has two nodes and two nodal displacements. Accordingly, we assume that, within an element, the displacement varies linearly from $u_1$ to $u_2$. For element number $e$, the displacement can then be expressed as

$$u = b_1 + b_2 s. \qquad (1.6.7)$$

The coefficients $b_1$ and $b_2$ can be related to nodal displacements $u_1$ and $u_2$ as follows:

At node 1, $s = 0$, and from Eq. (1.6.7):

$$u_1 = b_1 \quad \rightarrow \quad b_1 = u_1. \qquad (1.6.8)$$

At node 2, $s = 1$, and from Eqs (1.6.7) and (1.6.8):

$$u_2 = b_1 + b_2 = u_1 + b_2 \quad \rightarrow \quad b_2 = u_2 - u_1. \qquad (1.6.9)$$

Substituting Eqs (1.6.8) and (1.6.9) into Eq. (1.6.7) and rearranging yields

$$u = (1 - s)u_1 + s u_2. \qquad (1.6.10)$$

## 1.6.2  Construction of Element Stiffness Matrix

The mapping in Eq. (1.6.5) and the assumed displacement in Eq. (1.6.10) can be used to construct the element stiffness matrix. For an element:

$$\delta U_e = \int_{x=x_1}^{x=x_2} EA\delta\varepsilon_{xx}\varepsilon_{xx}dx. \qquad (1.6.11)$$

From the mapping in Eq. (1.6.5):

$$dx = (x_2 - x_1)ds = lds. \qquad (1.6.12)$$

From the assumed displacement in Eq. (1.6.10), the strain can be expressed as

$$\varepsilon_{xx} = \frac{\partial u}{\partial x} = \frac{\partial u}{\partial s}\frac{ds}{dx} = \frac{1}{l}(u_2 - u_1). \qquad (1.6.13)$$

In matrix form, the right-hand side of the above equation can be written as

$$\varepsilon_{xx} = \frac{1}{l}\lfloor -1 \quad 1 \rfloor \begin{Bmatrix} u_1 \\ u_2 \end{Bmatrix}, \qquad (1.6.14)$$

where

$$\begin{Bmatrix} u_1 \\ u_2 \end{Bmatrix} : 2 \times 1 \text{ element DOF vector.}$$

Alternatively:

$$\varepsilon_{xx} = \frac{1}{l} \lfloor u_1 \quad u_2 \rfloor \left\{ \begin{matrix} -1 \\ 1 \end{matrix} \right\} \tag{1.6.15}$$

and

$$\delta\varepsilon_{xx} = \frac{1}{l} \lfloor \delta u_1 \quad \delta u_2 \rfloor \left\{ \begin{matrix} -1 \\ 1 \end{matrix} \right\}. \tag{1.6.16}$$

Then

$$\delta\varepsilon_{xx}\varepsilon_{xx} = \frac{1}{l^2} \lfloor \delta u_1 \quad \delta u_2 \rfloor \left\{ \begin{matrix} -1 \\ 1 \end{matrix} \right\} \lfloor -1 \quad 1 \rfloor \left\{ \begin{matrix} u_1 \\ u_2 \end{matrix} \right\}$$

$$= \frac{1}{l^2} \lfloor \delta u_1 \quad \delta u_2 \rfloor \begin{bmatrix} 1 & -1 \\ -1 & 1 \end{bmatrix} \left\{ \begin{matrix} u_1 \\ u_2 \end{matrix} \right\}. \tag{1.6.17}$$

Substituting the above equation into Eq. (1.6.11):

$$\delta U_e = \int_{s=0}^{s=1} EA \frac{1}{l^2} \lfloor \delta u_1 \quad \delta u_2 \rfloor \begin{bmatrix} 1 & -1 \\ -1 & 1 \end{bmatrix} \left\{ \begin{matrix} u_1 \\ u_2 \end{matrix} \right\} l ds$$

$$= \left( \frac{1}{l} \int_{s=0}^{s=1} EA ds \right) \lfloor \delta u_1 \quad \delta u_2 \rfloor \begin{bmatrix} 1 & -1 \\ -1 & 1 \end{bmatrix} \left\{ \begin{matrix} u_1 \\ u_2 \end{matrix} \right\}$$

$$\rightarrow \delta U_e = \lfloor \delta u_1 \quad \delta u_2 \rfloor \begin{bmatrix} k_{11}^e & k_{12}^e \\ k_{21}^e & k_{22}^e \end{bmatrix} \left\{ \begin{matrix} u_1 \\ u_2 \end{matrix} \right\} = \lfloor \delta u_1 \quad \delta u_2 \rfloor \mathbf{k}^e \left\{ \begin{matrix} u_1 \\ u_2 \end{matrix} \right\}, \tag{1.6.18}$$

where

$$\mathbf{k}^e = \begin{bmatrix} k_{11}^e & k_{12}^e \\ k_{21}^e & k_{22}^e \end{bmatrix} = \left( \frac{1}{l} \int_{s=0}^{s=1} EA ds \right) \begin{bmatrix} 1 & -1 \\ -1 & 1 \end{bmatrix} : 2 \times 2 \text{ element stiffness matrix.} \tag{1.6.19}$$

In the above equation, superscript "*e*" stands for element number *e*. Here, the convenience enabled by the mapping in Eq. (1.6.5) is clear. In Eq. (1.6.11), the limit of integration is from $x_1$ to $x_2$, which varies from element to element. However, with the use of mapping, integration is carried out from $s = 0$ to $s = 1$ for every element.

For constant $E$ :

$$\int_{s=0}^{s=1} EA ds = E \int_{s=0}^{s=1} A ds = EA_m, \tag{1.6.20}$$

where

$$A_m = \int_{s=0}^{s=1} A ds \tag{1.6.21}$$

and thus

$$\begin{bmatrix} k_{11}^e & k_{12}^e \\ k_{21}^e & k_{22}^e \end{bmatrix} = \frac{EA_m}{l} \begin{bmatrix} 1 & -1 \\ -1 & 1 \end{bmatrix}. \tag{1.6.22}$$

Note that the element stiffness matrix is symmetric. At this point, we may compare the element stiffness matrix in Eq. (1.6.22) with Eq. (1.3.10). We can then observe that, with

$$k^{(e)} = \frac{EA_m}{l}, \tag{1.6.23}$$

the two matrices are identical. This reflects the fact that both the spring and the uniaxial element are structurally equivalent, with two nodal DOF and deformation under axial force.

## Example 1.3

Consider a slender body fixed at the root ($x = 0$). The cross-sectional area along the span is $A = A_0(1 - 0.5x/L)$, where $A_0$ and $L$ are given. The slender body is modeled using three elements of equal length. Construct the stiffness matrix of element #2.

## Solution:

For elements of equal length, $l = \frac{L}{3}$. Then, for element #2, $x_1 = \frac{L}{3}$, $x_2 = \frac{2L}{3}$ and the mapping is

$$x = (1 - s)x_1 + sx_2 = (1 - s)\frac{L}{3} + s\frac{2L}{3} = \frac{L}{3}(1 + s).$$

Using the mapping:

$$A = A_0\left(1 - \frac{x}{2L}\right) = A_0\left(1 - \frac{(1 + s)}{6}\right) = \frac{A_0}{6}(5 - s),$$

$$A_m = \int\limits_{s=0}^{s=1} A\,ds = \frac{A_0}{6} \int\limits_{s=0}^{s=1} (5 - s)\,ds = \frac{3}{4}A_0,$$

$$\mathbf{k}^2 = \frac{EA_m}{l} \begin{bmatrix} 1 & -1 \\ -1 & 1 \end{bmatrix} = \frac{9}{4}\left(\frac{EA_o}{L}\right) \begin{bmatrix} 1 & -1 \\ -1 & 1 \end{bmatrix}.$$

## 1.6.3 Construction of Element Load Vector

The assumed displacement for an element is given in Eq. (1.6.10). Accordingly:

$$\delta u = (1 - s)\delta u_1 + s\delta u_2. \tag{1.6.24}$$

Introducing Eq. (1.6.24) into Eq. (1.6.3) for an element:

$$\int\limits_{x=x_1}^{x=x_2} \delta u\, f\, dx = \int\limits_{s=0}^{s=1} [(1 - s)\delta u_1 + s\delta u_2] f l\, ds = \delta u_1 \int\limits_{s=0}^{s=1} (1 - s) f l\, ds + \delta u_2 \int\limits_{s=0}^{s=1} s f l\, ds. \tag{1.6.25}$$

The above equation can be expressed as

$$\int_{x=x_1}^{x=x_2} \delta u\, f dx = \delta u_1 Q_1^e + \delta u_2 Q_2^e, \tag{1.6.26}$$

where

$$Q_1^e = l \int_{s=0}^{s=1} (1-s) f ds, \quad Q_2^e = l \int_{s=0}^{s=1} s f ds. \tag{1.6.27}$$

Note that the FE formulation transforms the applied load distributed over the element into concentrated forces located at the nodes. In matrix form:

$$\int_{x=x_1}^{x=x_2} \delta u\, f dx = \lfloor \delta u_1 \quad \delta u_2 \rfloor \begin{Bmatrix} Q_1^e \\ Q_2^e \end{Bmatrix}, \tag{1.6.28}$$

where

$$\begin{Bmatrix} Q_1^e \\ Q_2^e \end{Bmatrix} : 2 \times 1 \text{ element nodal load vector due to } f.$$

One may consider $\delta u$, which is constant ($\delta u = c$, $\delta u_1 = \delta u_2 = c$) over the element. Then, from Eq. (1.6.26):

$$c \int_{x=x_1}^{x=x_2} f dx = c(Q_1^e + Q_2^e) \rightarrow \int_{x=x_1}^{x=x_2} f dx = Q_1^e + Q_2^e, \tag{1.6.29}$$

which shows that the sum of nodal loads is equal to the total force applied to the element.

## Example 1.4

Applying Eq. (1.6.27) for constant $f = c$ :

$$Q_1^e = l \int_{s=0}^{s=1} c(1-s)\, ds = cl \int_{s=0}^{s=1} (1-s)\, ds = \frac{cl}{2},$$

$$Q_2^e = l \int_{s=0}^{s=1} cs\, ds = cl \int_{s=0}^{s=1} s\, ds = \frac{cl}{2}.$$

Figure 1.13 shows the two nodal loads which have resulted from the FE formulation. Summing the nodal loads:

$$Q_1^e + Q_2^e = \frac{cl}{2} + \frac{cl}{2} = cl : \text{total load due to } f = c \text{ for element } i.$$

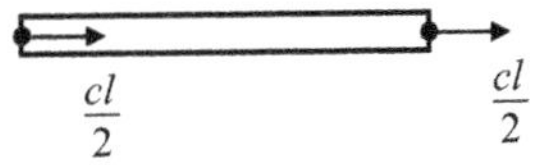

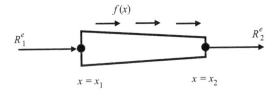

**Figure 1.13** Nodal loads due to $f = c$

### 1.6.4 Equilibrium Equation for an Element

Consider an element $e$ in static equilibrium as shown in Figure 1.14. In the figure:

$f$ : applied load per unit length
$R_1^e, R_2^e$ : reaction forces or applied forces at nodes 1 and 2.

Note that for element #1, $R_1^1 = R_o$, while for element #3, $R_2^3 = P$ for the three-element model of the problem described in Figures 1.6 and 1.9. Reaction forces at the element boundaries cancel out when elements are joined.

**Figure 1.14** Forces acting on an element

$f(x)$

$R_1^e$

$R_2^e$

$x = x_1$

$x = x_2$

Treating $R_1^e$ and $R_2^e$ as the "external forces":

$$\delta W_e = \int_{x=x_1}^{x=x_2} \delta u f dx + R_1^e \delta u_1 + R_2^e \delta u_2. \tag{1.6.30}$$

Introducing Eq. (1.6.26) into the above equation:

$$\delta W_e = \left(Q_1^e + R_1^e\right)\delta u_1 + \left(Q_2^e + R_2^e\right)\delta u_2 = P_1^e \delta u_1 + P_2^e \delta u_2 = \lfloor \delta u_1 \quad \delta u_2 \rfloor \begin{Bmatrix} P_1^e \\ P_2^e \end{Bmatrix}, \tag{1.6.31}$$

where

$$\begin{Bmatrix} P_1^e \\ P_2^e \end{Bmatrix} = \begin{Bmatrix} Q_1^e + R_1^e \\ Q_2^e + R_2^e \end{Bmatrix} : \text{element load vector.} \tag{1.6.32}$$

For an element in equilibrium:

$$\delta U_e - \delta W_e = \lfloor \delta u_1 \quad \delta u_2 \rfloor \left( \begin{bmatrix} k_{11}^e & k_{12}^e \\ k_{21}^e & k_{22}^e \end{bmatrix} \begin{Bmatrix} u_1 \\ u_2 \end{Bmatrix} - \begin{Bmatrix} P_1^e \\ P_2^e \end{Bmatrix} \right) = 0. \tag{1.6.33}$$

Accordingly:

$$\begin{bmatrix} k_{11}^e & k_{12}^e \\ k_{21}^e & k_{22}^e \end{bmatrix} \begin{Bmatrix} u_1 \\ u_2 \end{Bmatrix} - \begin{Bmatrix} P_1^e \\ P_2^e \end{Bmatrix} = \begin{Bmatrix} 0 \\ 0 \end{Bmatrix} \rightarrow \begin{bmatrix} k_{11}^e & k_{12}^e \\ k_{21}^e & k_{22}^e \end{bmatrix} \begin{Bmatrix} u_1 \\ u_2 \end{Bmatrix} = \begin{Bmatrix} P_1^e \\ P_2^e \end{Bmatrix}. \tag{1.6.34}$$

The above equation shows equilibrium between the elastic force and the "external" force at each node.

## 1.6.5  Finite Element Equilibrium Equation for Entire Structure

For the entire structure we need to introduce a global node numbering and thus a global DOF numbering system as shown in Figure 1.15 for the three-element model. The FE equilibrium equation for the entire structure can then be constructed from individual elements, utilizing the relationship between the element DOF and the global DOF. In the figure, 1, 2, 3, 4 are global node numbers, $q_1$, $q_2$, $q_3$, $q_4$ are global nodal DOF, and $F_1$, $F_2$, $F_3$, $F_4$ are global nodal loads.

**Figure 1.15** Global nodal DOF and global nodal loads

For the three-element model, we note that the relationship or connectivity between element DOF and global DOF is identical to that for the system of three springs discussed in Section 1.3. Accordingly, the element stiffness matrices and load vectors are assembled in the same manner into the global stiffness matrix and global load vector to construct the equilibrium equation for the entire structure as follows:

$$\begin{bmatrix} k_{11}^1 & k_{12}^1 & 0 & 0 \\ k_{21}^1 & k_{22}^1 + k_{11}^2 & k_{12}^2 & 0 \\ 0 & k_{21}^2 & k_{22}^2 + k_{11}^3 & k_{12}^3 \\ 0 & 0 & k_{21}^3 & k_{22}^3 \end{bmatrix} \begin{Bmatrix} q_1 \\ q_2 \\ q_3 \\ q_4 \end{Bmatrix} = \begin{Bmatrix} Q_1^1 + R_0 \\ Q_2^1 + Q_1^2 \\ Q_2^2 + Q_1^3 \\ Q_2^3 + P \end{Bmatrix} \rightarrow \mathbf{Kq} = \mathbf{F}, \qquad (1.6.35)$$

where

$$\mathbf{K} = \begin{bmatrix} k_{11}^1 & k_{12}^1 & 0 & 0 \\ k_{21}^1 & k_{22}^1 + k_{11}^2 & k_{12}^2 & 0 \\ 0 & k_{21}^2 & k_{22}^2 + k_{11}^3 & k_{12}^3 \\ 0 & 0 & k_{21}^3 & k_{22}^3 \end{bmatrix} : 4 \times 4 \text{ global stiffness matrix}, \qquad (1.6.36)$$

$$\mathbf{q} = \begin{Bmatrix} q_1 \\ q_2 \\ q_3 \\ q_4 \end{Bmatrix} : 4 \times 1 \text{ global DOF vector}, \qquad (1.6.37)$$

$$\mathbf{F} = \begin{Bmatrix} Q_1^1 + R_0 \\ Q_2^1 + Q_1^2 \\ Q_2^2 + Q_1^3 \\ Q_2^3 + P \end{Bmatrix} : 4 \times 1 \text{ global load vector,} \qquad (1.6.38)$$

where $R_0$ is the reaction force at $x = 0$ and $P$ is the applied load at $x = L$. Symbolically, the above equation can be written as

$$\begin{bmatrix} K_{11} & K_{12} & K_{13} & K_{14} \\ K_{21} & K_{22} & K_{23} & K_{24} \\ K_{31} & K_{32} & K_{33} & K_{34} \\ K_{41} & K_{42} & K_{43} & K_{44} \end{bmatrix} \begin{Bmatrix} q_1 \\ q_2 \\ q_3 \\ q_4 \end{Bmatrix} = \begin{Bmatrix} F_1 \\ F_2 \\ F_3 \\ F_4 \end{Bmatrix}. \qquad (1.6.39)$$

To complete the problem definition, we apply the geometric boundary condition at $x = 0$ and set $q_1 = 0$ in the above equation. We then delete the first equation with the reaction term in the load vector, corresponding to $q_1 = 0$, to match the three unknowns with the three equations as follows:

$$\begin{bmatrix} K_{22} & K_{23} & K_{24} \\ K_{32} & K_{33} & K_{34} \\ K_{42} & K_{43} & K_{44} \end{bmatrix} \begin{Bmatrix} q_2 \\ q_3 \\ q_4 \end{Bmatrix} = \begin{Bmatrix} F_2 \\ F_3 \\ F_4 \end{Bmatrix}. \qquad (1.6.40)$$

The above equation of reduced size can then be solved for $q_2, q_3,$ and $q_4$.

 **Example 1.5**

To appreciate the effectiveness of the finite element method, let us try the FE solution of the rotating slender body for which the exact solution was obtained in Example 1.2. We will then compare the FE solution with the exact solution. The slender body is modeled with three elements of equal length. In this exercise, we will do as follows:

(a) Assemble the global stiffness matrix.
(b) Construct the nodal load vector of each element due to the centrifugal force.
(c) Assemble the global load vector.
(d) Determine the unknown nodal displacements or degrees of freedom.
(e) Determine the reaction force at $x = 0$.
(f) Plot the non-dimensional axial displacement vs. $x/L$ and compare the FE solution with the exact solution. According to the solution given in Example 1.2, the maximum exact displacement occurs at $x = L$. Thus, we may non-dimensionalize the displacement as follows:

$$\bar{u} = \frac{u}{(\text{exact } u)_{x=L}} = \frac{u}{\left( \dfrac{\rho \Omega^2 L^3}{3E} \right)}.$$

(g) Determine the axial stress in each element.

(h) Plot the non-dimensional axial stress vs. $x/L$ and compare the FE solution with the exact solution. According to the solution given at the end of Example 1.2, the maximum exact stress occurs at $x = 0$. We may then non-dimensionalize the axial stress as follows:

$$\bar{\sigma}_{xx} = \frac{\sigma_{xx}}{(\text{exact } \sigma_{xx})_{x=0}} = \frac{\sigma_{xx}}{\left(\dfrac{\rho\Omega^2 L^2}{2}\right)}.$$

## Solution:

(a) For constant $A$:

$$A_m = \int_{s=0}^{s=1} A\,ds = A.$$

For element $e$, with $l = L/3$ :

$$\mathbf{k}^e = \frac{EA}{l}\begin{bmatrix} 1 & -1 \\ -1 & 1 \end{bmatrix} = \frac{3EA}{L}\begin{bmatrix} 1 & -1 \\ -1 & 1 \end{bmatrix}.$$

After assembly, the global stiffness matrix is

$$\mathbf{K} = \frac{3EA}{L}\begin{bmatrix} 1 & -1 & 0 & 0 \\ -1 & 1+1 & -1 & 0 \\ 0 & -1 & 1+1 & -1 \\ 0 & 0 & -1 & 1 \end{bmatrix} = \frac{3EA}{L}\begin{bmatrix} 1 & -1 & 0 & 0 \\ -1 & 2 & -1 & 0 \\ 0 & -1 & 2 & -1 \\ 0 & 0 & -1 & 1 \end{bmatrix}.$$

(b) Element load vectors due to the centrifugal force are determined using Eq. (1.6.27) as follows.

Element #1: The nodal coordinates are $x_1 = 0$, $x_2 = L/3$. Accordingly, the mapping is

$$x = (1 - s)x_1 + sx_2 = \frac{L}{3}s.$$

Using the mapping, the applied force per unit length can be expressed in terms of the non-dimensional coordinate $s$ as

$$f = cx = c\frac{L}{3}s, \text{where } c = m\Omega^2.$$

Then, according to Eq. (1.6.27), the nodal loads for the element are

$$Q_1^1 = l \int_{s=0}^{s=1} (1-s) f ds = c\left(\frac{L}{3}\right)^2 \int_{s=0}^{s=1} (1-s) s ds = \frac{1}{54} c L^2,$$

$$Q_2^1 = l \int_{s=0}^{s=1} s f ds = c\left(\frac{L}{3}\right)^2 \int_{s=0}^{s=1} s^2 ds = \frac{1}{27} c L^2.$$

Element #2: The nodal coordinates are $x_1 = L/3, x_2 = 2L/3,$ and the mapping is

$$x = (1-s)x_1 + s x_2 = (1-s)\frac{L}{3} + \frac{2L}{3}s = \frac{L}{3}(1+s) \rightarrow f = cx = c\frac{L}{3}(1+s).$$

Accordingly, the nodal loads for the element are

$$Q_1^2 = l \int_{s=0}^{s=1} (1-s) f ds = c\left(\frac{L}{3}\right)^2 \int_{s=0}^{s=1} (1-s)(1+s) ds = \frac{2}{27} c L^2,$$

$$Q_2^2 = l \int_{s=0}^{s=1} s f ds = c\left(\frac{L}{3}\right)^2 \int_{s=0}^{s=1} s(1+s) ds = \frac{5}{54} c L^2.$$

Element #3: The nodal coordinates are $x_1 = \frac{2L}{3}, x_2 = L,$ and the mapping is

$$x = (1-s)x_1 + s x_2 = (1-s)\frac{2L}{3} + Ls = \frac{L}{3}(2+s) \rightarrow f = cx = c\frac{L}{3}(2+s).$$

Accordingly, the nodal loads for the element are

$$Q_1^3 = l \int_{s=0}^{s=1} (1-s) f ds = c\left(\frac{L}{3}\right)^2 \int_{s=0}^{s=1} (1-s)(2+s) ds = \frac{7}{54} c L^2,$$

$$Q_2^3 = l \int_{s=0}^{s=1} s f ds = c\left(\frac{L}{3}\right)^2 \int_{s=0}^{s=1} s(2+s) ds = \frac{8}{54} c L^2.$$

(c) Assembling load vectors and reaction force, the global load vector is

$$\mathbf{F} = \frac{cL^2}{54} \left\{ \begin{array}{c} 1 + 54R_0/(cL^2) \\ 6 \\ 12 \\ 8 \end{array} \right\}.$$

(d) With the global stiffness matrix assembled in part (a) and the global load vector assembled in part (c), the equilibrium equation is

$$\frac{3EA}{L}\begin{bmatrix} 1 & -1 & 0 & 0 \\ -1 & 2 & -1 & 0 \\ 0 & -1 & 2 & -1 \\ 0 & 0 & -1 & 1 \end{bmatrix}\begin{Bmatrix} q_1 \\ q_2 \\ q_3 \\ q_4 \end{Bmatrix} = \frac{cL^2}{54}\begin{Bmatrix} 1 + 54R_0/(cL^2) \\ 6 \\ 12 \\ 8 \end{Bmatrix}.$$

Setting $q_1 = 0$ to apply the geometric boundary condition and deleting the first equation with the reaction term in the load vector, the discretized equilibrium equation is

$$\frac{3EA}{L}\begin{bmatrix} 2 & -1 & 0 \\ -1 & 2 & -1 \\ 0 & -1 & 1 \end{bmatrix}\begin{Bmatrix} q_2 \\ q_3 \\ q_4 \end{Bmatrix} = \frac{cL^2}{54}\begin{Bmatrix} 6 \\ 12 \\ 8 \end{Bmatrix}$$

or

$$\begin{bmatrix} 2 & -1 & 0 \\ -1 & 2 & -1 \\ 0 & -1 & 1 \end{bmatrix}\begin{Bmatrix} q_2 \\ q_3 \\ q_4 \end{Bmatrix} = \frac{cL^2}{162EA}\begin{Bmatrix} 6 \\ 12 \\ 8 \end{Bmatrix}.$$

Solving the above equation:

$$\begin{Bmatrix} q_2 \\ q_3 \\ q_4 \end{Bmatrix} = \frac{cL^3}{81EA}\begin{Bmatrix} 13 \\ 23 \\ 27 \end{Bmatrix} = \frac{m\Omega^2L^3}{81EA}\begin{Bmatrix} 13 \\ 23 \\ 27 \end{Bmatrix} = \frac{\rho A\Omega^2L^3}{81EA}\begin{Bmatrix} 13 \\ 23 \\ 27 \end{Bmatrix} = \frac{\rho\Omega^2L^3}{81E}\begin{Bmatrix} 13 \\ 23 \\ 27 \end{Bmatrix}.$$

(e) To determine the reaction force, we go back to the original equilibrium equation in part (d) involving all four DOF:

$$\frac{3EA}{L}\begin{bmatrix} 1 & -1 & 0 & 0 \\ -1 & 2 & -1 & 0 \\ 0 & -1 & 2 & -1 \\ 0 & 0 & -1 & 1 \end{bmatrix}\begin{Bmatrix} q_1 = 0 \\ q_2 \\ q_3 \\ q_4 \end{Bmatrix} = \frac{cL^2}{54}\begin{Bmatrix} 1 + 54R_0/(cL^2) \\ 6 \\ 12 \\ 8 \end{Bmatrix}.$$

Multiplying the first row of the stiffness matrix by the column vector of displacements:

$$\frac{3EA}{L}(-q_2) = \frac{cL^2}{54} + R_0 \rightarrow R_0 = \frac{3EA}{L}(-q_2) - \frac{cL^2}{54}.$$

Substituting $q_2$ obtained in part (d) into the above equation:

$$R_0 = \frac{3EA}{L}\left(-\frac{13cL^3}{81EA}\right) - \frac{cL^2}{54} = -\frac{1}{2}cL^2 = -\frac{1}{2}m\Omega^2L^2 = -\frac{1}{2}\rho A\Omega^2L^2.$$

The reaction force at $x = 0$ balances the total centrifugal force.

(f) The plot in Figure 1.16 shows that the exact displacement is approximated with the three straight-line segments. For this simple problem, the nodal displacements obtained by the three-element model match with the exact solution at the same location. However, there exist discrepancies between the three-element solution and the exact solution within each element. One may try a model with more elements to observe that these discrepancies get smaller as the number of elements increases.

(g) To calculate axial stress, recall that $\sigma_{xx} = E\varepsilon_{xx} = E\frac{\partial u}{\partial x}$.

For element #$e$, where $u_1$ and $u_2$ are the displacements of the nodes in the element, $\sigma^e_{xx} = \frac{E}{l}(u_2 - u_1) = \frac{3E}{L}(u_2 - u_1)$.

Note that the stress is constant within an element. Using the above equation:

$$\sigma^1_{xx} = \frac{3E}{L}(u_2 - u_1) = \frac{3E}{L}(q_2 - q_1) = \frac{13}{27}\rho\Omega^2 L^2,$$

$$\sigma^2_{xx} = \frac{3E}{L}(u_2 - u_1) = \frac{3E}{L}(q_3 - q_2) = \frac{10}{27}\rho\Omega^2 L^2,$$

$$\sigma^3_{xx} = \frac{3E}{L}(u_2 - u_1) = \frac{3E}{L}(q_4 - q_3) = \frac{4}{27}\rho\Omega^2 L^2.$$

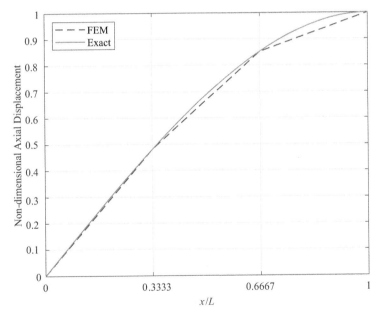

**Figure 1.16** Non-dimensional displacement of a rotating body: three-element model

(h) As shown in Figure 1.17, the FE stress solutions are not as accurate as the displacement solutions. This is because displacement is the primary variable which we assumed to be linear in each element, and it is necessary to take a derivative of the assumed displacement

**Figure 1.17** Non-dimensional stress of a rotating body: three-element model

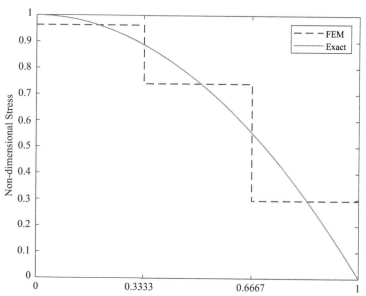

to determine stress in each element. We also note that axial stresses obtained from the three-element model are quite close to exact stresses at the element centroids. For this reason, the element centroid of the two-node element is called the "superconvergent stress point."

## 1.7   Additional Topics

### (1) Properties of Element Stiffness Matrix

For the two-node element:

$$\delta U_e = \lfloor \delta u_1 \quad \delta u_2 \rfloor \begin{bmatrix} k_{11}^e & k_{12}^e \\ k_{21}^e & k_{22}^e \end{bmatrix} \begin{Bmatrix} u_1 \\ u_2 \end{Bmatrix}, \tag{1.7.1}$$

where

$$\begin{bmatrix} k_{11}^e & k_{12}^e \\ k_{21}^e & k_{22}^e \end{bmatrix} \begin{Bmatrix} u_1 \\ u_2 \end{Bmatrix} : \text{nodal elastic force vector.} \tag{1.7.2}$$

Consider an element undergoing rigid-body translation  with $u = c$ where $c$ is a constant value. Then, substituting $u_1 = u_2 = c$ into the nodal elastic force vector:

$$\begin{bmatrix} k_{11}^e & k_{12}^e \\ k_{21}^e & k_{22}^e \end{bmatrix} \begin{Bmatrix} u_1 \\ u_2 \end{Bmatrix} = \begin{bmatrix} k_{11}^e & k_{12}^e \\ k_{21}^e & k_{22}^e \end{bmatrix} \begin{Bmatrix} c \\ c \end{Bmatrix} = c \begin{bmatrix} k_{11}^e & k_{12}^e \\ k_{21}^e & k_{22}^e \end{bmatrix} \begin{Bmatrix} 1 \\ 1 \end{Bmatrix}. \tag{1.7.3}$$

A rigid-body translation does not produce any elastic force in the element. Accordingly:

$$c \begin{bmatrix} k^e_{11} & k^e_{12} \\ k^e_{21} & k^e_{22} \end{bmatrix} \begin{Bmatrix} 1 \\ 1 \end{Bmatrix} = \begin{Bmatrix} 0 \\ 0 \end{Bmatrix} \rightarrow \begin{Bmatrix} k^e_{11} + k^e_{12} = 0 \\ k^e_{21} + k^e_{22} = 0 \end{Bmatrix}. \tag{1.7.4}$$

The sum of the entries in each row of the element stiffness matrix is equal to zero, which can be confirmed for the two-node element with the element stiffness matrix in Eq. (1.6.22).

## (2) Physical Meaning of the Individual Entries in the Stiffness Matrix

Consider again a three-element model of a slender body under axial loads as shown in Figure 1.15. For the three-element model:

$$\begin{bmatrix} K_{11} & K_{12} & K_{13} & K_{14} \\ K_{21} & K_{22} & K_{23} & K_{24} \\ K_{31} & K_{32} & K_{33} & K_{34} \\ K_{41} & K_{42} & K_{43} & K_{44} \end{bmatrix} \begin{Bmatrix} q_1 \\ q_2 \\ q_3 \\ q_4 \end{Bmatrix} = \begin{Bmatrix} F_1 \\ F_2 \\ F_3 \\ F_4 \end{Bmatrix}. \tag{1.7.5}$$

Now consider the following situation:

$$q_1 \neq 0, q_2 = q_3 = q_4 = 0. \tag{1.7.6}$$

Placing Eq. (1.7.6) into Eq. (1.7.5):

$$\begin{aligned} K_{11}q_1 &= F_1 & K_{11} &= F_1/q_1 \\ K_{21}q_1 &= F_2 & K_{21} &= F_2/q_1 \\ K_{31}q_1 &= F_3 & \rightarrow & K_{31} &= F_3/q_1 \\ K_{41}q_1 &= F_4 & K_{41} &= F_4/q_1 \end{aligned} \tag{1.7.7}$$

We observe that $K_{i1}$ $(i = 1, 2, 3, 4)$ is the nodal force at node $i$ due to a unit displacement at node 1 while all other nodal displacements are equal to zero. In general, $K_{ij}$ is the nodal force at node $i$ due to a unit displacement at node $j$ while keeping all other nodal displacements equal to zero.

## (3) Formulation using Strain Energy Expression

From Eqs (1.5.21) and (1.5.22), the strain energy can be expressed in terms of axial displacement as

$$U = \frac{1}{2} \int_{x=0}^{x=L} EA\varepsilon^2_{xx} dx = \frac{1}{2} \int_{x=0}^{x=L} EA \left( \frac{\partial u}{\partial x} \right)^2 dx. \tag{1.7.8}$$

For an element, recall that the axial strain is

$$\varepsilon_{xx} = \frac{1}{l} \lfloor -1 \quad 1 \rfloor \begin{Bmatrix} u_1 \\ u_2 \end{Bmatrix} \tag{1.7.9}$$

or

$$\varepsilon_{xx} = \frac{1}{l} \lfloor u_1 \quad u_2 \rfloor \begin{Bmatrix} -1 \\ 1 \end{Bmatrix}. \tag{1.7.10}$$

Then

$$\varepsilon_{xx}^2 = \frac{1}{l^2} \lfloor u_1 \quad u_2 \rfloor \begin{Bmatrix} -1 \\ 1 \end{Bmatrix} \lfloor -1 \quad 1 \rfloor \begin{Bmatrix} u_1 \\ u_2 \end{Bmatrix} = \frac{1}{l^2} \lfloor u_1 \quad u_2 \rfloor \begin{bmatrix} 1 & -1 \\ -1 & 1 \end{bmatrix} \begin{Bmatrix} u_1 \\ u_2 \end{Bmatrix}. \tag{1.7.11}$$

Using the above equation, the strain energy for the element can be expressed as

$$
\begin{aligned}
U_e &= \frac{1}{2} \int_{x=x_1}^{x=x_2} EA\varepsilon_{xx}^2 \, dx = \frac{1}{2} \int_{s=0}^{s=1} EA \frac{1}{l^2} \lfloor u_1 \quad u_2 \rfloor \begin{bmatrix} 1 & -1 \\ -1 & 1 \end{bmatrix} \begin{Bmatrix} u_1 \\ u_2 \end{Bmatrix} l\,ds \\
&= \frac{1}{2} \left\{ \frac{1}{l} \int_{s=0}^{s=1} EA\,ds \right\} \lfloor u_1 \quad u_2 \rfloor \begin{bmatrix} 1 & -1 \\ -1 & 1 \end{bmatrix} \begin{Bmatrix} u_1 \\ u_2 \end{Bmatrix} = \frac{1}{2} \lfloor u_1 \quad u_2 \rfloor \begin{bmatrix} k_{11}^e & k_{12}^e \\ k_{21}^e & k_{22}^e \end{bmatrix} \begin{Bmatrix} u_1 \\ u_2 \end{Bmatrix},
\end{aligned} \tag{1.7.12}
$$

where, for constant $E$:

$$\begin{bmatrix} k_{11}^e & k_{12}^e \\ k_{21}^e & k_{22}^e \end{bmatrix} = \frac{EA_m}{l} \begin{bmatrix} 1 & -1 \\ -1 & 1 \end{bmatrix}, \quad A_m = \int_{s=0}^{s=1} A\,ds \tag{1.7.13}$$

is the element stiffness matrix which is identical to that in Eq. (1.6.22). Summing over all elements:

$$U = \sum_e U_e = \frac{1}{2} \mathbf{q}^{\mathrm{T}} \mathbf{K} \mathbf{q}, \tag{1.7.14}$$

where the global stiffness matrix, assembled using the connectivity between element DOF and global DOF as previously described, is identical to that in Eq. (1.6.36). Then, noting that the transpose of a scalar is the scalar itself and the global stiffness matrix is symmetric:

$$
\begin{aligned}
\delta U &= \frac{1}{2} \delta\mathbf{q}^{\mathrm{T}} \mathbf{K}\mathbf{q} + \frac{1}{2} \mathbf{q}^{\mathrm{T}} \mathbf{K} \delta\mathbf{q} = \frac{1}{2} \delta\mathbf{q}^{\mathrm{T}} \mathbf{K}\mathbf{q} + \frac{1}{2} \left( \mathbf{q}^{\mathrm{T}} \mathbf{K} \delta\mathbf{q} \right)^{\mathrm{T}} \\
&= \frac{1}{2} \delta\mathbf{q}^{\mathrm{T}} \mathbf{K}\mathbf{q} + \frac{1}{2} \delta\mathbf{q}^{\mathrm{T}} \mathbf{K}^{\mathrm{T}}\mathbf{q} = \frac{1}{2} \delta\mathbf{q}^{\mathrm{T}} \mathbf{K}\mathbf{q} + \frac{1}{2} \delta\mathbf{q}^{\mathrm{T}} \mathbf{K}\mathbf{q} = \delta\mathbf{q}^{\mathrm{T}} \mathbf{K}\mathbf{q}.
\end{aligned} \tag{1.7.15}
$$

Alternatively, we can take increments at the element level. Then, from Eq. (1.7.12):

$$\delta U_e = \lfloor \delta u_1 \quad \delta u_2 \rfloor \begin{bmatrix} k_{11}^e & k_{12}^e \\ k_{21}^e & k_{22}^e \end{bmatrix} \begin{Bmatrix} u_1 \\ u_2 \end{Bmatrix} \tag{1.7.16}$$

The equation above is identical to Eq. (1.6.18).

## (4) Isoparametric Formulation

For the two-node element introduced in Section 1.6.1, the mapping is

$$x = (1 - s)x_1 + sx_2 = N_1 x_1 + N_2 x_2, \qquad (1.7.17)$$

where

$$N_1 = 1 - s, N_2 = s \qquad (1.7.18)$$

are called the mapping functions. The assumed displacement is

$$u = (1 - s)u_1 + su_2 = N_1 u_1 + N_2 u_2, \qquad (1.7.19)$$

where

$$N_1 = 1 - s, N_2 = s \qquad (1.7.20)$$

are called the shape functions. For the two-node element, the shape functions and the mapping functions are identical, which is called the "isoparametric formulation."

## (5) Higher-Order Elements

Throughout this chapter, we have considered the two-node element with linear displacement. However, we could construct an element in which the assumed displacement is a polynomial function of quadratic, cubic, or even higher order. For example, we may consider using the three-node element in which the displacement is assumed to be quadratic such that

$$u = N_1 u_1 + N_2 u_2 + N_3 u_3 = \sum_{i=1}^{3} N_i u_i, \qquad (1.7.21)$$

where $u_1$, $u_2$, $u_3$ are nodal displacements or DOF and $N_1$, $N_2$, $N_3$ are quadratic shape functions. The detailed derivations of the mapping and shape functions, element stiffness matrix, and load vector for the three-node element are left as a problem at the end of this chapter. In general, the assumed displacement of an element with $n$ nodes can be expressed as

$$u = \sum_{i=1}^{n} N_i u_i. \qquad (1.7.22)$$

## (6) Property of the Shape Functions

Consider now an element undergoing rigid-body translation with $u = c$ where $c$ is a constant value. Then, substituting $u_i = c$ into Eq. (1.7.22):

$$u = \sum_{i=1}^{n} N_i u_i \rightarrow c = c \sum_{i=1}^{n} N_i \rightarrow \sum_{i=1}^{n} N_i = 1. \qquad (1.7.23)$$

The sum of the shape functions is equal to 1, which can easily be confirmed for the two-node element.

## (7) Slender Body of Uniform Cross-section Subjected to Tip Force

Consider a slender body of uniform cross-section fixed at $x = 0$ and subjected to a tip force of $P$ at $x = L$, as shown in Figure 1.18. There is no other applied force.

**Figure 1.18** Slender body under a tip force

Following the approach described in Section 1.4, it can be shown that

$$u = \frac{Px}{EA}, \quad \sigma_{xx} = \frac{P}{A}. \tag{1.7.24}$$

The assumed displacement is linear for a two-node element. Accordingly, the FE model with one element produces exact solutions for axial displacement and stress as shown in the above equation.

## PROBLEMS

**1.1** Consider a slender column standing vertically under its own weight as shown in Figure 1.19. The column is of uniform cross-sectional area and is fixed against axial displacement at $x = 0$ and free at $x = L$ Note that $fdx = -(\rho Adx)g$ where $\rho$ is the mass density per volume and $g$ is the gravity constant. The geometric and material data are as follows: $L = 6$ m, $A = 0.05$ m$^2$, $E = 200$ GPa, $\rho = 8,000$ kg/m$^3$, $g = 9.81$ m/s$^2$.

**Figure 1.19** For Problem 1.1

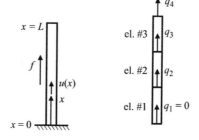

Following the procedure described in Section 1.4, the exact solutions for the axial displacement and axial stress can be determined as

$$u = -\frac{\rho g L^2}{2E}\left[2\left(\frac{x}{L}\right) - \left(\frac{x}{L}\right)^2\right], \quad \sigma_{xx} = -\rho g L\left(1 - \frac{x}{L}\right).$$

The column is modeled with three two-node elements of equal length. Do the following:
(a) Assemble the global stiffness matrix.

(b) Assemble the global load vector.

(c) Determine the unknown nodal displacements.

(d) Determine the reaction force at $x = 0$ following the procedure described in Example 1.5.

(e) Determine the axial stress in each element.

(f) Define the non-dimensional axial displacement as

$$\bar{u} = \frac{u}{(\text{exact } u)_{x=L}} = \frac{u}{\left(-\dfrac{\rho g L^2}{2E}\right)}.$$

Plot the non-dimensional axial displacement vs. $x/L$. Compare with the exact solution.

(g) Define the non-dimensional axial stress as

$$\bar{\sigma} = \frac{\sigma_{xx}}{(\text{exact } \sigma_{xx})_{x=0}} = \frac{\sigma_{xx}}{(-\rho g L)}.$$

Plot the non-dimensional axial stress vs. $x/L$. Compare with the exact solution.

**1.2** Repeat Problem 1.1 using a model with six two-node elements of equal length.

**1.3** A slender column of uniform cross-section is standing vertically under its own weight. The column is fixed against axial displacement at $x = 0$ and $x = L$, as shown in Figure 1.20. The column is modeled with three two-node elements of equal length. The geometric and material data are identical to those given in Problem 1.1.

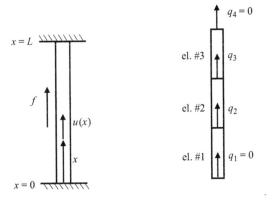

**Figure 1.20** For Problem 1.3

Do the following:

(a) Determine the unknown nodal displacements.

(b) Determine the reaction force at $x = 0$ and $x = L$.

(c) Determine the axial stress in each element.

(d) Define the non-dimensional axial displacement as

$$\bar{u} = \frac{u}{(\text{exact } u)_{x=L/2}}.$$

Plot the non-dimensional axial displacement vs. $x/L$. Compare with the exact solution.

(e) Define the non-dimensional axial stress as

$$\bar{\sigma}_{xx} = \frac{\sigma_{xx}}{(\text{exact } \sigma_{xx})_{x=L}}.$$

Plot the non-dimensional axial stress vs. $x/L$. Compare with the exact solution.

*Note:* Following the procedure described in Section 1.4, the exact solutions for the axial displacement and axial stress can be determined as

$$u = -\frac{\rho g}{2E}\left(Lx - x^2\right), \quad \sigma_{xx} = \rho g\left(x - \frac{L}{2}\right).$$

**1.4** Repeat Problem 1.3 using a model with six elements of equal length.

**1.5** Repeat the problem of the rotating body described in Example 1.5 using a model with six elements of equal length.

**1.6** Figure 1.21 shows a slender body of length $L$ constrained against axial displacement at $x = 0$ and subjected to a force $P$ applied at $x = L$. The cross-sectional area along the length is $A = A_0(1 - 0.5x/L)$ where $A_0$ is given. The geometric and material data are as follows: $L = 7$ m, $A_0 = 0.1$ m², $E = 72$ GPa. Do the following using a model with four two-node uniaxial elements of equal length:

(a) Determine the $2 \times 2$ stiffness matrix for each element.

(b) Construct the global stiffness matrix.

(c) Determine the nodal displacements. Compare with the exact solution.

(d) Determine the axial stress in the element. Compare with the exact stress at the element centroid.

**Figure 1.21** For Problem 1.6

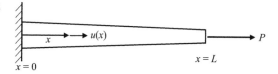

**1.7** Consider a slender column standing vertically under its own weight as shown in Figure 1.19. The column is fixed against axial displacement at $x = 0$ and is free at $x = L$ The cross-sectional area along the body length is $A = A_0(1 - 0.5x/L)$ The geometric and material data are $L = 6$ m, $A_0 = 0.1$ m², $E = 200$ GPa, $\rho = 8{,}000\,\text{kg/m}^3$.

The column is modeled with three two-node elements of equal length:

(a) Determine the $2 \times 2$ stiffness matrix for each element.

(b) Determine the $2 \times 1$ load vector due to gravity for each element.

(c) Construct the global stiffness matrix and the global load vector.

(d) Determine the nodal displacements.

(e) Determine the reaction force at the bottom.

(f) Determine the axial stress in each element.

**1.8** Repeat Problem 1.7 with a model with six two-node elements of equal length.

**1.9** Consider a uniaxial element with three nodes as shown in Figure 1.22. The mapping and assumed displacement can be expressed as

$$x = a_1 + a_2 s + a_3 s^2 \text{ or } x = N_1 x_1 + N_2 x_2 + N_3 x_3,$$

$$u = b_1 + b_2 s + b_3 s^2 \text{ or } u = N_1 u_1 + N_2 u_2 + N_3 u_3,$$

where $x_1, x_2, x_3$ are nodal coordinates and $u_1, u_2, u_3$ are nodal displacements.
(a) Express $N_1, N_2, N_3$ as functions of $s$.
(b) Confirm that $N_1 + N_2 + N_3 = 1$.
(c) Set $x_2 = (x_1 + x_3)/2$ and observe that the mapping reduces to be linear.
(d) Determine the $3 \times 3$ element stiffness matrix for constant cross-sectional area.

Note: $k_{11} = \frac{7}{3} \left( \frac{EA}{l} \right)$, $k_{22} = \frac{16}{3} \left( \frac{EA}{l} \right)$.

(e) Determine the element load vector for constant $f = c$

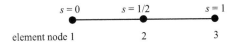

$s = 0$    $s = 1/2$    $s = 1$    **Figure 1.22** For Problem 1.9

element node 1    2    3

**1.10** Consider the rotating slender body described in Section 1.4. The slender body is now modeled with two three-node elements (described in Problem 1.9) of equal length. Do the following:
(a) Construct the global stiffness matrix.
(b) Construct the nodal load vector of each element due to the centrifugal force.
(c) Construct the global load vector.
(d) Determine the unknown nodal displacements.
(e) Determine the axial stress in each element.
(f) Determine the reaction force at $x = 0$.
(g) Plot non-dimensional axial displacement vs. $x/L$. Compare with the exact solution.
(h) Plot non-dimensional axial stress vs. $x/L$. Compare with the exact solution.
**1.11** Repeat Problem 1.10 using a model with four elements of equal length.

# 2 Truss, Temperature Effect, and Torsion

What we have learned in the previous chapter can be extended to model truss structures. A truss is a structure built up from individual slender body members connected at common joints. For simplicity, we may assume that the members are connected through hinge joints which are free to rotate and thus cannot transmit moment. Accordingly, individual members carry only axial tensile or compressive force. For a member of constant cross-section, axial stress is then constant along the member length and thus axial strain is also constant. Within the context of FE formulation, each member is naturally treated as a uniaxial element – as introduced in the previous chapter. However, in order to construct the global stiffness matrix of a truss structure in 3D, it is necessary to construct the element stiffness matrices with 3 DOF at each node, corresponding to three displacement components in the Cartesian coordinate system.

Subsequently, we consider the effect of temperature change on slender bodies and truss structures. When a member or element subjected to temperature change is not free to expand or contract, axial stress can develop in other members as well as the member itself. It can be shown that the effect of temperature change manifests as the element load vector dependent on temperature change and the coefficient of thermal expansion of the member material.

In this chapter we also consider FE modeling of slender bodies undergoing torsional deformation. We will observe that the problem of a slender body undergoing torsional deformation subjected to applied torque is mathematically equivalent to that of the slender body undergoing uniaxial deformation under axially applied loads. This equivalence allows us to apply what we have learned in Chapter 1 to FE modeling of torsional deformation.

## 2.1 Stiffness Matrix of Truss Member

Consider a truss structure with hinge joints. The loads are applied at the joints. For each member, the cross-sectional area is uniform along the length. The hinge joints are free to rotate and thus cannot transmit moment. Accordingly, individual members carry only axial tensile or compressive load.

Each element has its own axis oriented in an arbitrary direction in 3D space as shown in Figure 2.1. In the figure:

$x$, $y$, $z$: global coordinates
$x_1$, $y_1$, $z_1$: coordinates of node 1
$x_2$, $y_2$, $z_2$: coordinates of node 2

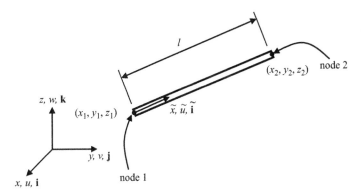

**Figure 2.1** Global coordinate system and the element coordinate system

$u, v, w$: displacements in the $x, y, z$ directions
$\tilde{x}$: coordinate along the member axis
$\tilde{u}$: displacement in the $\tilde{x}$ direction
$l$: element length.

Axial force is related to axial stress in an element or a member as

$$F = \sigma_{\tilde{x}\tilde{x}}A,\tag{2.1.1}$$

where $A$ is the cross-sectional area and

$$\sigma_{\tilde{x}\tilde{x}} = E\varepsilon_{\tilde{x}\tilde{x}}.\tag{2.1.2}$$

For a uniform cross-section, axial stress is constant along the element length and thus axial strain is also constant. Accordingly, from Eq. (1.5.12) in Chapter 1:

$$\delta U_e = \int_e \delta\varepsilon_{\tilde{x}\tilde{x}}\sigma_{\tilde{x}\tilde{x}}Ad\tilde{x}.\tag{2.1.3}$$

Introducing the stress–strain relation:

$$\delta U_e = \int_e EA\delta\varepsilon_{\tilde{x}\tilde{x}}\varepsilon_{\tilde{x}\tilde{x}}d\tilde{x}.\tag{2.1.4}$$

To assemble the element stiffness matrix of individual members or elements into the global stiffness matrix, we need an element with 3 DOF per node as shown in Figure 2.2. In the figure:

$u_1, v_1, w_1$: displacements of node 1
$u_2, v_2, w_2$: displacements of node 2.

Referring to Eq. (1.6.14) in Chapter 1:

$$\varepsilon_{\tilde{x}\tilde{x}} = \frac{1}{l}\lfloor -1 \quad 1\rfloor\begin{Bmatrix} \tilde{u}_1 \\ \tilde{u}_2 \end{Bmatrix}.\tag{2.1.5}$$

**Figure 2.2** Element with 6 DOF in 3D space

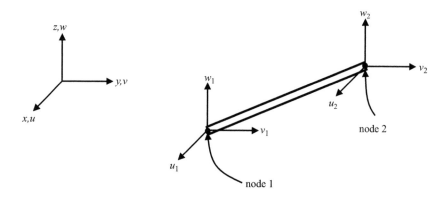

To express axial strain in the above equation in terms of the six nodal DOF, we first express displacement of a point in the body axis as

$$\mathbf{u} = u\mathbf{i} + v\mathbf{j} + w\mathbf{k}. \tag{2.1.6}$$

The displacement in the axial direction is then

$$\tilde{u} = \tilde{\mathbf{i}} \cdot \mathbf{u} = \tilde{\mathbf{i}} \cdot (u\mathbf{i} + v\mathbf{j} + w\mathbf{k}) = u(\tilde{\mathbf{i}} \cdot \mathbf{i}) + v(\tilde{\mathbf{i}} \cdot \mathbf{j}) + w(\tilde{\mathbf{i}} \cdot \mathbf{k}). \tag{2.1.7}$$

The vector drawn from element node 1 to node 2 is

$$l\tilde{\mathbf{i}} = \Delta x \mathbf{i} + \Delta y \mathbf{j} + \Delta z \mathbf{k} \rightarrow \tilde{\mathbf{i}} = \frac{\Delta x}{l}\mathbf{i} + \frac{\Delta y}{l}\mathbf{j} + \frac{\Delta z}{l}\mathbf{k}, \tag{2.1.8}$$

where

$$l = \sqrt{(\Delta x)^2 + (\Delta y)^2 + (\Delta z)^2} : \text{element length}, \tag{2.1.9}$$

$$\Delta x = x_2 - x_1, \quad \Delta y = y_2 - y_1, \quad \Delta z = z_2 - z_1. \tag{2.1.10}$$

From Eq. (2.1.8):

$$\tilde{\mathbf{i}} \cdot \mathbf{i} = \frac{\Delta x}{l}, \tilde{\mathbf{i}} \cdot \mathbf{j} = \frac{\Delta y}{l}, \text{ and } \tilde{\mathbf{i}} \cdot \mathbf{k} = \frac{\Delta z}{l}, \tag{2.1.11}$$

which are the direction cosines. Introducing the above equation into Eq. (2.1.7):

$$\tilde{u} = \frac{\Delta x}{l}u + \frac{\Delta y}{l}v + \frac{\Delta z}{l}w. \tag{2.1.12}$$

Accordingly:

$$\tilde{u}_1 = \frac{\Delta x}{l}u_1 + \frac{\Delta y}{l}v_1 + \frac{\Delta z}{l}w_1, \quad \tilde{u}_2 = \frac{\Delta x}{l}u_2 + \frac{\Delta y}{l}v_2 + \frac{\Delta z}{l}w_2. \tag{2.1.13}$$

Written in matrix form:

$$
\left\{ \begin{array}{c} \tilde{u}_1 \\ \tilde{u}_2 \end{array} \right\} = \frac{1}{l} \begin{bmatrix} \Delta x & \Delta y & \Delta z & 0 & 0 & 0 \\ 0 & 0 & 0 & \Delta x & \Delta y & \Delta z \end{bmatrix} \left\{ \begin{array}{c} u_1 \\ v_1 \\ w_1 \\ u_2 \\ v_2 \\ w_2 \end{array} \right\}.
\tag{2.1.14}
$$

Substituting the above equation into Eq. (2.1.5) and carrying out multiplication:

$$
\varepsilon_{\tilde{x}\tilde{x}} = \frac{1}{l^2} \lfloor -\Delta x \quad -\Delta y \quad -\Delta z \quad \Delta x \quad \Delta y \quad \Delta z \rfloor \left\{ \begin{array}{c} u_1 \\ v_1 \\ w_1 \\ u_2 \\ v_2 \\ w_2 \end{array} \right\} = \frac{1}{l^2} \hat{\mathbf{B}} \mathbf{d},
\tag{2.1.15}
$$

where

$$
\hat{\mathbf{B}} = \lfloor -\Delta x \quad -\Delta y \quad -\Delta z \quad \Delta x \quad \Delta y \quad \Delta z \rfloor,
\tag{2.1.16}
$$

$$
\mathbf{d} = \left\{ \begin{array}{c} u_1 \\ v_1 \\ w_1 \\ u_2 \\ v_2 \\ w_2 \end{array} \right\} : \text{element DOF vector.}
\tag{2.1.17}
$$

Alternatively:

$$
\varepsilon_{\tilde{x}\tilde{x}} = \frac{1}{l^2} \mathbf{d}^{\mathrm{T}} \hat{\mathbf{B}}^{\mathrm{T}}
\tag{2.1.18}
$$

and

$$
\delta\varepsilon_{\tilde{x}\tilde{x}} = \frac{1}{l^2} \delta\mathbf{d}^{\mathrm{T}} \hat{\mathbf{B}}^{\mathrm{T}} = \frac{1}{l^2} \lfloor \delta u_1 \quad \delta v_1 \quad \delta w_1 \quad \delta u_2 \quad \delta v_2 \quad \delta w_2 \rfloor \left\{ \begin{array}{c} -\Delta x \\ -\Delta y \\ -\Delta z \\ \Delta x \\ \Delta y \\ \Delta z \end{array} \right\}.
\tag{2.1.19}
$$

Then

$$\delta U_e = \int_e EA\delta\varepsilon_{\tilde{x}\tilde{x}}\varepsilon_{\tilde{x}\tilde{x}}\,d\tilde{x} = \frac{1}{l^4}\int_{s=0}^{s=1} EA\delta\mathbf{d}^\mathrm{T}\hat{\mathbf{B}}^\mathrm{T}\hat{\mathbf{B}}\mathbf{d}l\,ds$$

$$= \delta\mathbf{d}^\mathrm{T}\left(\frac{1}{l^3}\int_{s=0}^{s=1} EA\hat{\mathbf{B}}^\mathrm{T}\hat{\mathbf{B}}\,ds\right)\mathbf{d} = \delta\mathbf{d}^\mathrm{T}\mathbf{k}^e\mathbf{d}, \qquad (2.1.20)$$

where

$$\mathbf{k}^e = \frac{1}{l^3}\int_{s=0}^{s=1} EA\hat{\mathbf{B}}^\mathrm{T}\hat{\mathbf{B}}\,ds \qquad (2.1.21)$$

is the $6\times 6$ element stiffness matrix. For constant $EA$:

$$\mathbf{k}^e = \frac{EA}{l^3}\hat{\mathbf{B}}^\mathrm{T}\hat{\mathbf{B}}. \qquad (2.1.22)$$

Expressed in expanded form:

$$\mathbf{k}^e = \frac{EA}{l^3}\begin{Bmatrix} -\Delta x \\ -\Delta y \\ -\Delta z \\ \Delta x \\ \Delta y \\ \Delta z \end{Bmatrix}\lfloor -\Delta x \quad -\Delta y \quad -\Delta z \quad \Delta x \quad \Delta y \quad \Delta z\rfloor. \qquad (2.1.23)$$

## 2.2   Planar Truss

For a two-dimensional (2D) or planar truss in the $xy$-plane as shown in Figure 2.3, $\Delta z = 0$ and $w = 0$. Then, Eq. (2.1.15) is simplified to

$$\varepsilon_{\tilde{x}\tilde{x}} = \frac{1}{l^2}\lfloor -\Delta x \quad -\Delta y \quad \Delta x \quad \Delta y\rfloor\begin{Bmatrix} u_1 \\ v_1 \\ u_2 \\ v_2 \end{Bmatrix}. \qquad (2.2.1)$$

**Figure 2.3** A planar truss structure

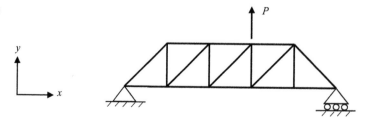

Similarly, the element stiffness matrix in Eq. (2.1.23) is reduced to

$$\mathbf{k}^e = \frac{EA}{l^3} \begin{Bmatrix} -\Delta x \\ -\Delta y \\ \Delta x \\ \Delta y \end{Bmatrix} \lfloor -\Delta x \quad -\Delta y \quad \Delta x \quad \Delta y \rfloor. \tag{2.2.2}$$

## Example 2.1

Consider a truss structure with five elements as shown in Figure 2.4. Each member has the same constant cross-sectional area and Young's modulus:

(a) Write down the $4 \times 4$ stiffness matrix of each element.
(b) Construct the $8 \times 8$ global stiffness matrix and the $8 \times 1$ global load vector.
(c) Solve for the unknown nodal DOF.
(d) Determine the axial stress in each element.
(e) Determine the reaction forces $R_1$, $R_2$, $R_7$, $R_8$ at the fixed hinges.

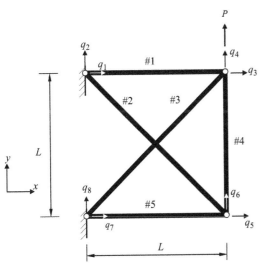

Figure 2.4 2D truss with five members

## Solution:

(a) One may use the element numbering shown in Figure 2.5.

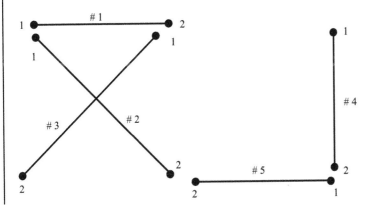

Figure 2.5 Element numbers and element node numbers

Using Eq. (2.2.2) and Table 2.1, the element stiffness matrices are obtained as

$$
\mathbf{k}^1 = \frac{EA}{L}
\begin{array}{c}
(1)\\(2)\\(3)\\(4)
\end{array}
\begin{bmatrix}
\overset{(1)}{1} & \overset{(2)}{0} & \overset{(3)}{-1} & \overset{(4)}{0}\\
0 & 0 & 0 & 0\\
-1 & 0 & 1 & 0\\
0 & 0 & 0 & 0
\end{bmatrix},
\qquad
\mathbf{k}^2 = \frac{EA}{2\sqrt{2}L}
\begin{array}{c}
(1)\\(2)\\(5)\\(6)
\end{array}
\begin{bmatrix}
\overset{(1)}{1} & \overset{(2)}{-1} & \overset{(5)}{-1} & \overset{(6)}{1}\\
-1 & 1 & 1 & -1\\
-1 & 1 & 1 & -1\\
1 & -1 & -1 & 1
\end{bmatrix},
$$

$$
\mathbf{k}^3 = \frac{EA}{2\sqrt{2}L}
\begin{array}{c}
(3)\\(4)\\(7)\\(8)
\end{array}
\begin{bmatrix}
\overset{(3)}{1} & \overset{(4)}{1} & \overset{(7)}{-1} & \overset{(8)}{-1}\\
1 & 1 & -1 & -1\\
-1 & -1 & 1 & 1\\
-1 & -1 & 1 & 1
\end{bmatrix},
\qquad
\mathbf{k}^4 = \frac{EA}{L}
\begin{array}{c}
(3)\\(4)\\(5)\\(6)
\end{array}
\begin{bmatrix}
\overset{(3)}{0} & \overset{(4)}{0} & \overset{(5)}{0} & \overset{(6)}{0}\\
0 & 1 & 0 & -1\\
0 & 0 & 0 & 0\\
0 & -1 & 0 & 1
\end{bmatrix},
$$

$$
\mathbf{k}^5 = \frac{EA}{L}
\begin{array}{c}
(5)\\(6)\\(7)\\(8)
\end{array}
\begin{bmatrix}
\overset{(5)}{1} & \overset{(6)}{0} & \overset{(7)}{-1} & \overset{(8)}{0}\\
0 & 0 & 0 & 0\\
-1 & 0 & 1 & 0\\
0 & 0 & 0 & 0
\end{bmatrix}.
$$

Note that the numbers in parentheses are shown to indicate the corresponding row and column numbers in the global stiffness matrix. One may confirm how they are obtained from the connectivity information in the second column of Table 2.1.

**Table 2.1 Element connectivity and geometry for Example 2.1**

| Element # | Connectivity | $\Delta x$ | $\Delta y$ | Element length |
|---|---|---|---|---|
| 1 | 1, 2, 3, 4 | $L$ | 0 | $L$ |
| 2 | 1, 2, 5, 6 | $L$ | $-L$ | $\sqrt{2}L$ |
| 3 | 3, 4, 7, 8 | $-L$ | $-L$ | $\sqrt{2}L$ |
| 4 | 3, 4, 5, 6 | 0 | $-L$ | $L$ |
| 5 | 5, 6, 7, 8 | $-L$ | 0 | $L$ |

(b) After assembly, the global $8 \times 8$ stiffness matrix is

$$
\mathbf{K} = \frac{EA}{L}
\begin{bmatrix}
1+\dfrac{1}{2\sqrt{2}} & -\dfrac{1}{2\sqrt{2}} & -1 & 0 & -\dfrac{1}{2\sqrt{2}} & \dfrac{1}{2\sqrt{2}} & 0 & 0 \\[2ex]
-\dfrac{1}{2\sqrt{2}} & \dfrac{1}{2\sqrt{2}} & 0 & 0 & \dfrac{1}{2\sqrt{2}} & -\dfrac{1}{2\sqrt{2}} & 0 & 0 \\[2ex]
-1 & 0 & 1+\dfrac{1}{2\sqrt{2}} & \dfrac{1}{2\sqrt{2}} & 0 & 0 & -\dfrac{1}{2\sqrt{2}} & -\dfrac{1}{2\sqrt{2}} \\[2ex]
0 & 0 & \dfrac{1}{2\sqrt{2}} & 1+\dfrac{1}{2\sqrt{2}} & 0 & -1 & -\dfrac{1}{2\sqrt{2}} & -\dfrac{1}{2\sqrt{2}} \\[2ex]
-\dfrac{1}{2\sqrt{2}} & \dfrac{1}{2\sqrt{2}} & 0 & 0 & 1+\dfrac{1}{2\sqrt{2}} & -\dfrac{1}{2\sqrt{2}} & -1 & 0 \\[2ex]
\dfrac{1}{2\sqrt{2}} & -\dfrac{1}{2\sqrt{2}} & 0 & -1 & -\dfrac{1}{2\sqrt{2}} & 1+\dfrac{1}{2\sqrt{2}} & 0 & 0 \\[2ex]
0 & 0 & -\dfrac{1}{2\sqrt{2}} & -\dfrac{1}{2\sqrt{2}} & -1 & 0 & 1+\dfrac{1}{2\sqrt{2}} & \dfrac{1}{2\sqrt{2}} \\[2ex]
0 & 0 & -\dfrac{1}{2\sqrt{2}} & -\dfrac{1}{2\sqrt{2}} & 0 & 0 & \dfrac{1}{2\sqrt{2}} & \dfrac{1}{2\sqrt{2}}
\end{bmatrix}.
$$

For the point loads acting on the joints, the incremental work is expressed as

$$
\delta W = F_1 \delta q_1 + F_2 \delta q_2 + \cdots + F_8 \delta q_8 = \delta \mathbf{q}^{\mathrm{T}} \mathbf{F},
$$

where

$$
\mathbf{F} =
\begin{Bmatrix}
F_1 = R_1 \\
F_2 = R_2 \\
F_3 = 0 \\
F_4 = P \\
F_5 = 0 \\
F_6 = 0 \\
F_7 = R_7 \\
F_8 = R_8
\end{Bmatrix}.
$$

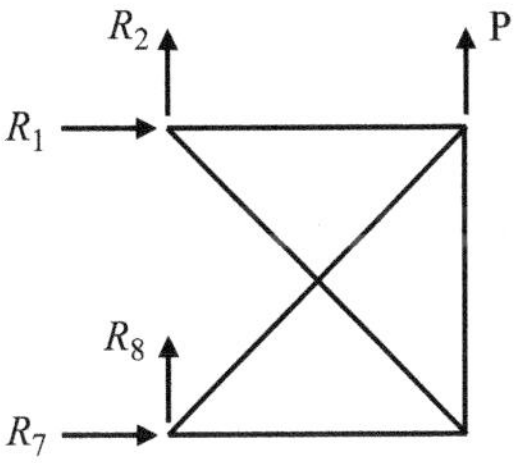

**Figure 2.6** The applied force and unknown reaction forces

The global load vector includes the reaction forces $R_1$, $R_2$, $R_7$, and $R_8$ as "external" forces, as shown in Figure 2.6.

(c) Applying the geometric boundary conditions ($q_1 = q_2 = q_7 = q_8 = 0$), and deleting the equations with unknown reactions in the global load vector

$$\frac{EA}{L}\begin{bmatrix} 1+\dfrac{1}{2\sqrt{2}} & \dfrac{1}{2\sqrt{2}} & 0 & 0 \\[2ex] \dfrac{1}{2\sqrt{2}} & 1+\dfrac{1}{2\sqrt{2}} & 0 & -1 \\[2ex] 0 & 0 & 1+\dfrac{1}{2\sqrt{2}} & -\dfrac{1}{2\sqrt{2}} \\[2ex] 0 & -1 & -\dfrac{1}{2\sqrt{2}} & 1+\dfrac{1}{2\sqrt{2}} \end{bmatrix}\begin{Bmatrix} q_3 \\ q_4 \\ q_5 \\ q_6 \end{Bmatrix}=\begin{Bmatrix} 0 \\ P \\ 0 \\ 0 \end{Bmatrix}.$$

Solving the above equation:

$$\begin{Bmatrix} q_3 \\ q_4 \\ q_5 \\ q_6 \end{Bmatrix}=\begin{Bmatrix} -0.5578 \\ 2.1353 \\ 0.4422 \\ 1.6931 \end{Bmatrix}\frac{PL}{EA}.$$

(d) $\sigma_{\tilde{x}\tilde{x}} = E\varepsilon_{\tilde{x}\tilde{x}}$, where

$$\varepsilon_{\tilde{x}\tilde{x}} = \frac{1}{l^2}\lfloor -\Delta x \quad -\Delta y \quad \Delta x \quad \Delta y \rfloor\begin{Bmatrix} u_1 \\ v_1 \\ u_2 \\ v_2 \end{Bmatrix}.$$

Element 1:

$$\sigma_{\tilde{x}\tilde{x}}^1 = \frac{E}{L^2}\lfloor -L \quad 0 \quad L \quad 0\rfloor\begin{Bmatrix} q_1 \\ q_2 \\ q_3 \\ q_4 \end{Bmatrix} = \frac{E}{L^2}\lfloor -L \quad 0 \quad L \quad 0\rfloor\begin{Bmatrix} 0 \\ 0 \\ -0.5578 \\ 2.1353 \end{Bmatrix}\frac{PL}{EA} = -0.5578\frac{P}{A}.$$

Element 2:

$$\sigma_{\tilde{x}\tilde{x}}^2 = \frac{E}{(\sqrt{2}L)^2}\lfloor -L \quad L \quad L \quad -L\rfloor\begin{Bmatrix} q_1 \\ q_2 \\ q_5 \\ q_6 \end{Bmatrix} = \frac{E}{(\sqrt{2}L)^2}\lfloor -L \quad L \quad L \quad -L\rfloor\begin{Bmatrix} 0 \\ 0 \\ 0.4422 \\ 1.6931 \end{Bmatrix}\frac{PL}{EA}$$

$$\rightarrow \quad \sigma_{\tilde{x}\tilde{x}}^2 = -0.6254\frac{P}{A}.$$

Element 3:

$$\sigma_{\tilde{x}\tilde{x}}^3 = \frac{E}{(\sqrt{2}L)^2} \lfloor L \quad L \quad -L \quad -L \rfloor \begin{Bmatrix} q_3 \\ q_4 \\ q_7 \\ q_8 \end{Bmatrix}$$

$$\rightarrow \sigma_{\tilde{x}\tilde{x}}^3 = 0.7888 \frac{P}{A}.$$

Element 4:

$$\sigma_{\tilde{x}\tilde{x}}^4 = \frac{E}{L^2} \lfloor 0 \quad L \quad 0 \quad -L \rfloor \begin{Bmatrix} q_3 \\ q_4 \\ q_5 \\ q_6 \end{Bmatrix}$$

$$\rightarrow \sigma_{\tilde{x}\tilde{x}}^4 = 0.4422 \frac{P}{A}.$$

Element 5:

$$\sigma_{\tilde{x}\tilde{x}}^5 = \frac{E}{L^2} \lfloor L \quad 0 \quad -L \quad 0 \rfloor \begin{Bmatrix} q_5 \\ q_6 \\ q_7 \\ q_8 \end{Bmatrix}$$

$$\rightarrow \sigma_{\tilde{x}\tilde{x}}^5 = 0.4422 \frac{P}{A}.$$

(e) From $\mathbf{Kq} = \mathbf{F}$ involving all 8 DOF, one can determine the reaction forces as $R_1 = P$, $R_2 = -0.4422P$, $R_7 = -P$, $R_8 = -0.5578P$, as shown in Figure 2.7.

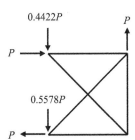

Figure 2.7 The applied force and reaction forces in equilibrium

## 2.3 Effect of Temperature Change: A Slender Body

In order to appreciate the effect of temperature change within the context of the FE formulation, consider a slender body of constant cross-sectional area. The bar is unconstrained and initially stress free. Suppose the body is subjected to a uniform temperature change of $\Delta T$ and uniform axial stress. Under these conditions, the total strain experienced by the body is

**Figure 2.8** A bar constrained by rigid walls and subjected to temperature change

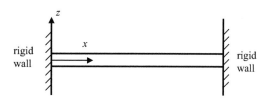

$$\varepsilon_{xx} = \frac{1}{E}\sigma_{xx} + \alpha\Delta T, \qquad (2.3.1)$$

where

$\varepsilon_{xx}$: total strain
$\frac{1}{E}\sigma_{xx}$: stress-induced strain
$\alpha$: coefficient of thermal expansion, a material constant
$\alpha\Delta T$: thermally induced strain.

From Eq. (2.3.1), the axial stress is

$$\sigma_{xx} = E(\varepsilon_{xx} - \alpha\Delta T). \qquad (2.3.2)$$

For a structural component with geometric constraints, a temperature change induces stress because the body cannot freely expand or contract. To appreciate this, consider a uniaxial bar constrained by rigid walls as shown in Figure 2.8. The temperature of the bar is raised uniformly by $\Delta T(> 0)$.

The bar is allowed to expand freely in the $y$ and $z$ directions, but it cannot expand in the axial direction. Accordingly:

$$\varepsilon_{xx} = \frac{1}{E}\sigma_{xx} + \alpha\Delta T = 0,$$

$$\sigma_{xx} = -E\alpha\Delta T: \text{compressive stress for } \Delta T \ (> 0).$$

**Element Load Vector Due to Temperature Change**

From Eq. (1.5.12):

$$\delta U_e = \int_e \delta\varepsilon_{xx}\sigma_{xx}A\,dx \qquad (2.3.3)$$

for an element. Introducing Eq. (2.3.2) into Eq. (2.3.3):

$$\delta U_e = \int_e EA\delta\varepsilon_{xx}(\varepsilon_{xx} - \alpha\Delta T)dx = \int_e EA\delta\varepsilon_{xx}\varepsilon_{xx}dx - \int_e \delta\varepsilon_{xx}(EA\alpha\Delta T)dx. \qquad (2.3.4)$$

The first term in Eq. (2.3.4) leads to the element stiffness matrix which was shown in Eq. (1.6.19). The second term in the same equation results in a load vector due to temperature change. Introducing Eq. (1.6.16) in Chapter 1:

$$\int_e \delta\varepsilon_{xx}(EA\alpha\Delta T)dx = \int_{s=0}^{s=1} \frac{1}{l}\lfloor \delta u_1 \quad \delta u_2\rfloor\begin{Bmatrix} -1 \\ 1 \end{Bmatrix}(EA\alpha\Delta T)lds$$

$$= \lfloor \delta u_1 \quad \delta u_2\rfloor\begin{Bmatrix} -1 \\ 1 \end{Bmatrix}\int_{s=0}^{s=1} EA\alpha\Delta Tds = \lfloor \delta u_1 \quad \delta u_2\rfloor\begin{Bmatrix} F_{1T}^e \\ F_{2T}^e \end{Bmatrix},$$

(2.3.5)

where

$$\begin{Bmatrix} F_{1T}^e \\ F_{2T}^e \end{Bmatrix} = \begin{Bmatrix} -1 \\ 1 \end{Bmatrix}\int_{s=0}^{s=1} EA\alpha\Delta Tds$$

(2.3.6)

is the element load vector. We observe that the effect of the temperature change manifests as the element load vector. Note that this element load vector is self-equilibrating in that the sum of the two nodal forces is equal in magnitude but opposite in direction.

## Example 2.2

A long slender body of length $L$ is constrained against axial displacement via two rigid walls placed at the two ends ($x = 0$ and $x = L$). The temperature in the section of the body from $x = L/3$ to $x = 2L/3$ is now raised by $\Delta T = C$, where $C$ is a given constant. Assume constant $E$, $A$, and $\alpha$ for the body. The body is modeled with three two-node uniaxial elements of equal length:

(a) Construct the global load vector.
(b) Determine the unknown nodal displacements.
(c) Determine the axial stress in each element.
(d) Determine the reaction forces at $x = 0$ and $x = L$.

## Solution:

(a) For elements #1 and #3, $\Delta T = 0$. For element #2, $\Delta T = C$. Then from Eq. (2.3.6):

$$\begin{Bmatrix} F_{1T}^1 \\ F_{2T}^1 \end{Bmatrix} = \begin{Bmatrix} 0 \\ 0 \end{Bmatrix}, \quad \begin{Bmatrix} F_{1T}^2 \\ F_{2T}^2 \end{Bmatrix} = \begin{Bmatrix} -1 \\ 1 \end{Bmatrix}EA\alpha C, \quad \begin{Bmatrix} F_{1T}^3 \\ F_{2T}^3 \end{Bmatrix} = \begin{Bmatrix} 0 \\ 0 \end{Bmatrix}.$$

Accordingly, the global load vector is

$$\mathbf{F} = \begin{Bmatrix} R_1 \\ -EA\alpha C \\ EA\alpha C \\ R_4 \end{Bmatrix},$$

where $R_1$ and $R_4$ are the reaction forces.

(b) With the global stiffness matrix assembled in Example 1.5 of Chapter 1, the FE equilibrium equation is

$$\frac{3EA}{L} \begin{bmatrix} 1 & -1 & 0 & 0 \\ -1 & 2 & -1 & 0 \\ 0 & -1 & 2 & -1 \\ 0 & 0 & -1 & 1 \end{bmatrix} \begin{Bmatrix} q_1 \\ q_2 \\ q_3 \\ q_4 \end{Bmatrix} = \begin{Bmatrix} R_1 \\ -EA\alpha C \\ EA\alpha C \\ R_4 \end{Bmatrix}.$$

Applying geometric constraints ($q_1 = q_4 = 0$) and deleting the first and fourth equations with unknown reaction forces:

$$\frac{3EA}{L} \begin{bmatrix} 2 & -1 \\ -1 & 2 \end{bmatrix} \begin{Bmatrix} q_2 \\ q_3 \end{Bmatrix} = EA\alpha C \begin{Bmatrix} -1 \\ 1 \end{Bmatrix} \rightarrow \begin{bmatrix} 2 & -1 \\ -1 & 2 \end{bmatrix} \begin{Bmatrix} q_2 \\ q_3 \end{Bmatrix} = \frac{\alpha CL}{3} \begin{Bmatrix} -1 \\ 1 \end{Bmatrix}.$$

Solving the above equation:

$$q_2 = -\frac{1}{9}\alpha CL, \quad q_3 = \frac{1}{9}\alpha CL.$$

(c) Use Eq. (2.3.2) to determine axial stress in each element. For element #e:

$$\sigma^e_{xx} = E(\varepsilon_{xx} - \alpha\Delta T) = E\left[\frac{1}{l}(u_2 - u_1) - \alpha\Delta T\right].$$

Accordingly:

$$\sigma^1_{xx} = \frac{E}{l}(u_2 - u_1) = \frac{3E}{L}(q_2 - q_1) = -\frac{1}{3}E\alpha C,$$

$$\sigma^2_{xx} = E\left[\frac{1}{l}(u_2 - u_1) - \alpha C\right] = E\left[\frac{3}{L}(q_3 - q_2) - \alpha C\right] = -\frac{1}{3}E\alpha C,$$

$$\sigma^3_{xx} = \frac{E}{l}(u_2 - u_1) = \frac{3E}{L}(q_4 - q_3) = -\frac{1}{3}E\alpha C.$$

(c) Substituting the nodal displacements in the equilibrium equation in part (b), the reactions are determined as

$$R_1 = \frac{1}{3}EA\alpha C, \quad R_4 = -\frac{1}{3}EA\alpha C.$$

Note that, for this simple problem, axial stress is uniform along the body. According to Eq. (2.3.1), axial strain is uniform within each element and the displacement is linear within the element. Accordingly, the FE solution is the exact solution.

## 2.4    Effect of Temperature Change: Truss Structures

When a truss member or element subjected to temperature change is not free to expand or contract, axial stress can develop in other members as well as the member itself. As in the case of the slender body in the last section, it will be shown that the effect of temperature change manifests as an element load vector that depends on temperature change and the coefficient of thermal expansion of the member material.

For an element subjected to temperature change of $\Delta T$, the axial stress is

$$\sigma_{\tilde{x}\tilde{x}} = E(\varepsilon_{\tilde{x}\tilde{x}} - \alpha\Delta T). \tag{2.4.1}$$

Introducing the above equation into Eq. (2.1.3):

$$\delta U_e = \int_e EA\delta\varepsilon_{\tilde{x}\tilde{x}}(\varepsilon_{\tilde{x}\tilde{x}} - \alpha\Delta T)d\tilde{x} = \int_e EA\delta\varepsilon_{\tilde{x}\tilde{x}}\varepsilon_{\tilde{x}\tilde{x}}d\tilde{x} - \int_e \delta\varepsilon_{\tilde{x}\tilde{x}}(EA\alpha\Delta T)d\tilde{x}. \tag{2.4.2}$$

Recall from Eq. (2.1.19) that

$$\delta\varepsilon_{\tilde{x}\tilde{x}} = \frac{1}{l^2}\delta\mathbf{d}^{\mathrm{T}}\hat{\mathbf{B}}^{\mathrm{T}}. \tag{2.4.3}$$

The first term in Eq. (2.4.2) leads to the element stiffness matrix derived in Section 2.1, while the second term in the same equation results in an element load vector due to temperature change. Substituting Eq. (2.4.3) into Eq. (2.4.2):

$$\int_e \delta\varepsilon_{\tilde{x}\tilde{x}}(EA\alpha\Delta T)d\tilde{x} = \int_{s=0}^{s=1} \frac{1}{l^2}\delta\mathbf{d}^{\mathrm{T}}\hat{\mathbf{B}}^{\mathrm{T}}(EA\alpha\Delta T)l\,ds = \delta\mathbf{d}^{\mathrm{T}}\hat{\mathbf{B}}^{\mathrm{T}}\int_{s=0}^{s=1} \frac{EA\alpha\Delta T}{l}\,ds$$

$$= \delta\mathbf{d}^{\mathrm{T}}\hat{\mathbf{B}}^{\mathrm{T}}\frac{EA\alpha\Delta T}{l} = \delta\mathbf{d}^{\mathrm{T}}\mathbf{F}^e_{thermal}, \tag{2.4.4}$$

where

$$\mathbf{F}^e_{thermal} = \hat{\mathbf{B}}^{\mathrm{T}}\frac{EA\alpha\Delta T}{l} \tag{2.4.5}$$

is the $6\times1$ element load vector due to temperature change. Writing in expanded form:

$$\mathbf{F}^i_{thermal} = \frac{EA\alpha\Delta T}{l}\begin{Bmatrix} -\Delta x \\ -\Delta y \\ -\Delta z \\ \Delta x \\ \Delta y \\ \Delta z \end{Bmatrix}. \tag{2.4.6}$$

## Example 2.3

Consider the truss structure identical to the one in Example 2.1, but with no externally applied loads. The truss is initially stress free. The temperature of element #3 is raised uniformly by $\Delta T$.

(a) Determine the load vector of element #3 corresponding to $\Delta T$.
(b) Construct the $8 \times 1$ global load vector before reduction.
(c) Reduce to the final equation and solve for $\mathbf{q}$.
(d) Determine the axial stress in each element.
(e) Determine the reaction forces $R_1$, $R_2$, $R_7$, $R_8$ at the fixed hinges.

## Solution:

(a) For a planar truss:

$$\mathbf{F}^e_{thermal} = \hat{\mathbf{B}}^{\mathrm{T}} \frac{EA\alpha\Delta T}{l} = \frac{EA\alpha\Delta T}{l} \begin{Bmatrix} -\Delta x \\ -\Delta y \\ \Delta x \\ \Delta y \end{Bmatrix}.$$

For element 3:
$$\Delta x = -L, \ \Delta y = -L, \ l = \sqrt{2}L,$$

$$\mathbf{F}^3_{thermal} = \frac{EA\alpha\Delta T}{\sqrt{2}} \begin{Bmatrix} 1 \\ 1 \\ -1 \\ -1 \end{Bmatrix}.$$

(b) The global load vector is

$$\mathbf{F} = \mathbf{F}_{thermal} + \mathbf{F}_{reactions} = \begin{Bmatrix} R_1 \\ R_2 \\ \frac{\sqrt{2}}{2}EA\alpha\Delta T \\ \frac{\sqrt{2}}{2}EA\alpha\Delta T \\ 0 \\ 0 \\ R_7 - \frac{\sqrt{2}}{2}EA\alpha\Delta T \\ R_8 - \frac{\sqrt{2}}{2}EA\alpha\Delta T \end{Bmatrix}.$$

(c) Applying the geometric boundary conditions and deleting the equations with unknown reactions:

$$\frac{EA}{L} \begin{bmatrix} 1+\frac{1}{2\sqrt{2}} & \frac{1}{2\sqrt{2}} & 0 & 0 \\ \frac{1}{2\sqrt{2}} & 1+\frac{1}{2\sqrt{2}} & 0 & -1 \\ 0 & 0 & 1+\frac{1}{2\sqrt{2}} & -\frac{1}{2\sqrt{2}} \\ 0 & -1 & -\frac{1}{2\sqrt{2}} & 1+\frac{1}{2\sqrt{2}} \end{bmatrix} \begin{Bmatrix} q_3 \\ q_4 \\ q_5 \\ q_6 \end{Bmatrix} = \frac{\sqrt{2}}{2}EA\alpha\Delta T \begin{Bmatrix} 1 \\ 1 \\ 0 \\ 0 \end{Bmatrix}.$$

Solving the above equation:

$$\begin{Bmatrix} q_3 \\ q_4 \\ q_5 \\ q_6 \end{Bmatrix} = \begin{Bmatrix} 0.2310 \\ 1.1154 \\ 0.2310 \\ 0.8843 \end{Bmatrix} (\alpha \Delta T) L.$$

(d)

$$\sigma_{\tilde{x}\tilde{x}} = E(\varepsilon_{\tilde{x}\tilde{x}} - \alpha \Delta T), \; \varepsilon_{\tilde{x}\tilde{x}} = \frac{1}{l^2} \lfloor -\Delta x \quad -\Delta y \quad \Delta x \quad \Delta y \rfloor \begin{Bmatrix} u_1 \\ v_1 \\ u_2 \\ v_2 \end{Bmatrix}.$$

For elements 1, 2, 4, and 5, $\Delta T = 0$, while for element 3, $\Delta T \neq 0$. It can be shown that

$$\sigma_{\tilde{x}\tilde{x}}^1 = 0.2310 E \alpha \Delta T, \quad \sigma_{\tilde{x}\tilde{x}}^2 = -0.3267 E \alpha \Delta T,$$

$$\sigma_{\tilde{x}\tilde{x}}^3 = -0.3267 E \alpha \Delta T, \quad \sigma_{\tilde{x}\tilde{x}}^4 = 0.2310 E \alpha \Delta T,$$

$$\sigma_{\tilde{x}\tilde{x}}^5 = 0.2310 E \alpha \Delta T.$$

(e) To determine $R_1$, $R_2$, $R_7$, and $R_8$, use $\mathbf{Kq} = \mathbf{F}$ before reduction involving all DOF. As shown in Figure 2.9:

$$R_1 = 0,$$
$$R_2 = -0.2310 E A \alpha \Delta T,$$
$$R_7 = 0,$$
$$R_8 = 0.2310 E A \alpha \Delta T.$$

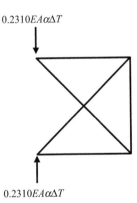

$0.2310 E A \alpha \Delta T$

$0.2310 E A \alpha \Delta T$

**Figure 2.9** Thermally induced reaction forces in equilibrium

## 2.5     Inclined Support

Consider a planar truss structure as shown in Figure 2.10. The joint at the bottom right is supported by a roller hinge which is free to slide along the $\hat{x}$-axis inclined at an angle with respect to the $x$-axis. But it is not allowed to move in the $\hat{y}$ direction normal to the inclined surface. To accommodate this geometric constraint or boundary condition, we first express the displacement vector in terms of the unit vectors in the reference $xy$-coordinate system as well as the $\hat{x}\hat{y}$ system of the inclined constraint as

$$\mathbf{u} = u\mathbf{i} + v\mathbf{j} = \hat{u}\hat{\mathbf{i}} + \hat{v}\hat{\mathbf{j}}. \qquad (2.5.1)$$

Accordingly:

$$u = \mathbf{i} \cdot \mathbf{u} = \mathbf{i} \cdot \left( \hat{u}\hat{\mathbf{i}} + \hat{v}\hat{\mathbf{j}} \right) = \left( \mathbf{i} \cdot \hat{\mathbf{i}} \right)\hat{u} + \left( \mathbf{i} \cdot \hat{\mathbf{j}} \right)\hat{v},$$
$$v = \mathbf{j} \cdot \mathbf{u} = \mathbf{j} \cdot \left( \hat{u}\hat{\mathbf{i}} + \hat{v}\hat{\mathbf{j}} \right) = \left( \mathbf{j} \cdot \hat{\mathbf{i}} \right)\hat{u} + \left( \mathbf{j} \cdot \hat{\mathbf{j}} \right)\hat{v}. \qquad (2.5.2)$$

Written in matrix form:

$$\left\{ \begin{matrix} u \\ v \end{matrix} \right\} = \begin{bmatrix} \mathbf{i}\cdot\hat{\mathbf{i}} & \mathbf{i}\cdot\hat{\mathbf{j}} \\ \mathbf{j}\cdot\hat{\mathbf{i}} & \mathbf{j}\cdot\hat{\mathbf{j}} \end{bmatrix} \left\{ \begin{matrix} \hat{u} \\ \hat{v} \end{matrix} \right\} = \begin{bmatrix} d_{x\hat{x}} & d_{x\hat{y}} \\ d_{y\hat{x}} & d_{y\hat{y}} \end{bmatrix} \left\{ \begin{matrix} \hat{u} \\ \hat{v} \end{matrix} \right\}, \qquad (2.5.3)$$

where $d_{x\hat{y}}$, etc. are the direction cosines. For the problem at hand, Eq. (2.5.3) can be expressed as

$$\left\{ \begin{matrix} q_5 \\ q_6 \end{matrix} \right\} = \begin{bmatrix} d_{x\hat{x}} & d_{x\hat{y}} \\ d_{y\hat{x}} & d_{y\hat{y}} \end{bmatrix} \left\{ \begin{matrix} \hat{q}_5 \\ \hat{q}_6 \end{matrix} \right\}. \qquad (2.5.4)$$

**Figure 2.10** Truss structure with a support on an inclined surface

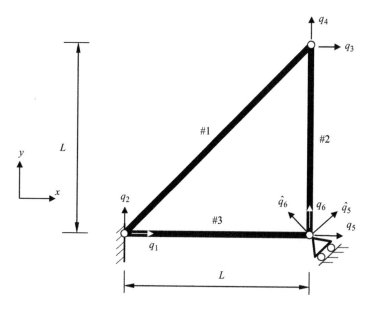

We can then establish the following relationship between the DOF vector in the original $xy$-coordinate system and the new system involving the DOF in the inclined coordinate system such that

$$\begin{Bmatrix} q_1 \\ q_2 \\ q_3 \\ q_4 \\ q_5 \\ q_6 \end{Bmatrix} = \begin{Bmatrix} 1 & 0 & 0 & 0 & 0 & 0 \\ 0 & 1 & 0 & 0 & 0 & 0 \\ 0 & 0 & 1 & 0 & 0 & 0 \\ 0 & 0 & 0 & 1 & 0 & 0 \\ 0 & 0 & 0 & 0 & d_{x\hat{x}} & d_{x\hat{y}} \\ 0 & 0 & 0 & 0 & d_{y\hat{x}} & d_{y\hat{y}} \end{Bmatrix} \begin{Bmatrix} q_1 \\ q_2 \\ q_3 \\ q_4 \\ \hat{q}_5 \\ \hat{q}_6 \end{Bmatrix} \rightarrow \mathbf{q}_{old} = \mathbf{T}\mathbf{q}_{new}. \tag{2.5.5}$$

Introducing the above relationship:

$$\delta U = \delta\mathbf{q}_{old}^T\mathbf{K}_{old}\mathbf{q}_{old} = \delta\mathbf{q}_{new}^T\mathbf{T}^T\mathbf{K}_{old}\mathbf{T}\mathbf{q}_{new} = \delta\mathbf{q}_{new}^T\mathbf{K}_{new}\mathbf{q}_{new}. \tag{2.5.6}$$

It follows that

$$\mathbf{K}_{new} = \mathbf{T}^T\mathbf{K}_{old}\mathbf{T}. \tag{2.5.7}$$

Similarly:

$$\delta W = \delta\mathbf{q}_{old}^T\mathbf{F}_{old} = \delta\mathbf{q}_{new}^T\mathbf{T}^T\mathbf{F}_{old} = \delta\mathbf{q}_{new}^T\mathbf{F}_{new}. \tag{2.5.8}$$

Thus

$$\mathbf{F}_{new} = \mathbf{T}^T\mathbf{F}_{old}. \tag{2.5.9}$$

The equilibrium equation for the model can now be expressed as

$$\mathbf{K}_{new}\mathbf{q}_{new} = \mathbf{F}_{new}. \tag{2.5.10}$$

For the model shown in Figure 2.10, the $\hat{q}_6 = 0$ condition can now be applied to Eq. (2.5.10). Alternatively, Eq. (2.5.3) can be applied to Eq. (2.1.15) or Eq. (2.1.20) to construct the element stiffness matrix in terms of the DOF in the inclined coordinate system. The reader may try the problems at the end of this chapter to appreciate the procedures described in this section.

## 2.6 Slender Body under Torque

Consider a slender body undergoing torsional deformation subjected to applied torque as shown in Figure 2.11. In the figure:

$\phi(x)$: twist angle of the cross-section located at $x$
$f_T(x)$: applied torque per unit length
$\bar{T}$: torque applied at $x = L$.

For equilibrium, let's consider a free-body diagram of an infinitesimal segment of length $dx$ as shown in Figure 2.12.

**Figure 2.11** Slender body under applied torques

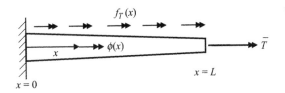

**Figure 2.12** Free-body diagram for torsion

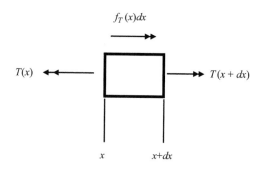

Summing torques acting on the free body:

$$T(x + dx) - T(x) + f_T(x)dx = 0$$
$$\rightarrow T(x) + \frac{\partial T}{\partial x}dx - T(x) + f_T(x)dx = 0 \rightarrow \left(\frac{\partial T}{\partial x} + f_T\right)dx = 0. \qquad (2.6.1)$$

Thus

$$\frac{\partial T}{\partial x} + f_T = 0. \qquad (2.6.2)$$

According to the elementary theory of torsion:

$$T = GJ\frac{\partial \phi}{\partial x}, \qquad (2.6.3)$$

where $G$ is the shear modulus and $J$ is the torsion constant of the cross-section. The boundary conditions are:

(1) geometric boundary condition at $x = 0$, $\phi = 0$;
(2) torque or moment boundary condition at $x = L$, $T = \bar{T}$.

By comparing Eqs (2.6.2), (2.6.3), and the boundary conditions of the slender body under applied torque with Eqs (1.4.2), (1.4.6), and the boundary conditions of the same body under axial force described in Section 1.4 of Chapter 1, one can observe a mathematical equivalence between the two physically different problems – as shown in Table 2.2.

**Table 2.2 Mathematical equivalence between axial deformation and torsion**

| Axial | Torsional |
| --- | --- |
| $F$ | $T$ |
| $u$ | $\phi$ |
| $EA$ | $GJ$ |
| $P$ | $\bar{T}$ |
| $f$ | $f_T$ |

Based on Eq. (1.5.22), one can then construct the strain–energy expression for a slender body under applied torque as

$$U = \frac{1}{2} \int_{x=0}^{x=L} GJ \left(\frac{\partial \phi}{\partial x}\right)^2 dx. \tag{2.6.4}$$

The infinitesimal increment in work done by the applied loads is

$$\delta W = \int_{x=0}^{x=L} \delta \phi f_T dx + (\bar{T}\delta\phi)_{x=L}. \tag{2.6.5}$$

Replacing the fixed boundary condition at $x = 0$ with reaction torque $R_0$:

$$\delta W = \int_{x=0}^{x=L} \delta \phi f_T dx + (\bar{T}\delta\phi)_{x=L} + (R_0\delta\phi)_{x=0}. \tag{2.6.6}$$

### Finite Element Modeling

Consider the two-node torsional element as shown in Figure 2.13. The twist angle over the element is assumed to be linear as

$$\phi = (1-s)\phi_1 + s\phi_2, \tag{2.6.7}$$

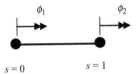

$\phi_1$   $\phi_2$

$s = 0$   $s = 1$

**Figure 2.13** Two-node torsional element

where $\phi_1$ and $\phi_2$ are nodal twist angles. The element stiffness matrix for constant $G$ is then

$$\mathbf{k}^e = \frac{GJ_m}{l}\begin{bmatrix} 1 & -1 \\ -1 & 1 \end{bmatrix}, \quad J_m = \int_{s=0}^{s=1} J ds. \tag{2.6.8}$$

This element stiffness matrix is analogous to that in Eq. (1.6.22). For the element load vector due to the distributed applied torque per unit length:

$$\int_{x=x_1}^{x=x_2} \delta \phi f_T \, dx = \lfloor \delta\phi_1 \quad \delta\phi_2 \rfloor \begin{Bmatrix} Q_1^e \\ Q_2^e \end{Bmatrix}, \tag{2.6.9}$$

where

$$Q_1^e = l \int_{s=0}^{s=1} (1-s)f_T ds, \quad Q_2^e = l \int_{s=0}^{s=1} sf_T ds. \tag{2.6.10}$$

The element nodal loads in the above equation are analogous to the nodal loads in Eq. (1.6.27).

## 2.7    Additional Topics

### (1) Nonzero Prescribed Geometric Boundary Condition

A structure may be subjected to known displacements prescribed over a part of the body in lieu of applied forces. In this case, the effect of the prescribed displacements manifests as the applied loads. To appreciate this, consider a slender body undergoing torsional deformation. For the three-element model with 4 DOF as shown in Figure 2.14, the FE equilibrium equation can be symbolically expressed as

$$\begin{bmatrix} K_{11} & K_{12} & K_{13} & K_{14} \\ K_{21} & K_{22} & K_{23} & K_{24} \\ K_{31} & K_{32} & K_{33} & K_{34} \\ K_{41} & K_{42} & K_{43} & K_{44} \end{bmatrix} \begin{Bmatrix} q_1 \\ q_2 \\ q_3 \\ q_4 \end{Bmatrix} = \begin{Bmatrix} F_1 \\ F_2 \\ F_3 \\ F_4 \end{Bmatrix}. \tag{2.7.1}$$

Suppose the left end is fixed ($q_1 = 0$), while a nonzero displacement is applied at global node 3 such that $q_3 = C$, where $C$ is a given nonzero quantity. We note that $q_3$ multiplies with the entries in the third column of the global stiffness matrix. Accordingly, from Eq. (2.7.1):

$$\begin{bmatrix} K_{11} & K_{12} & K_{14} \\ K_{21} & K_{22} & K_{24} \\ K_{31} & K_{32} & K_{34} \\ K_{41} & K_{42} & K_{44} \end{bmatrix} \begin{Bmatrix} q_1 \\ q_2 \\ q_4 \end{Bmatrix} + \begin{Bmatrix} K_{13} \\ K_{23} \\ K_{33} \\ K_{43} \end{Bmatrix} q_3 = \begin{Bmatrix} F_1 \\ F_2 \\ F_3 \\ F_4 \end{Bmatrix}. \tag{2.7.2}$$

Moving the $q_3$ term to the right side:

$$\begin{bmatrix} K_{11} & K_{12} & K_{14} \\ K_{21} & K_{22} & K_{24} \\ K_{31} & K_{32} & K_{34} \\ K_{41} & K_{42} & K_{44} \end{bmatrix} \begin{Bmatrix} q_1 \\ q_2 \\ q_4 \end{Bmatrix} = \begin{Bmatrix} F_1 \\ F_2 \\ F_3 \\ F_4 \end{Bmatrix} - \begin{Bmatrix} K_{13} \\ K_{23} \\ K_{33} \\ K_{43} \end{Bmatrix} q_3 = \begin{Bmatrix} \hat{F}_1 = F_1 - K_{13}q_3 \\ \hat{F}_2 = F_2 - K_{23}q_3 \\ \hat{F}_3 = F_3 - K_{33}q_3 \\ \hat{F}_4 = F_4 - K_{43}q_3 \end{Bmatrix}. \tag{2.7.3}$$

We observe that the nonzero prescribed torsional displacement manifests as an effective load vector. Setting $q_1 = 0$, the above equation reduces to

$$\begin{bmatrix} K_{12} & K_{14} \\ K_{22} & K_{24} \\ K_{32} & K_{34} \\ K_{42} & K_{44} \end{bmatrix} \begin{Bmatrix} q_2 \\ q_4 \end{Bmatrix} = \begin{bmatrix} F_1 - K_{13}q_3 \\ F_2 - K_{23}q_3 \\ F_3 - K_{33}q_3 \\ F_4 - K_{43}q_3 \end{bmatrix}. \tag{2.7.4}$$

Nodal loads $F_1$ and $F_3$ contain reaction torques corresponding to the applied geometric boundary conditions. Accordingly, deleting the first and third rows in the above equation:

**Figure 2.14** Three-element model for a body undergoing torsional deformation

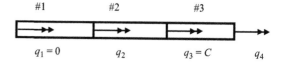

$$q_1 = 0 \qquad q_2 \qquad q_3 = C \qquad q_4$$

$$\begin{bmatrix} K_{22} & K_{24} \\ K_{42} & K_{44} \end{bmatrix} \begin{Bmatrix} q_2 \\ q_4 \end{Bmatrix} = \begin{Bmatrix} F_2 - K_{23}q_3 \\ F_4 - K_{43}q_3 \end{Bmatrix}, \tag{2.7.5}$$

which can be solved for the two unknown nodal DOF.

## Example 2.4

A slender body of length $L$ and uniform cross-section is constrained against torsion at $x = 0$ while the twist angle at $x = 2L/3$ is given as $C$. The body is modeled with three elements of equal length. Note that there is no applied torque per unit length ($f_T = 0$).

(a) Determine the load vector due to the prescribed displacement at $x = 2L/3$.
(b) Determine the unknown nodal displacement.
(c) Determine the torsional moment in each element.
(d) Determine the reaction moments or torques.

## Solution:

(a) For the three-element model, the equilibrium equation is

$$\frac{3GJ}{L} \begin{bmatrix} 1 & -1 & 0 & 0 \\ -1 & 2 & -1 & 0 \\ 0 & -1 & 2 & -1 \\ 0 & 0 & -1 & 1 \end{bmatrix} \begin{Bmatrix} q_1 \\ q_2 \\ q_3 \\ q_4 \end{Bmatrix} = \begin{Bmatrix} F_1 = R_1 \\ F_2 = 0 \\ F_3 = R_3 \\ F_4 = 0 \end{Bmatrix},$$

$$\frac{3GJ}{L} \begin{bmatrix} 1 & -1 & 0 \\ -1 & 2 & 0 \\ 0 & -1 & -1 \\ 0 & 0 & 1 \end{bmatrix} \begin{Bmatrix} q_1 \\ q_2 \\ q_4 \end{Bmatrix} = \begin{Bmatrix} R_1 \\ 0 \\ R_3 \\ 0 \end{Bmatrix} - \frac{3GJ}{L} \begin{Bmatrix} 0 \\ -1 \\ 2 \\ -1 \end{Bmatrix} q_3,$$

where the second term on the right-hand side is the load vector due to the prescribed nonzero displacement.

(b) Setting $q_1 = 0$ and deleting the first row and the third row with unknown reactions on the right-hand side:

$$\frac{3GJ}{L} \begin{bmatrix} 2 & 0 \\ 0 & 1 \end{bmatrix} \begin{Bmatrix} q_2 \\ q_4 \end{Bmatrix} = \frac{3GJ}{L} \begin{Bmatrix} 1 \\ 1 \end{Bmatrix} q_3 \rightarrow \begin{bmatrix} 2 & 0 \\ 0 & 1 \end{bmatrix} \begin{Bmatrix} q_2 \\ q_4 \end{Bmatrix} = \begin{Bmatrix} 1 \\ 1 \end{Bmatrix} C \rightarrow \begin{Bmatrix} q_2 \\ q_4 \end{Bmatrix} = \begin{Bmatrix} 1/2 \\ 1 \end{Bmatrix} C.$$

The DOF vector involving all 4 DOF is

$$\begin{Bmatrix} q_1 \\ q_2 \\ q_3 \\ q_4 \end{Bmatrix} = \begin{Bmatrix} 0 \\ 1/2 \\ 1 \\ 1 \end{Bmatrix} C.$$

(c) According to Eq. (2.6.3), for the two-node element:

$$T^e = GJ\frac{\partial\phi}{\partial s}\frac{ds}{dx} = GJ\frac{1}{l}(\phi_2 - \phi_1) = \frac{3GJ}{L}(\phi_2 - \phi_1).$$

Using the above equation:

$$T^1 = \frac{3GJ}{L}(\phi_2 - \phi_1) = \frac{3GJ}{L}(q_2 - q_1) = \frac{3}{2}\left(\frac{GJC}{L}\right),$$

$$T^2 = \frac{3GJ}{L}(\phi_2 - \phi_1) = \frac{3GJ}{L}(q_3 - q_2) = \frac{3}{2}\left(\frac{GJC}{L}\right),$$

$$T^3 = \frac{3GJ}{L}(\phi_2 - \phi_1) = \frac{3GJ}{L}(q_4 - q_3) = 0.$$

(d) From part (a):

$$\frac{3GJ}{L}\begin{bmatrix} 1 & -1 & 0 & 0 \\ -1 & 2 & -1 & 0 \\ 0 & -1 & 2 & -1 \\ 0 & 0 & -1 & 1 \end{bmatrix}\begin{Bmatrix} q_1 \\ q_2 \\ q_3 \\ q_4 \end{Bmatrix} = \begin{Bmatrix} R_1 \\ 0 \\ R_3 \\ 0 \end{Bmatrix}$$

$$\rightarrow \frac{3GJ}{L}\begin{bmatrix} 1 & -1 & 0 & 0 \\ -1 & 2 & -1 & 0 \\ 0 & -1 & 2 & -1 \\ 0 & 0 & -1 & 1 \end{bmatrix}\begin{Bmatrix} 0 \\ 1/2 \\ 1 \\ 1 \end{Bmatrix}C = \begin{Bmatrix} R_1 \\ 0 \\ R_3 \\ 0 \end{Bmatrix}.$$

From the above equation, we find that

$$R_1 = -\frac{3}{2}\left(\frac{GJC}{L}\right), \quad R_3 = \frac{3}{2}\left(\frac{GJC}{L}\right),$$

where $R_1$ is the reaction at the fixed boundary and $R_3$ is the torque corresponding to the applied twist angle. Note that the FE solutions are the exact solutions for this simple problem.

## (2) Alternative Way of Handling the Geometric Boundary Condition

Applying the geometric boundary conditions to the global stiffness matrix and the global load vector results in an equation with the global stiffness matrix and load vector of a reduced size. However, there is an alternative way of handling the geometric boundary conditions while maintaining the original sizes of the global stiffness matrix and the global load vector. As an illustration, consider an FE equilibrium equation for a model with 4 DOF as shown in Eq. (2.7.1) and suppose the geometric boundary conditions are $q_1 = 0$ and $q_3 = C$. We may then express these boundary conditions as

$$K_{11}q_1 = 0,$$
$$K_{33}q_3 = K_{33}C. \tag{2.7.6}$$

The above equation and Eq. (2.7.5) can be combined into a single matrix equation such that

$$\begin{bmatrix} K_{11} & 0 & 0 & 0 \\ 0 & K_{22} & 0 & K_{24} \\ 0 & 0 & K_{33} & 0 \\ 0 & K_{42} & 0 & K_{44} \end{bmatrix} \begin{Bmatrix} q_1 \\ q_2 \\ q_3 \\ q_4 \end{Bmatrix} = \begin{Bmatrix} 0 \\ F_2 - K_{23}C \\ K_{33}C \\ F_4 - K_{43}C \end{Bmatrix}. \tag{2.7.7}$$

We observe that the above equation can be constructed from the original global stiffness matrix shown in Eq. (2.7.1) and the global load vector in Eq. (2.7.3) as follows:

(1) For $q_1 = 0$, delete all off-diagonal entries in the first column and the first row of the global stiffness matrix in Eq. (2.7.1), and set $\hat{F}_1 = 0$ in Eq. (2.7.3).
(2) For $q_3 = C$, delete all off-diagonal entries in the third column and the third row of the global stiffness matrix in Eq. (2.7.1), and set $\hat{F}_3 = K_{33}C$ in Eq. (2.7.3).

We can then solve the above equation to recover the geometric boundary conditions and determine the unknown nodal DOF, while maintaining the original size of the global stiffness matrix and the global load vector. This approach can easily be generalized to problems with other types of finite elements.

## PROBLEMS

**2.1** Consider a slender body of constant cross-section as shown in Figure 2.15. The body is subjected to a force of $F_3 = -P$. The geometric and material data are given as $L = 300$ cm, $A = 64$ cm$^2$, $E = 200$ GPa. Determine the displacement $q_3$ in terms of $P$.

**2.2** Consider a truss structure with two elements as shown in Figure 2.16. Each member has the same values of $A = 0.05$ m$^2$, $E = 200$ GPa, $\alpha = 13 \times 10^{-6}/°$C, and $L = 5.0$ m. The truss is initially stress free. Now the temperature of element #1 is lowered by $\Delta T$.

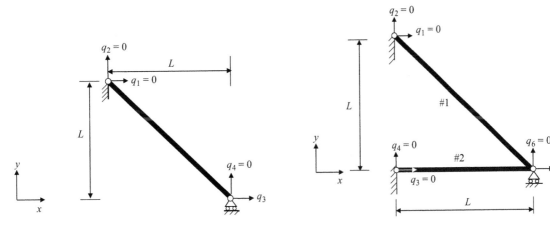

**Figure 2.15** For Problem 2.1        **Figure 2.16** For Problem 2.2

(a) Determine the load vector of element #1 corresponding to $\Delta T$.

(b) Construct the global stiffness matrix and the global load vector.

(c) Determine the displacement $q_5$.

(d) Determine the axial stress in each element.

**2.3** Consider a truss structure with three elements as shown in Figure 2.17. Each member has the same cross-sectional area of $A = 0.05 \text{ m}^2$, $E = 200 \text{ GPa}$, $\alpha = 13 \times 10^{-6}/{}^\circ\text{C}$, and $L = 5.0 \text{ m}$. The truss is initially stress free. Now a load $P$ is applied such that $F_8 = P$.

(a) Construct element stiffness matrices.

(b) Construct the global stiffness matrix.

(c) Determine nodal displacements $q_7$ and $q_8$.

(d) Determine the axial stress in each element.

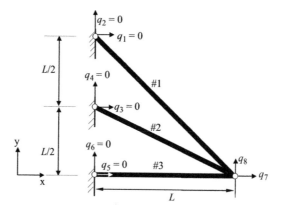

**Figure 2.17** For Problem 2.3

**2.4** The truss in Problem 2.3 is initially stress free. Now the temperature of element #2 is raised by $\Delta T$.

(a) Determine the load vector of element #2 corresponding to $\Delta T$.

(b) Determine nodal displacements $q_7$ and $q_8$

(c) Determine the axial stress in each element.

**2.5** Consider a simple 3D truss structure with four members as shown in Figure 2.18. The four hinges at the base are fixed and only 3 DOF at the top are free. Each member has the same constant cross-sectional area $A = 0.05 \text{ m}^2$, Young's modulus $E = 200 \text{ GPa}$, and $L = 5.0 \text{ m}$. The truss is initially stress free. Now a load $P$ is applied such that $F_3 = -P$.

(a) Construct the element stiffness matrices and assemble them into the global stiffness matrix.

(b) Determine nodal displacements $q_1$, $q_2$, and $q_3$ in terms of $P$.

(c) Determine the axial stress in each element in terms of $P$.

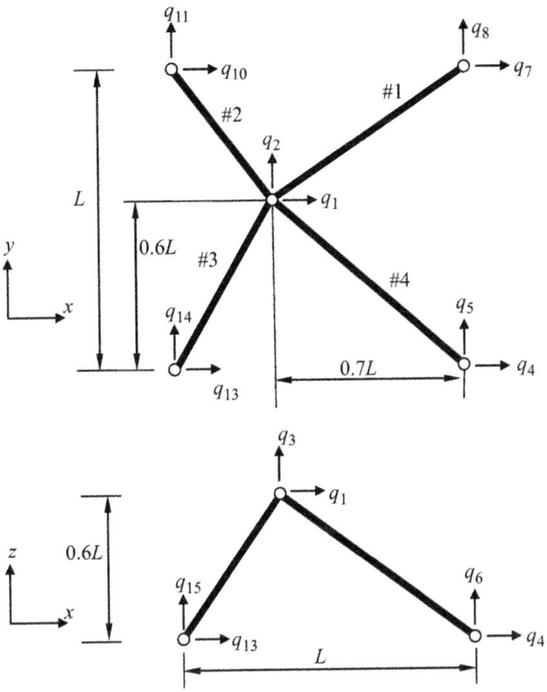

**Figure 2.18** For Problem 2.5

For convenience, do as follows:
(1) Place node 1 of each element at the joint at which all elements meet.
(2) Place the origin of the global coordinate system at the joint where $q_{13}$, $q_{14}$, $q_{15}$ are located.

**2.6** The truss in Problem 2.5 is now loaded such that $F_1 = Q$ There is no other applied load:
   (a) Determine unknown nodal displacements $q_1$, $q_2$, and $q_3$ in terms of $Q$.
   (b) Determine the axial stress in each element in terms of $Q$.

**2.7** The truss in Problem 2.5 is initially stress free. Now the temperature of element #2 is raised by $\Delta T$. The coefficient of thermal expansion is given as $\alpha = 13 \times 10^{-6}/°C$.
   (a) Determine the load vector of element #2 corresponding to $\Delta T$.
   (b) Determine nodal displacements in terms of $\Delta T$.
   (c) Determine the axial stress in each element in terms of $\Delta T$.

**2.8** For the truss shown in Figure 2.10, the applied load is $F_3 = P$.
   (a) Determine the unknown nodal displacements in terms of $P$.
   (b) Determine the axial stress in each element in terms of $P$.
   (c) Determine the reaction forces.

**2.9** The truss shown in Figure 2.4 is free of applied force, but a displacement is prescribed such that $q_4 = C$ where $C$ is a given value.
   (a) Determine the unknown nodal displacements in terms of $C$.
   (b) Determine the axial stress in each element in terms of $C$.
   (c) Determine the reaction forces at the fixed hinges.

**2.10** Consider the truss structure in Figure 2.19 in which global node numbers are indicated. The truss is subjected to applied force at global node #3 as shown in the figure. Geometric and material data are as follows: $L = 127$ cm, $A = 12.9$ cm², $E = 200$ GPa.

(a) Determine the unknown nodal displacements in terms of $P$.

(b) Determine the axial stress in each element in terms of $P$.

(c) Determine the reaction forces.

**Figure 2.19** For Problem 2.10

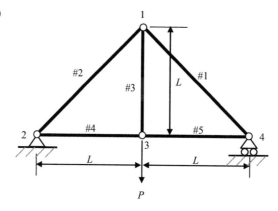

**2.11** Equation (2.5.3) can be introduced at the element level. To appreciate this, apply Eq. (2.5.3) to Eq. (2.2.1) to construct the element stiffness matrices of elements 2, 4, and 5 with the DOF in the inclined coordinate system as shown in Figure 2.10.

**2.12** A slender body of length $L$ is constrained against torsional rotation at $x = 0$ and $x = L$, while torsional rotation at $x = L/3$ is given as $C$. Assume constant $GJ$ for the body. The body is modeled with three elements of equal length. There is no applied torque per unit length ($f_T = 0$).

(a) Determine the load vector due to the prescribed rotation at $x = L/3$.

(b) Determine the torsional rotation at $x = 2L/3$.

(c) Determine the torsional moment in each element.

(d) Determine the reaction torques at $x = 0$, $x = L/3$ and $x = L$.

Note that the FE solutions are the exact solutions for this simple problem.

# 3  Beams and Frames

In this chapter we first consider FE modeling of slender bodies undergoing bending deformation. This will be followed by a discussion on frame structures which can be modeled as an assemblage of slender bodies rigidly connected.

For the description of bending, we may use the Bernoulli–Euler (B–E) theory or the Timoshenko theory. In the B–E theory, cross-sections remain always normal to the beam axis even in the deformed configuration, and thus the transverse shear strain is assumed zero, while this assumption is not made in the Timoshenko theory. In this chapter, we will consider the FE formulation based on the B–E theory. The Timoshenko beam bending theory and the corresponding FE formulation will be considered in Chapters 9 and 10.

First, we will introduce the B–E theory of beam bending as a review and extension of what is typically covered in an undergraduate sophomore-level course on mechanics of materials. The notations and sign conventions that are used in this chapter and the following chapters will be defined, under the assumption that bending occurs in the $xz$-plane. For the FE formulation, the displacement field is assumed to be a cubic polynomial for the two-node bending element. The assumed displacement is introduced into the incremental forms of strain energy and work done by applied loads to construct the element stiffness matrix and the element load vector. They are then assembled into the global stiffness matrix and the global load vector, using the relationship between the element DOF and the global DOF. An example problem is solved to demonstrate the effectiveness of the FE modeling and analysis. We will then consider bending of beams on elastic supports, which is represented as a spring or distributed springs. This will be followed by a discussion on beam bending in the $xy$-plane as a prelude to construction of the frame element.

We will then introduce the frame element which can be used to model frame structures deforming in the 2D plane and 3D space. First, the planar frame element is constructed via assembly of the two-node uniaxial element developed in Chapter 1 and the two-node bending element. Subsequently, the two-node 3D frame element is constructed by combining the two-node torsional element introduced in Chapter 2 and the two-node element for bending in the $xy$-plane.

## 3.1  Slender Body under Bending Loads

As an example of a slender body undergoing bending deformation, consider a cantilevered beam under applied loads as shown in Figure 3.1.

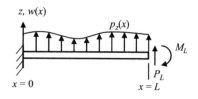

In the figure:

$w(x)$: transverse displacement in the $z$ direction
$p_z(x)$: applied force per unit length
$P_L$: force applied at $x = L$
$M_L$: moment applied at $x = L$.

**Figure 3.1** Cantilevered beam under bending loads

## 3.1.1   Beam Bending Theory

Figure 3.2 shows the cross-section of a slender body located at position $x$. The $x$-axis or beam axis is along the centroid of the cross-section. The origin of the $y$ and $z$-axes is located at the centroid of the cross-section and the orientation of the two axes is chosen such that $\int yzdA = 0$. They are then called the principal axes, and it can be shown that the bending in the $xz$-plane can be decoupled from the bending in the $xy$-plane. For sections that are symmetric, such as rectangular or circular, the principal axes can easily be identified by inspection.

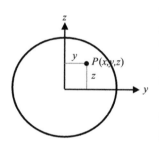

**Figure 3.2** The $y$ and $z$ coordinates on the cross-section located at position $x$

In the bending theory of a slender body, it is assumed that the cross-section located at position $x$ translates and rotates around the $y$ and $z$-axes as a rigid plane. For simplicity, we will assume that bending deformation occurs only in the $xz$-plane (Figure 3.3) and rotation of the cross-section is small. In Figure 3.3, $u_o$ is the displacement of the cross-section in the $x$ direction, $w_0$ is the displacement of the cross-section in the transverse or $z$ direction, and $\theta_y$ is the rotation of the cross-section around the $y$-axis, which is normal to the page, following the right-hand rule.

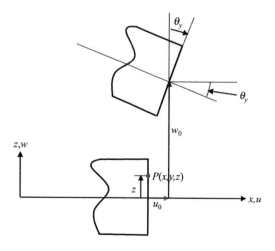

**Figure 3.3** Translation and rotation of the cross-section following the right-hand rule

**Kinematics of Deformation and Stress–Strain Relation**

For pure bending with no axial displacement ($u_o = 0$) the displacement of a point $P(x, y, z)$ on the cross-section located at position $x$ can be expressed as

$$u(x, y, z) = z\theta_y(x),$$
$$w(x, y, z) = w_0(x), \tag{3.1.1}$$

where $z\theta_y(x)$ is the horizontal displacement of point $P$ due to a small rotation $\theta_y$ (defined positive clockwise) of the cross-section around the $y$-axis. The bending displacement $w$ is positive in the upward direction along positive $z$. Then, for axial strain:

$$\varepsilon_{xx} = \frac{\partial u}{\partial x} = z\frac{\partial \theta_y}{\partial x}, \tag{3.1.2}$$

which shows that axial strain is proportional to $z$ for a given curvature.

According to the B–E beam bending theory, the bending rotation is related to the deflection of the beam axis by

$$\theta_y = -\frac{\partial w_0}{\partial x}. \tag{3.1.3}$$

The above equation is based on the assumption that the cross-section remains always normal to the beam axis, even in the deformed configuration, and thus the transverse shear strain is equal to zero. Substituting Eq. (3.1.3) in Eq. (3.1.2), we can relate the axial strain to the curvature of the beam axis such that

$$\varepsilon_{xx} = -z\frac{\partial^2 w_0}{\partial x^2} = -z\hat{\varepsilon}, \tag{3.1.4}$$

where

$$\hat{\varepsilon} = \frac{\partial^2 w_0}{\partial x^2}. \tag{3.1.5}$$

**Stress–Strain Relationship**

Axial stress $\sigma_{xx}$ normal to the cross-section is related to axial strain $\varepsilon_{xx}$ through Young's modulus $E$ such that

$$\sigma_{xx} = E\varepsilon_{xx}. \tag{3.1.6}$$

The moment around the $y$-axis is defined as

$$M_y(x) = \int_A z\sigma_{xx}dA, \tag{3.1.7}$$

where $\sigma_{xx}dA$ is the force normal to the infinitesimal surface area of $dA$ and $z$ is the moment arm measured from the $y$-axis. Then, using Eqs (3.1.6) and (3.1.2):

$$M_y(x) = \int_A zE\varepsilon_{xx}dA = \int_A zE\left(z\frac{\partial\theta_y}{\partial x}\right)dA = \frac{\partial\theta_y}{\partial x}E\int_A z^2 dA, \qquad (3.1.8)$$

which results in the moment–curvature relation

$$M_y(x) = EI_y\frac{\partial\theta_y}{\partial x}, \qquad (3.1.9)$$

where

$$I_y = \int_A z^2 dA : \text{area moment of inertia.} \qquad (3.1.10)$$

Introducing Eq. (3.1.3) into Eq. (3.1.9):

$$M_y(x) = -EI_y\frac{\partial^2 w_0}{\partial x^2} \qquad (3.1.11)$$

for the B–E beam bending theory. From now on, for convenience, we will use $w$ instead of $w_0$ for bending displacement in the $xz$-plane.

## Equilibrium Equation

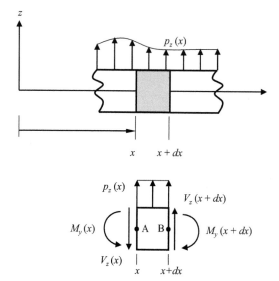

**Figure 3.4** Positive forces and moments acting on a beam segment of infinitesimal length

A slender body undergoes bending deformation when the load is applied. To be precise, equilibrium is a condition that must be satisfied by the body in the *deformed* configuration. However, for bending deformation with small displacements and rotations, the differences between the original undeformed configuration and the deformed configuration are very small and negligible, and the undeformed configuration can be used when equilibrium is considered. In this chapter we will adopt this simplifying assumption. To consider equilibrium, we may use a free body of infinitesimal length or a free body of finite length.

Consider a slender body subjected to a lateral load per unit length as shown in Figure 3.4, and introduce cuts at $x$ and $x + dx$ to isolate a free body of infinitesimal length. Then, for the equilibrium of all forces in the $z$ direction:

$$V_z(x + dx) - V_z(x) + p_z(x)dx = 0. \tag{3.1.12}$$

Introducing

$$V_z(x + dx) = V_z(x) + \frac{\partial V_z}{\partial x} dx \tag{3.1.13}$$

into Eq. (3.1.12):

$$\left(\frac{\partial V_z}{\partial x} + p_z\right) dx = 0 \rightarrow \frac{\partial V_z}{\partial x} + p_z = 0. \tag{3.1.14}$$

Thus

$$\frac{\partial V_z}{\partial x} = -p_z. \tag{3.1.15}$$

Summing all moments about point $B$ in Figure 3.4 for the moment equilibrium:

$$-M_y(x) + M_y(x + dx) - V_z(x)dx + p_z(x)dx\frac{dx}{2} = 0. \tag{3.1.16}$$

Introducing

$$M_y(x + dx) = M_y(x) + \frac{\partial M_y}{\partial x} dx \tag{3.1.17}$$

into Eq. (3.1.16):

$$\left(\frac{\partial M_y}{\partial x} - V_z\right) dx + \frac{p_z(x)}{2}(dx)^2 = 0. \tag{3.1.18}$$

In the above equation, $\frac{p_z(x)}{2}(dx)^2$ is a higher-order term and can be dropped as $dx \rightarrow 0$. Accordingly:

$$\left(\frac{\partial M_y}{\partial x} - V_z\right) dx = 0 \rightarrow \frac{\partial M_y}{\partial x} - V_z = 0. \tag{3.1.19}$$

Thus

$$\frac{\partial M_y}{\partial x} = V_z. \tag{3.1.20}$$

Introducing Eq. (3.1.15) into Eq. (3.1.20):

$$\frac{\partial^2 M_y}{\partial x^2} + p_z = 0. \tag{3.1.21}$$

Equations (3.1.11), (3.1.15), and (3.1.20) can be used to determine shear force, moment, and displacement distributions along the body length by applying proper boundary conditions for given problems. Alternatively, we may use a free body of finite length.

## 3.1.2   Exact Solution Example

We will now use these equations to determine shear force, moment, and displacement for a simple example problem, which can then be used as a reference solution later.

### Example 3.1

Consider a cantilevered slender body of constant $EI_y$ and length $L$ as shown in Figure 3.5. The body is clamped (i.e. no displacement and no rotation of the cross-section) at $x = 0$, and subjected to a force $P$ at $x = L$.

(a) Find the shear force and moment distribution.
(b) Find the bending displacement.

**Figure 3.5** Cantilevered beam under tip load

### Solution:

For this problem, the boundary conditions are as follows:

$$\text{At } x = 0, \ w = 0, \text{ and } \theta_y = 0.$$
$$\text{At } x = L, \ V_z = P, \text{ and } M_y = 0.$$

(a) To find the shear force and moment distribution, we may use Eq. (3.1.15) and (3.1.20) or we may consider a free body of finite size. We will try both approaches as an illustration.
Approach 1: Setting $p_z = 0$ in Eq. (3.1.15) and integrating:

$$\frac{\partial V_z}{\partial x} = 0 \rightarrow V_z = c_1,$$

where $c_1$ is a constant of integration. For load $P$ applied at $x = L$, $V_z = P$ and thus

$$c_1 = P \rightarrow V_z = P.$$

Then, setting $V_z = P$ in Eq. (3.1.20) and integrating:

$$\frac{\partial M_y}{\partial x} = P \rightarrow M_y = P(x + c_2),$$

where $c_2$ is another constant of integration. The moment is zero at $x = L$. Accordingly:

$$0 = P(L + c_2) \rightarrow L + c_2 = 0 \rightarrow c_2 = -L.$$

Thus

$$M_y = -P(L - x).$$

Approach 2: In this approach, we will work with a free body of finite length. Let's introduce an imaginary cut as shown in Figure 3.6.

Consider the free body on the right-hand side of the cut.

(the sum of vertical forces) $= 0 \rightarrow -V_z(x) + P = 0 \rightarrow V_z(x) = P$.

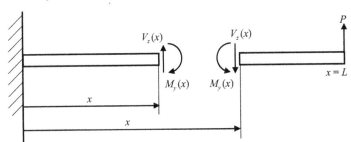

Figure 3.6 Free body of finite size on the right side

(the sum of moments) $= 0 \rightarrow -M_y(x) - P(L - x) = 0 \rightarrow M_y(x) = -P(L - x)$.

(b) To find the displacement, we use Eq. (3.1.11) as follows:

$$-EI_y \frac{\partial^2 w}{\partial x^2} = M_y = -P(L - x) \rightarrow \frac{\partial^2 w}{\partial x^2} = \frac{P}{EI_y}(L - x).$$

Integrating the above equation:

$$\frac{\partial w}{\partial x} = \frac{P}{EI_y}\left(Lx - \frac{1}{2}x^2 + c_3\right) = -\theta_y.$$

At $x = 0$, $\theta_y = 0$. Accordingly, $c_3 = 0$. Integrating again:

$$w = \frac{P}{EI_y}\left(\frac{1}{2}Lx^2 - \frac{1}{6}x^3 + c_4\right).$$

Since $w = 0$ at $x = 0$, $c_4 = 0$ and

$$w = \frac{P}{EI_y}\left(\frac{1}{2}Lx^2 - \frac{1}{6}x^3\right) = \frac{PL^3}{EI_y}\left[\frac{1}{2}\left(\frac{x}{L}\right)^2 - \frac{1}{6}\left(\frac{x}{L}\right)^3\right].$$

### 3.1.3  Virtual Work, Incremental Work, and Strain Energy

For a slender body in equilibrium:

$$\delta U - \delta W = 0, \tag{3.1.22}$$

where $\delta U$ is the internal virtual work or the increment in the strain energy and $\delta W$ is the external virtual work or the increment in the work done by applied loads. To derive the expressions for $\delta U$ and $\delta W$, we may follow the approach described in Chapter 1 for a slender body undergoing axial deformation and introduce a virtual displacement $\delta w(x)$ to construct an integral such that

$$\int\limits_{x=0}^{x=L} \left( \frac{\partial^2 M_y}{\partial x^2} + p_z \right) \delta w \, dx = 0, \tag{3.1.23}$$

which is equivalent to Eq. (3.1.21). The above integral can then be transformed into a virtual work form by performing the integration by parts twice on the first term. The details are left as a problem at the end of Chapter 6. Instead, we will take a different approach, utilizing the strain energy expression developed in Chapter 1 as follows.

## Incremental Work Done

For the applied loads shown in Figure 3.1, an infinitesimal increment in work done by applied loads can be expressed as

$$\delta W = \int\limits_{x=0}^{x=L} \delta w p_z \, dx + \delta W_B, \tag{3.1.24}$$

where

$$\delta W_B = (P_L \delta w)_{x=L} + (M_L \delta \theta)_{x=L} \tag{3.1.25}$$

is the contribution from the loads applied at the boundary. For the cantilever beam example shown in Figure 3.1, one may remove the geometric constraints (or boundary conditions) at $x = 0$ and replace them with reaction force $R_1$ and moment $R_2$ as shown in Figure 3.7.

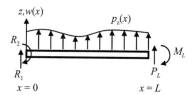

**Figure 3.7** Applied and reaction forces and moments

Then:

$$\delta W_B = (P_L \delta w)_{x=L} + (M_L \delta \theta)_{x=L} + (R_1 \delta w)_{x=0} + (R_2 \delta \theta)_{x=0}. \tag{3.1.26}$$

## Strain Energy

In the B–E beam bending theory, the only nonzero strain is the axial strain $\varepsilon_{xx}$. Based on Eq. (1.5.20) in Chapter 1, the strain energy of infinitesimal volume $dV = dA dx$ under uniaxial strain is

$$(U)_{dV} = \frac{1}{2} E \varepsilon_{xx}^2 \, dA dx, \tag{3.1.27}$$

where $dA = dydz$ is an infinitesimal area over the cross-section located at position $x$. Introducing Eq. (3.1.4) into the above equation:

$$(U)_{dV} = \frac{1}{2} E \varepsilon_{xx}^2 dA dx = \frac{1}{2} E(-z\hat{\varepsilon})^2 dydzdx = \frac{1}{2} Ez^2(\hat{\varepsilon})^2 dydzdx. \tag{3.1.28}$$

According to Eq. (3.1.5):

$$\hat{\varepsilon} = \frac{\partial^2 w}{\partial x^2}. \tag{3.1.29}$$

Integrating over the cross-sectional area, the strain energy for infinitesimal length $dx$ is

$$(U)_{dx} = \int_A (U)_{dV} = \frac{1}{2} \left( E \int_A z^2 dydz \right) \hat{\varepsilon}^2 dx = \frac{1}{2} EI_y \hat{\varepsilon}^2 dx. \tag{3.1.30}$$

Then, for the entire body:

$$U = \int_{x=0}^{x=L} (U)_{dx} = \frac{1}{2} \int_{x=0}^{x=L} EI_y \hat{\varepsilon}^2 dx. \tag{3.1.31}$$

The infinitesimal increment of the strain energy is then

$$\delta U = \int_{x=0}^{x=L} EI_y \hat{\varepsilon} \delta \hat{\varepsilon} dx. \tag{3.1.32}$$

Either Eq. (3.1.31) or Eq. (3.1.32) can be used to construct the element stiffness matrix in the FE formulation.

## 3.2  Finite Element Formulation for Beam Bending

The primary variable in the B–E beam bending theory presented in the previous section is the transverse displacement $w$. In the FE formulation, the body is divided into many elements and an assumed displacement is introduced in each element as follows.

### 3.2.1  Assumed Displacement

Noting that $w$, the transverse displacement, and $\theta$, the rotational angle of cross-section, must be continuous at the element boundaries, we consider a two-node element with $w$ and $\theta$ as the nodal DOF, as shown in Figure 3.8, where $w_1$, $\theta_1$, $w_2$ and $\theta_2$ are the four nodal DOF. Recall that for the mapping:

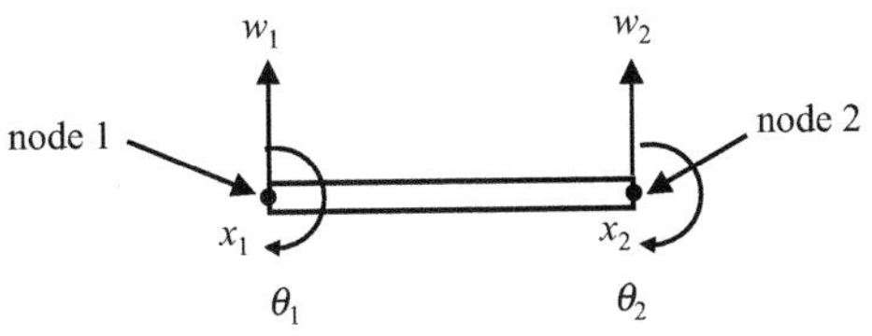

Figure 3.8 Two-node beam bending element

$$x = (1 - s)x_1 + sx_2, \tag{3.2.1}$$

where $s = 0$ at node 1 and $s = 1$ at node 2. Since the element has 4 DOF, the displacement $w$ is assumed to be a cubic function with four coefficients such that

$$w = b_1 + b_2 s + b_3 s^2 + b_4 s^3. \tag{3.2.2}$$

Noting that the assumed displacement in the element is expressed as a function of the $s$ coordinate:

$$\theta = -\frac{\partial w}{\partial x} = -\frac{\partial w}{\partial s}\frac{\partial s}{\partial x} = -\frac{1}{l}\left(b_2 + 2b_3 s + 3b_4 s^2\right), \tag{3.2.3}$$

where $l = x_2 - x_1$ is the element length. Coefficients $b_1, b_2, b_3, b_4$ can be expressed in terms of $w_1, \theta_1, w_2, \theta_2$ by matching Eqs (3.2.2) and (3.2.3) with the element DOF at the two nodes as follows.

At node 1, $s = 0$. Then from Eqs (3.2.2) and (3.2.3):

$$w_1 = b_1, \tag{3.2.4}$$

$$\theta_1 = -\frac{1}{l}b_2. \tag{3.2.5}$$

At node 2, $s = 1$. Then from Eqs (3.2.2) and (3.2.3):

$$w_2 = b_1 + b_2 + b_3 + b_4, \tag{3.2.6}$$

$$\theta_2 = -\frac{1}{l}(b_2 + 2b_3 + 3b_4). \tag{3.2.7}$$

From the above four equations, we determine the coefficients for the polynomial in Eq. (3.2.2) to be

$$\begin{aligned} b_1 &= w_1, \quad b_2 = -l\theta_1, \quad b_3 = -3w_1 + 3w_2 + 2l\theta_1 + l\theta_2, \\ b_4 &= 2w_1 - 2w_2 - l\theta_1 - l\theta_2. \end{aligned} \tag{3.2.8}$$

Substituting Eq. (3.2.8) into Eq. (3.2.2) and rearranging:

$$w = N_1 w_1 + N_2 l\theta_1 + N_3 w_2 + N_4 l\theta_2, \tag{3.2.9}$$

where

$$\begin{aligned} N_1 &= 1 - 3s^2 + 2s^3, \quad N_2 = -s + 2s^2 - s^3, \\ N_3 &= 3s^2 - 2s^3, \quad N_4 = s^2 - s^3. \end{aligned} \tag{3.2.10}$$

In matrix form, Eq. (3.2.9) can be expressed as

$$w = \lfloor N_1 \ N_2 l \ N_3 \ N_4 l \rfloor \begin{Bmatrix} w_1 \\ \theta_1 \\ w_2 \\ \theta_2 \end{Bmatrix} = \mathbf{Nd}, \tag{3.2.11}$$

where

$$\mathbf{N} = \lfloor N_1 \quad N_2 l \quad N_3 \quad N_4 l \rfloor, \tag{3.2.12}$$

$$\mathbf{d} = \begin{Bmatrix} d_1 \\ d_2 \\ d_3 \\ d_4 \end{Bmatrix} = \begin{Bmatrix} w_1 \\ \theta_1 \\ w_2 \\ \theta_2 \end{Bmatrix} : \text{element DOF vector.} \tag{3.2.13}$$

## 3.2.2  Element Stiffness Matrix and Element Load Vector

The assumed displacement described in the previous section can now be used to construct the element stiffness matrix and the element load vector due to an applied load per unit length.

### Element Stiffness Matrix

According to Eq. (3.1.32), the strain energy increment for element "$e$" is expressed as

$$\delta U_e = \int_{x=x_1}^{x=x_2} EI_y \delta\hat{\varepsilon}\hat{\varepsilon}dx. \tag{3.2.14}$$

Noting that the assumed displacement is expressed as a function of non-dimensional $s$ coordinate:

$$\hat{\varepsilon} = \frac{\partial^2 w}{\partial x^2} = \frac{\partial}{\partial s}\left(\frac{\partial w}{\partial s}\frac{\partial s}{\partial x}\right)\frac{\partial s}{\partial x} = \frac{1}{l^2}\frac{\partial^2 w}{\partial s^2}. \tag{3.2.15}$$

Introducing the assumed displacement in Eq. (3.2.9):

$$\hat{\varepsilon} = \frac{1}{l^2}\left(\frac{\partial^2 N_1}{\partial s^2}w_1 + \frac{\partial^2 N_2}{\partial s^2}l\theta_1 + \frac{\partial^2 N_3}{\partial s^2}w_2 + \frac{\partial^2 N_4}{\partial s^2}l\theta_2\right)$$

$$= \frac{1}{l^2}\lfloor -6 + 12s \quad (4 - 6s)l \quad 6 - 12s \quad (2 - 6s)\,l \rfloor \begin{Bmatrix} w_1 \\ \theta_1 \\ w_2 \\ \theta_2 \end{Bmatrix}. \tag{3.2.16}$$

Symbolically:

$$\hat{\varepsilon} = \frac{1}{l^2}\mathbf{Bd}, \tag{3.2.17}$$

where

$$\mathbf{B} = \lfloor -6 + 12s \quad (4 - 6s)l \quad 6 - 12s \quad (2 - 6s)l \rfloor. \tag{3.2.18}$$

Alternatively:

$$\hat{\varepsilon} = \frac{1}{l^2} \lfloor w_1 \quad \theta_1 \quad w_2 \quad \theta_2 \rfloor \left\{ \begin{array}{c} -6 + 12s \\ (4 - 6s)l \\ 6 - 12s \\ (2 - 6s)l \end{array} \right\} = \frac{1}{l^2} \mathbf{d}^\mathrm{T} \mathbf{B}^\mathrm{T} \tag{3.2.19}$$

and

$$\delta\hat{\varepsilon} = \frac{1}{l^2} \lfloor \delta w_1 \quad \delta\theta_1 \quad \delta w_2 \quad \delta\theta_2 \rfloor \left\{ \begin{array}{c} -6 + 12s \\ (4 - 6s)l \\ 6 - 12s \\ (2 - 6s)l \end{array} \right\} = \frac{1}{l^2} \delta\mathbf{d}^\mathrm{T} \mathbf{B}^\mathrm{T}. \tag{3.2.20}$$

Introducing Eqs (3.2.17) and (3.2.20) into Eq. (3.2.14):

$$\delta U_e = \frac{1}{l^4} \int_{s=0}^{s=1} EI_y \delta\mathbf{d}^\mathrm{T} \mathbf{B}^\mathrm{T} \mathbf{B}\mathbf{d} \, l \, ds = \delta\mathbf{d}^\mathrm{T} \left( \frac{1}{l^3} \int_{s=0}^{s=1} EI_y \mathbf{B}^\mathrm{T} \mathbf{B} \, ds \right) \mathbf{d} = \delta\mathbf{d}^\mathrm{T} \mathbf{k}^e \mathbf{d}, \tag{3.2.21}$$

where

$$\mathbf{k}^e = \frac{1}{l^3} \int_{s=0}^{s=1} EI_y \mathbf{B}^\mathrm{T} \mathbf{B} \, ds : 4 \times 4 \text{ element stiffness matrix.} \tag{3.2.22}$$

For constant $EI_y$, we can show that

$$\mathbf{k}^e = \frac{2EI_y}{l^3} \begin{bmatrix} 6 & -3l & -6 & -3l \\ -3l & 2l^2 & 3l & l^2 \\ -6 & 3l & 6 & 3l \\ -3l & l^2 & 3l & 2l^2 \end{bmatrix}. \tag{3.2.23}$$

Note that the entries in the above element stiffness matrix are not of the same unit because $w$ and $\theta$ have dissimilar units.

Alternatively, we can construct the element stiffness matrix using the strain energy. From Eq. (3.1.31), the strain energy for an element can be expressed as

$$U_e = \frac{1}{2} \int_{x=x_1}^{x=x_2} EI_y \hat{\varepsilon}^2 \, dx. \tag{3.2.24}$$

Substituting Eqs (3.2.17) and (3.2.19) into the above equation:

$$U_e = \frac{1}{2l^4} \int_{s=0}^{s=1} EI_y \mathbf{d}^T \mathbf{B}^T \mathbf{B} \mathbf{d} l ds = \frac{1}{2} \mathbf{d}^T \left( \frac{1}{l^3} \int_{s=0}^{s=1} EI_y \mathbf{B}^T \mathbf{B} ds \right) \mathbf{d} = \frac{1}{2} \mathbf{d}^T \mathbf{k}^e \mathbf{d}, \qquad (3.2.25)$$

where

$$\mathbf{k}^e = \frac{1}{l^3} \int_{s=0}^{s=1} EI_y \mathbf{B}^T \mathbf{B} ds \qquad (3.2.26)$$

is the element stiffness matrix identical to that in Eq. (3.2.22). We note that the incremental form of Eq. (3.2.25) is identical to Eq. (3.2.21).

**Element Load Vector due to Applied Load per Unit Length**

From Eq. (3.2.11):

$$\delta w = \mathbf{N} \delta \mathbf{d} = \delta \mathbf{d}^T \mathbf{N}^T \qquad (3.2.27)$$

for an element. Then, introducing the above equation into the first term on the right-hand side of Eq. (3.1.24):

$$\int_{x=x_1}^{x=x_2} \delta w p_z dx = \int_{s=0}^{s=1} \delta \mathbf{d}^T \mathbf{N}^T p_z l ds = \delta \mathbf{d}^T \mathbf{Q}^e, \qquad (3.2.28)$$

where

$$\delta \mathbf{d}^T = \lfloor \delta w_1 \quad \delta \theta_1 \quad \delta w_2 \quad \delta \theta_2 \rfloor \qquad (3.2.29)$$

and

$$\mathbf{Q}^e = l \int_{s=0}^{s=1} \mathbf{N}^T p_z ds : \text{element load vector due to } p_z \qquad (3.2.30)$$

or

$$\mathbf{Q}^e = \left\{ \begin{array}{c} Q_1^e \\ Q_2^e \\ Q_3^e \\ Q_4^e \end{array} \right\} = l \int_{s=0}^{s=1} \left\{ \begin{array}{c} N_1 \\ N_2 l \\ N_3 \\ N_4 l \end{array} \right\} p_z ds. \qquad (3.2.31)$$

For example, for constant $p_z = c$, one can show that

$$Q_1^e = cl \int_0^1 N_1 ds = \frac{cl}{2}, \quad Q_2^e = cl \int_0^1 N_2 l ds = -\frac{cl^2}{12},$$

$$Q_3^e = ct \int_0^1 N_3 ds = \frac{cl}{2}, \quad Q_4^e = cl \int_0^1 N_4 l ds = \frac{cl^2}{12}. \qquad (3.2.32)$$

### Equilibrium Equation at the Element Level

Consider a single element as shown in Figure 3.9. In the figure:

$$\left\{\begin{array}{c} F_1^e \\ M_{y1}^e \\ F_2^e \\ M_{y2}^e \end{array}\right\} = \left\{\begin{array}{c} Q_1^e + R_1^e \\ Q_2^e + R_2^e \\ Q_3^e + R_3^e \\ Q_4^e + R_4^e \end{array}\right\}. \tag{3.2.33}$$

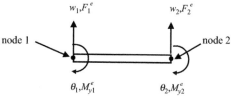

node 1       node 2

Figure 3.9 Element DOF, and nodal forces and moments

In the above equation, $Q_1^e$, $Q_2^e$, $Q_3^e$, and $Q_4^e$ are the nodal loads due to $p_z(x)$, applied force per unit length, and $R_1^e$, $R_2^e$, $R_3^e$, and $R_4^e$ are either reaction forces and moments or applied nodal forces and moments. The element is in equilibrium and thus

$$\delta U_e - \delta W_e = \delta \mathbf{d}^T \mathbf{k}^e \mathbf{d} - \delta \mathbf{d}^T \mathbf{f}^e = \delta \mathbf{d}^T (\mathbf{k}^e \mathbf{d} - \mathbf{f}^e) = 0 \rightarrow \mathbf{k}^e \mathbf{d} - \mathbf{f}^e = \mathbf{0}, \tag{3.2.34}$$

where

$$\mathbf{f}^e = \left\{\begin{array}{c} F_1^e \\ M_{y1}^e \\ F_2^e \\ M_{y2}^e \end{array}\right\} : \text{element load vector}. \tag{3.2.35}$$

For an element of uniform cross-section, the equilibrium equation at the element level can be expressed as

$$\left\{\begin{array}{c} F_1^e \\ M_{y1}^e \\ F_2^e \\ M_{y2}^e \end{array}\right\} = \frac{2EI_y}{l^3} \begin{bmatrix} 6 & -3l & -6 & -3l \\ -3l & 2l^2 & 3l & l^2 \\ -6 & 3l & 6 & 3l \\ -3l & l^2 & 3l & 2l^2 \end{bmatrix} \left\{\begin{array}{c} w_1 \\ \theta_1 \\ w_2 \\ \theta_2 \end{array}\right\}. \tag{3.2.36}$$

## 3.2.3  Global Stiffness Matrix and Global Load Vector

Element stiffness matrices and element load vectors are assembled into the global stiffness matrix and global load vector using the relationship or connectivity between the element DOF and the global DOF. As an example, consider a three-element model as shown in Figure 3.10.

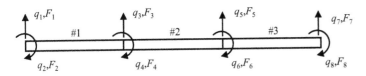

**Figure 3.10** Three-element model for a beam

For element #2:

$$d_1 = q_3, \quad d_2 = q_4, \quad d_3 = q_5, \quad d_4 = q_6. \tag{3.2.37}$$

The element stiffness matrix is then assembled into the global stiffness matrix as shown below:

$$
\begin{array}{cccc}
 & (3) & (4) & (5) & (6) \\
(3) & \begin{bmatrix} k_{11}^2 & k_{12}^2 & k_{13}^2 & k_{14}^2 \\
(4) & k_{21}^2 & k_{22}^2 & k_{23}^2 & k_{24}^2 \\
(5) & k_{31}^2 & k_{32}^2 & k_{33}^2 & k_{34}^2 \\
(6) & k_{41}^2 & k_{42}^2 & k_{43}^2 & k_{44}^2 \end{bmatrix}.
\end{array}
$$

Note that numbers 3, 4, 5, and 6 in parentheses indicate the row number and the column number of the global stiffness matrix into which each entry is added. For example, $k_{24}^2$ sums to $K_{46}$.

Similarly, for element #2, the element load vector due to an applied force per unit length is assembled into the global load vector as

$$
\begin{array}{c}
(3) \\
(4) \\
(5) \\
(6)
\end{array}
\begin{Bmatrix}
Q_1^2 \\
Q_2^2 \\
Q_3^2 \\
Q_4^2
\end{Bmatrix}.
$$

Note that numbers 3, 4, 5, and 6 in parentheses indicate the row number of the global load vector into which each entry above is added. For example, $Q_3^2$ sums to $F_5$ in the global load vector. In addition, for the problem described in Figure 3.7, force $P_L$ applied at $x = L$ sums to $F_7$, while moment $M_L$ applied at $x = L$ sums to $F_8$ in the global load vector.

## 3.3 Alternate Formulation of Beam Bending Element

In Figure 3.8 we see that $w_1$ and $w_2$ are not of the same units as $\theta_1$ and $\theta_2$ – the former have units of displacement (length) while the latter are rotations. At times, it may be convenient to use an alternate formulation in which all 4 DOF have the same unit of length. Accordingly, for a slender body modeled with elements of equal length, we may introduce

$$\hat{\theta} = l\theta, \tag{3.3.1}$$

where $l$ is the element length. From Eq. (3.2.9), the assumed displacement can then be expressed as

$$w = N_1 w_1 + N_2 \hat{\theta}_1 + N_3 w_2 + N_4 \hat{\theta}_2 = \hat{\mathbf{N}} \hat{\mathbf{d}}, \qquad (3.3.2)$$

where

$$\hat{\mathbf{N}} = \lfloor N_1 \quad N_2 \quad N_3 \quad N_4 \rfloor, \qquad (3.3.3)$$

$$\hat{\mathbf{d}} = \begin{Bmatrix} w_1 \\ \hat{\theta}_1 \\ w_2 \\ \hat{\theta}_2 \end{Bmatrix} : \text{element DOF vector.} \qquad (3.3.4)$$

**Element Stiffness Matrix**

Using the assumed displacement expressed in Eq. (3.3.2):

$$\hat{\varepsilon} = \frac{\partial^2 w}{\partial x^2} = \frac{1}{l^2} \frac{\partial^2 w}{\partial s^2} = \frac{1}{l^2} \hat{\mathbf{B}} \hat{\mathbf{d}}, \qquad (3.3.5)$$

where

$$\hat{\mathbf{B}} = \lfloor -6 + 12s \quad 4 - 6s \quad 6 - 12x \quad 2 - 6s \rfloor. \qquad (3.3.6)$$

Also:

$$\delta\hat{\varepsilon} = \frac{1}{l^2} \delta\hat{\mathbf{d}}^{\mathrm{T}} \hat{\mathbf{B}}^{\mathrm{T}}. \qquad (3.3.7)$$

Then the incremental strain energy of an element can be expressed as

$$\delta U_e = \int_{x=x_1}^{x=x_2} EI_y \delta\hat{\varepsilon}\hat{\varepsilon} \, dx = \frac{1}{l^4} \int_{s=0}^{s=1} EI_y \delta\hat{\mathbf{d}}^{\mathrm{T}} \hat{\mathbf{B}}^{\mathrm{T}} \hat{\mathbf{B}} \hat{\mathbf{d}} l \, ds$$

$$= \delta\hat{\mathbf{d}}^{\mathrm{T}} \left( \frac{1}{l^3} \int_{s=0}^{s=1} EI_y \hat{\mathbf{B}}^{\mathrm{T}} \hat{\mathbf{B}} \, ds \right) \hat{\mathbf{d}} = \delta\hat{\mathbf{d}}^{\mathrm{T}} \hat{\mathbf{k}}^e \hat{\mathbf{d}}, \qquad (3.3.8)$$

where

$$\hat{\mathbf{k}}^e = \frac{1}{l^3} \int_{s=0}^{s=1} EI_y \hat{\mathbf{B}}^{\mathrm{T}} \hat{\mathbf{B}} \, ds : 4 \times 4 \text{ element stiffness matrix.} \qquad (3.3.9)$$

For constant $EI_y$, it can be shown that

$$\hat{\mathbf{k}}^e = \frac{2EI_y}{l^3} \begin{bmatrix} 6 & -3 & -6 & -3 \\ -3 & 2 & 3 & 1 \\ -6 & 3 & 6 & 3 \\ -3 & 1 & 3 & 2 \end{bmatrix}. \qquad (3.3.10)$$

**Element Load Vector due to Applied Load per Unit Length**

From Eq. (3.3.2):

$$\delta w = \hat{\mathbf{N}}\delta\hat{\mathbf{d}} = \delta\hat{\mathbf{d}}^{\mathrm{T}}\hat{\mathbf{N}}^{\mathrm{T}}, \tag{3.3.11}$$

where

$$\delta\hat{\mathbf{d}}^{\mathrm{T}} = \lfloor \delta w_1 \quad \delta\hat{\theta}_1 \quad \delta w_2 \quad \delta\hat{\theta}_2 \rfloor. \tag{3.3.12}$$

Accordingly:

$$\int_{x=x_1}^{x=x_2} \delta w p_z dx = \int_{s=0}^{s=1} \delta\hat{\mathbf{d}}^{\mathrm{T}}\hat{\mathbf{N}}^{\mathrm{T}}p_z l ds = \delta\hat{\mathbf{d}}^{\mathrm{T}}\hat{\mathbf{Q}}^e, \tag{3.3.13}$$

where

$$\hat{\mathbf{Q}}^e = l \int_{s=0}^{s=1} \hat{\mathbf{N}}^{\mathrm{T}}p_z ds \tag{3.3.14}$$

or

$$\mathbf{Q}^e = \begin{Bmatrix} \hat{Q}_1^e \\ \hat{Q}_2^e \\ \hat{Q}_3^e \\ \hat{Q}_4^e \end{Bmatrix} = l \int_{s=0}^{s=1} \begin{Bmatrix} N_1 \\ N_2 \\ N_3 \\ N_4 \end{Bmatrix} p_z ds : \text{element load vector due to } p_z. \tag{3.3.15}$$

For constant $p_z = c$, one can show that

$$\hat{Q}_1^e = cl \int_{s=0}^{s=1} N_1 ds = \frac{cl}{2}, \quad \hat{Q}_4^e = cl \int_{s=0}^{s=1} N_2 ds = -\frac{cl}{12},$$

$$\hat{Q}_3^e = cl \int_{s=0}^{s=1} N_3 ds = \frac{cl}{2}, \quad \hat{Q}_4^e = cl \int_{s=0}^{s=1} N_4 ds = \frac{cl}{12}. \tag{3.3.16}$$

## 3.4  Finite Element Solution Examples

To appreciate the effectiveness of the FE formulation for beam bending, we will consider two simple example problems for which the exact solutions are also available for comparison.

## Example 3.2

Consider a cantilevered body clamped at $x = 0$ and subjected to a tip force $P$ applied at $x = L$, as described in Example 3.1. The body is of uniform cross-section with constant $EI_y$. Use a model with one element to determine (a) the bending displacement and (b) the reaction at $x = 0$.

## Solution:

(a) For the one-element model, $l = L$, and then from Eq. (3.2.36), the FE equilibrium equation is

$$\frac{2EI_y}{L^3} \begin{bmatrix} 6 & -3L & -6 & -3L \\ -3L & 2L^2 & 3L & L^2 \\ -6 & 3L & 6 & 3L \\ -3L & L^2 & 3L & 2L^2 \end{bmatrix} \begin{Bmatrix} w_1 \\ \theta_1 \\ w_2 \\ \theta_2 \end{Bmatrix} = \begin{Bmatrix} R_1 \\ R_2 \\ P \\ 0 \end{Bmatrix},$$

where $R_1$ and $R_2$ are the reaction force and moment at the clamped end $(x = 0)$. Applying the geometric boundary conditions $(w_1 = 0, \theta_1 = 0)$ and deleting the first two equations with reactions in the load vector, the above equation reduces to

$$\frac{2EI_y}{L^3} \begin{bmatrix} 6 & 3L \\ 3L & 2L^2 \end{bmatrix} \begin{Bmatrix} w_2 \\ \theta_2 \end{Bmatrix} = \begin{Bmatrix} P \\ 0 \end{Bmatrix}.$$

Solving the above equation:

$$w_2 = \frac{PL^3}{3EI_y}, \quad \theta_2 = -\frac{PL^2}{2EI_y}.$$

Substituting $w_2$ and $\theta_2$ into Eq. (3.2.9), we obtain

$$w = \frac{PL^3}{EI_y}\left(\frac{1}{2}s^2 - \frac{1}{6}s^3\right).$$

For the one-element model, the mapping is

$$x = (1 - s)x_1 + sx_2 = Ls \rightarrow x = Ls \rightarrow s = \frac{x}{L}.$$

Using the mapping:

$$w = \frac{PL^3}{EI_y}\left[\frac{1}{2}\left(\frac{x}{L}\right)^2 - \frac{1}{6}\left(\frac{x}{L}\right)^3\right].$$

We note that this solution matches the exact solution for this simple problem. This is because the exact solution is cubic and the "assumed" displacement for the element is also cubic. Accordingly, the one-element model produces the exact solution.

(b) Substituting the nodal displacement and rotation determined in part (a) into the original equilibrium equation involving all 4 DOF:

$$\frac{2EI_y}{L^3} \begin{bmatrix} 6 & -3L & -6 & -3L \\ -3L & 2L^2 & 3L & L^2 \\ -6 & 3L & 6 & 3L \\ -3L & L^2 & 3L & 2L^2 \end{bmatrix} \begin{Bmatrix} w_1 = 0 \\ \theta_1 = 0 \\ w_2 \\ \theta_2 \end{Bmatrix} = \begin{Bmatrix} R_1 \\ R_2 \\ P \\ 0 \end{Bmatrix}.$$

From the first two rows in the above matrix equation:

$$R_1 = \frac{2EI_y}{L^3}(-6w_2 - 3L\theta_2) = \frac{2EI_y}{L^3}\left(-\frac{2PL^3}{EI_y} + \frac{3PL^3}{2EI_y}\right) = -P,$$

$$R_{21} = \frac{2EI_y}{L^3}(3Lw_2 + L^2\theta_2) = \frac{2EI_y}{L^3}\left(\frac{PL^4}{EI_y} - \frac{PL^4}{2EI_y}\right) = PL.$$

 **Example 3.3**

Consider a slender body of constant $EI_y$ and length $L$ as shown in Figure 3.11. The body is clamped at $x = 0$ and constrained against displacement but free to rotate at $x = L$. The body is subjected to an applied force per unit length expressed as $p_z(x) = -p_o$, where $p_o$ is a given constant value.

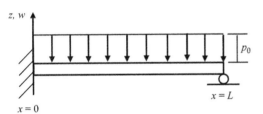

**Figure 3.11** A beam clamped at one end and supported by a roller at the other end

Use a model with two elements of equal length as shown in Figure 3.12 to do as follows:

(a) Construct the element stiffness matrix and the load vector for each element.
(b) Assemble the global stiffness matrix and the global load vector.
(c) Apply the geometric boundary conditions and set up the reduced equilibrium equation.
(d) Solve for the nodal DOF vector.
(e) Compare $w$, $\theta$ at $x = L/2$ and $\theta$ at $x = L$ with the exact solutions.
(f) Determine the reaction force and moment at $x = 0$ and the reaction force at $x = L$. Compare with the exact solutions.
(g) Determine the shear and moment distributions.

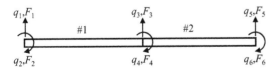

**Figure 3.12** Two-element model for beam bending

*Note:* Exact solutions are given as

$$V_z = p_0 \left( x - \frac{5}{8} L \right),$$

$$M_y = \frac{3}{8} p_0 L (x - L) + p_0 \left( \frac{1}{2} x^2 - Lx + \frac{1}{2} L^2 \right),$$

$$w = \frac{3p_0 L}{8EI_y} \left( \frac{1}{2} Lx^2 - \frac{1}{6} x^3 \right) - \frac{p_0}{EI_y} \left( \frac{1}{24} x^4 - \frac{1}{6} Lx^3 + \frac{1}{4} L^2 x^2 \right).$$

## Solution:

(a) For each element, $l = L/2$, and introducing $\hat{\theta} = l\theta$, the element stiffness matrix is

$$\mathbf{k}^e = \frac{2EI_y}{l^3} \begin{bmatrix} 6 & -3 & -6 & -3 \\ -3 & 2 & 3 & 1 \\ -6 & 3 & 6 & 3 \\ -3 & 1 & 3 & 2 \end{bmatrix} = \frac{16EI_y}{L^3} \begin{bmatrix} 6 & -3 & -6 & -3 \\ -3 & 2 & 3 & 1 \\ -6 & 3 & 6 & 3 \\ -3 & 1 & 3 & 2 \end{bmatrix}$$

and the element load vector due to the distributed load is

$$\hat{\mathbf{Q}}^e = l \int_{s=0}^{s=1} \begin{Bmatrix} N_1 \\ N_2 \\ N_3 \\ N_4 \end{Bmatrix} p_z ds = l \int_{s=0}^{s=1} \begin{Bmatrix} N_1 \\ N_2 \\ N_3 \\ N_4 \end{Bmatrix} cds = \begin{Bmatrix} cl/2 \\ -cl/12 \\ cl/2 \\ cl/12 \end{Bmatrix} = \frac{cl}{12} \begin{Bmatrix} 6 \\ -1 \\ 6 \\ 1 \end{Bmatrix} = \frac{cL}{24} \begin{Bmatrix} 6 \\ -1 \\ 6 \\ 1 \end{Bmatrix},$$

where $c = -p_0$.

(b) The global stiffness matrix and global load vector are assembled using the connectivity between the element DOF and the global DOF. The FE equilibrium equation is then

$$\frac{16EI_y}{L^3} \begin{bmatrix} 6 & -3 & -6 & -3 & 0 & 0 \\ -3 & 2 & 3 & 1 & 0 & 0 \\ -6 & 3 & 12 & 0 & -6 & -3 \\ -3 & 1 & 0 & 4 & 3 & 1 \\ 0 & 0 & -6 & 3 & 6 & 3 \\ 0 & 0 & -3 & 1 & 3 & 2 \end{bmatrix} \begin{Bmatrix} q_1 \\ q_2 \\ q_3 \\ q_4 \\ q_5 \\ q_6 \end{Bmatrix} = \frac{cL}{24} \begin{Bmatrix} 6 \\ -1 \\ 12 \\ 0 \\ 6 \\ 1 \end{Bmatrix} + \begin{Bmatrix} R_1 \\ \hat{R}_2 \\ 0 \\ 0 \\ \hat{R}_5 \\ 0 \end{Bmatrix}.$$

(c) Applying the geometric boundary conditions ($q_1 = q_2 = q_5 = 0$) and deleting the rows including the unknown reactions (i.e., deleting columns 1, 2, 5 and rows 1, 2, 5):

$$\frac{16EI_y}{L^3} \begin{bmatrix} 12 & 0 & -3 \\ 0 & 4 & 1 \\ -3 & 1 & 2 \end{bmatrix} \begin{Bmatrix} q_3 \\ q_4 \\ q_6 \end{Bmatrix} = \frac{cL}{24} \begin{Bmatrix} 12 \\ 0 \\ 1 \end{Bmatrix}.$$

(d) Solving the equation in part (c):

$$\begin{Bmatrix} q_3 \\ q_4 \\ q_6 \end{Bmatrix} = \frac{-p_0 L^4}{384 E I_y} \begin{Bmatrix} 2 \\ -1 \\ 4 \end{Bmatrix} \rightarrow \begin{Bmatrix} q_3 = w_{x=L/2} \\ q_4 = \hat{\theta}_{x=L/2} \\ q_6 = \hat{\theta}_{x=L} \end{Bmatrix} = \frac{-p_0 L^4}{384 E I_y} \begin{Bmatrix} 2 \\ -1 \\ 4 \end{Bmatrix}.$$

Noting that $\hat{\theta} = l\theta = (L/2)\theta$:

$$\begin{Bmatrix} w_{x=L/2} \\ \theta_{x=L/2} \\ \theta_{x=L} \end{Bmatrix} = -\frac{p_0 L^3}{384 E I_y} \begin{Bmatrix} 2L \\ -2 \\ 8 \end{Bmatrix} = -\frac{p_0 L^3}{192 E I_y} \begin{Bmatrix} L \\ -1 \\ 4 \end{Bmatrix}.$$

(e) The exact solution is as follows:

$$\begin{Bmatrix} w_{x=L/2} \\ \theta_{x=L/2} \\ \theta_{x=L} \end{Bmatrix}_{exact} = -\frac{p_0 L^3}{192 E I_y} \begin{Bmatrix} L \\ -1 \\ 4 \end{Bmatrix}.$$

We observe that the FE solution matches the exact solution at the nodes.

(f) Substituting the geometric boundary conditions and the solution obtained in part (d) into the original equation in part (a):

$$\frac{16 E I_y}{L^3} \begin{bmatrix} 6 & -3 & -6 & -3 & 0 & 0 \\ -3 & 2 & 3 & 1 & 0 & 0 \\ -6 & 3 & 12 & 0 & -6 & -3 \\ -3 & 1 & 0 & 4 & 3 & 1 \\ 0 & 0 & -6 & 3 & 6 & 3 \\ 0 & 0 & -3 & 1 & 3 & 2 \end{bmatrix} \begin{Bmatrix} q_1 = 0 \\ q_2 = 0 \\ q_3 \\ q_4 \\ q_5 = 0 \\ q_6 \end{Bmatrix} = \frac{cL}{24} \begin{Bmatrix} 6 \\ -1 \\ 12 \\ 0 \\ 6 \\ 1 \end{Bmatrix} + \begin{Bmatrix} R_1 \\ \hat{R}_2 \\ 0 \\ 0 \\ \hat{R}_5 \\ 0 \end{Bmatrix}.$$

From the first row:

$$\frac{16 E I_y}{L^3}(-6 \times q_3 - 3 \times q_4) = cL/4 + R_1 \rightarrow \frac{16 E I_y}{L^3} \times \frac{p_0 L^4}{384 E I_y}(6 \times 2 - 3) = \frac{-p_0 L}{4} + R_1$$

$$\rightarrow R_1 = \frac{5}{8} p_0 L.$$

Following a similar procedure:

$$\hat{R}_2 = -\frac{1}{4} p_0 L, \quad R_5 = \frac{3}{8} p_0 L.$$

Noting that $\hat{R}_2 \delta\hat{\theta} = R_2 \delta\theta$ and $\delta\hat{\theta} = l\delta\theta$:

$$\hat{R}_2 l\delta\theta = R_2 \delta\theta \rightarrow R_2 = \hat{R}_2 l = -\frac{1}{4} p_0 L \times \frac{L}{2} \rightarrow R_2 = -\frac{1}{8} p_0 L^2.$$

These reactions match the exact reactions.

(g) We may attempt to find the moment distribution first using Eq. (3.1.11). However, this requires taking two derivatives of the assumed displacement, which degrades the accuracy of approximation. To find the shear force using Eq. (3.1.20) requires taking one more derivative, further degrading the accuracy of approximation. However, for 1D problems such as beam bending, it would be preferable to determine shear force and moment distributions by considering the equilibrium of a free body of finite size once all nodal reaction forces and moments are known. For the present problem, this results in the shear force and moment distributions given at the end of the problem statement.

## 3.5    Beams on Elastic Supports

Beams may be supported by rollers or bearings exhibiting elastic behavior. In such cases, these supports can be modeled as a spring. For beams resting on soil, a system of distributed springs may be used to represent the effect of the soil foundation.

### 3.5.1    Elastic Support Modeled as a Spring

As an example, consider a slender body clamped at one end and supported by a linear spring as shown in Figure 3.13. The strain energy of the spring is

$$U = \frac{1}{2}k(w_{x=L})^2,$$

(3.5.1)

which will contribute to the total strain energy of the entire system and the global stiffness matrix.

**Figure 3.13** A beam clamped at one end and supported by a spring at the other end

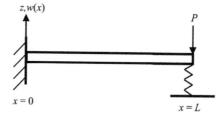

## Example 3.4

The slender body in Figure 3.13 is subjected to a tip vertical force of $P$. Use a model of one bending element to determine the nodal displacement and rotational angle. Introduce

$$\bar{k} = \frac{k}{\left(\dfrac{EI_y}{L^3}\right)} : \text{non-dimensional spring constant relative to bending rigidity.}$$

## Solution:

The FE equilibrium equation for the one-element model is

$$\frac{2EI_y}{L^3}\begin{bmatrix} 6 & -3L & -6 & -3L \\ -3L & 2L^2 & 3L & L^2 \\ -6 & 3L & 6+\bar{k}/2 & 3L \\ -3L & L^2 & 3L & 2L^2 \end{bmatrix}\begin{Bmatrix} w_1 \\ \theta_1 \\ w_2 \\ \theta_2 \end{Bmatrix} = \begin{Bmatrix} R_1 \\ R_2 \\ -P \\ 0 \end{Bmatrix}.$$

Applying the geometric boundary conditions ($w_1 = 0, \theta_1 = 0$) and deleting the equations (or rows) with unknown reactions:

$$\frac{2EI_y}{L^3}\begin{bmatrix} 6+\bar{k}/2 & 3L \\ 3L & 2L^2 \end{bmatrix}\begin{Bmatrix} w_2 \\ \theta_2 \end{Bmatrix} = \begin{Bmatrix} -P \\ 0 \end{Bmatrix}.$$

For $\bar{k} = 2$, it can be shown that

$$w_2 = -\frac{1}{5}\left(\frac{PL^3}{EI_y}\right), \quad \theta_2 = 10\left(\frac{PL^2}{EI_y}\right).$$

The determination of the reactions, shear force, and moment distributions is left as a problem for the reader.

### 3.5.2 Elastic Foundation

Rail tracks on the ground or floating slender structures may be modeled as a beam on an elastic foundation. As shown in Figure 3.14, the elastic foundation can be represented by an array of distributed springs. Introducing $k$, the spring constant per unit length (N/m/m), the strain energy of the elastic foundation for a segment of length $dx$ can be expressed as

$$dU_{EF} = \frac{1}{2}(kdx)w^2. \tag{3.5.2}$$

Then, for the entire body:

$$U_{EF} = \int dU_{EF} = \frac{1}{2}\int_{x=0}^{x=L} kw^2 dx. \tag{3.5.3}$$

Introducing the assumed displacement in Eq. (3.2.11) into an elastic foundation element:

**Figure 3.14** Elastic foundation modeled as a distributed spring

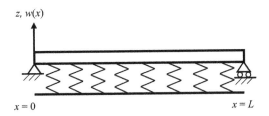

$z, w(x)$

$x = 0$        $x = L$

$$(U_{EF})_e = \frac{1}{2} \int_{x=x_1}^{x=x_2} kw^2 dx = \frac{1}{2} \int_{s=0}^{s=1} k\mathbf{d}^T\mathbf{N}^T\mathbf{N}\mathbf{d}l\,ds = \frac{1}{2}\mathbf{d}^T\left(l\int_{s=0}^{s=1} k\mathbf{N}^T\mathbf{N}ds\right)\mathbf{d} = \frac{1}{2}\mathbf{d}^T\mathbf{k}_{EF}^e\mathbf{d}, \quad (3.5.4)$$

where

$$\mathbf{k}_{EF}^e = l\int_{s=0}^{s=1} k\mathbf{N}^T\mathbf{N}ds \tag{3.5.5}$$

is the element stiffness matrix of the elastic foundation. For the entire body:

$$U_{EF} = \frac{1}{2}\mathbf{q}^T\mathbf{K}_{EF}\mathbf{q} \tag{3.5.6}$$

and

$$\delta U_{EF} = \delta\mathbf{q}^T\mathbf{K}_{EF}\mathbf{q}, \tag{3.5.7}$$

where the global $\mathbf{K}_{EF}$ matrix is assembled from the element $\mathbf{k}_{EF}^e$ matrices. For constant $k$, it can be shown that

$$\mathbf{k}_{EF}^e = \frac{kl}{420}\begin{bmatrix} 156 & -22l & 54 & 13l \\ -22l & 4l^2 & -13l & -3l^2 \\ 54 & -13l & 156 & 22l \\ 13l & -3l^2 & 22l & 4l^2 \end{bmatrix}. \tag{3.5.8}$$

Alternatively, introducing $\hat{\theta} = l\theta$, one can show that

$$\hat{\mathbf{k}}_{EF}^e = \frac{kl}{420}\begin{bmatrix} 156 & -22 & 54 & 13 \\ -22 & 4 & -13 & -3 \\ 54 & -13 & 156 & 22 \\ 13 & -3 & 22 & 4 \end{bmatrix}. \tag{3.5.9}$$

## Example 3.5

Consider the bending of a cantilevered beam under a tip load as shown in Figure 3.15. The beam is placed on an elastic foundation modeled as a distributed spring with spring constant $k$ per unit length. Use a model with one bending element to determine the nodal displacement and rotation angle.

**Figure 3.15** Cantilevered beam on elastic foundation under a tip load

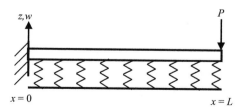

## Solution:

For the one-element model, $l = L$ and

$$\hat{\mathbf{k}}_B^e + \hat{\mathbf{k}}_{EF}^e = \frac{2EI_y}{L^3} \begin{bmatrix} 6 & -3 & -6 & -3 \\ -3 & 2 & 3 & 1 \\ -6 & 3 & 6 & 3 \\ -3 & 1 & 3 & 2 \end{bmatrix} + \frac{kL}{420} \begin{bmatrix} 156 & -22 & 54 & 13 \\ -22 & 4 & -13 & -3 \\ 54 & -13 & 156 & 22 \\ 13 & -3 & 22 & 4 \end{bmatrix}.$$

We may introduce a non-dimensional stiffness parameter $\bar{k}$ defined as

$$\bar{k} = \frac{kL}{\left(\dfrac{EI_y}{L^3}\right)},$$

which is a non-dimensional measure of the elastic foundation stiffness relative to the beam bending rigidity.

Then

$$\hat{\mathbf{k}}_B^e + \hat{\mathbf{k}}_{EF}^e = \frac{EI_y}{L^3} \left( 2\begin{bmatrix} 6 & -3 & -6 & -3 \\ -3 & 2 & 3 & 1 \\ -6 & 3 & 6 & 3 \\ -3 & 1 & 3 & 2 \end{bmatrix} + \frac{\bar{k}}{420} \begin{bmatrix} 156 & -22 & 54 & 13 \\ -22 & 4 & -13 & -3 \\ 54 & -13 & 156 & 22 \\ 13 & -3 & 22 & 4 \end{bmatrix} \right).$$

For the one-element model, the equilibrium equation is

$$\frac{EI_y}{L^3} \left( 2\begin{bmatrix} 6 & -3 & -6 & -3 \\ -3 & 2 & 3 & 1 \\ -6 & 3 & 6 & 3 \\ -3 & 1 & 3 & 2 \end{bmatrix} + \frac{\bar{k}}{420} \begin{bmatrix} 156 & -22 & 54 & 13 \\ -22 & 4 & -13 & -3 \\ 54 & -13 & 156 & 22 \\ 13 & -3 & 22 & 4 \end{bmatrix} \right) \begin{Bmatrix} w_1 \\ \hat{\theta}_1 \\ w_2 \\ \hat{\theta}_2 \end{Bmatrix} = \begin{Bmatrix} R_1 \\ \hat{R}_2 \\ P \\ 0 \end{Bmatrix}.$$

Applying the geometric boundary conditions ($w_1 = 0, \theta_1 = 0$) and deleting the equations (or rows) with unknown reactions, the equation is reduced to

$$\frac{EI_y}{L^3} \left( \begin{bmatrix} 12 & 6 \\ 6 & 4 \end{bmatrix} + \frac{\bar{k}}{420} \begin{bmatrix} 156 & 22 \\ 22 & 4 \end{bmatrix} \right) \begin{Bmatrix} w_2 \\ \hat{\theta}_2 \end{Bmatrix} = \begin{Bmatrix} -P \\ 0 \end{Bmatrix}.$$

For a given $\bar{k}$ value, the above equation can be solved to determine the two nodal DOF. For example, for $\bar{k} = 4.2$:

$$\begin{Bmatrix} w_2 \\ \hat{\theta}_2 \end{Bmatrix} = -\frac{PL^3}{EI_y} \begin{Bmatrix} 0.2510 \\ -0.3865 \end{Bmatrix} \rightarrow \begin{Bmatrix} w_2 \\ L\theta_2 \end{Bmatrix} = -\frac{PL^3}{EI_y} \begin{Bmatrix} 0.2510 \\ -0.3865 \end{Bmatrix}.$$

Accordingly:

$$w_2 = -0.2510\frac{PL^3}{EI_y}, \quad \theta_2 = 0.3865\frac{PL^2}{EI_y}.$$

The determination of the reactions, shear force, and moment distributions is left as a problem for the reader.

## 3.6  Bending in the Other Plane

As a prelude to construction of the frame element which can undergo bending deformation in 3D space, we will consider pure bending in the $xy$-plane with no axial displacement. Figure 3.16 shows an example of a cantilevered slender body under bending loads.

**Figure 3.16** Cantilevered beam bending in the $xy$-plane

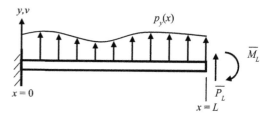

In the figure:

$v$: transverse displacement in the $y$ direction
$p_y(x)$: applied force per unit length in the $y$ direction
$\bar{P}_L$: force applied at $x = L$
$\bar{M}_L$: moment applied at $x = L$.

Following the procedures described in the previous sections for bending in the $xz$-plane, we may consider the two-node element as shown in Figure 3.17.
For uniform cross-section, the element stiffness matrix corresponding to Figure 3.17 is

**Figure 3.17** Two-node element for bending in the $xy$-plane, following the left-hand (LH) rule for the rotational angle

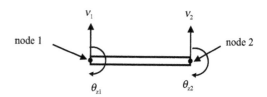

$$\mathbf{k}^e = \frac{2EI_z}{l^3} \begin{bmatrix} 6 & -3l & -6 & -3l \\ -3l & 2l^2 & 3l & l^2 \\ -6 & 3l & 6 & 3l \\ -3l & l^2 & 3l & 2l^2 \end{bmatrix}, \tag{3.6.1}$$

where

$$I_z = \int_A y^2 dA : \text{area moment of inertia with respect to the } z\text{-axis.} \tag{3.6.2}$$

However, we note that the rotational angles in Figure 3.17 follow the left-hand (LH) rule. For the rotational angle following the right-hand (RH) rule as shown in Figure 3.18:

$$(\theta_z)_{RH} = -(\theta_z)_{LH}, \ (M_z)_{RH} = -(M_z)_{LH}. \tag{3.6.3}$$

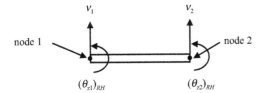

**Figure 3.18** Two-node element for bending in the $xy$-plane, following the right-hand (RH) rule for the rotational angle

Based on the FE formulation described in Section 3.2, the equilibrium equation for the element shown in Figure 3.17 is

$$\begin{Bmatrix} F_{y1} \\ (M_{z1})_{LH} \\ F_{y2} \\ (M_{z2})_{LH} \end{Bmatrix} = \frac{2EI_z}{l^3} \begin{bmatrix} 6 & -3l & -6 & -3l \\ -3l & 2l^2 & 3l & l^2 \\ -6 & 3l & 6 & 3l \\ -3l & l^2 & 3l & 2l^2 \end{bmatrix} \begin{Bmatrix} v_1 \\ (\theta_{z1})_{LH} \\ v_2 \\ (\theta_{z2})_{LH} \end{Bmatrix}. \tag{3.6.4}$$

Introducing the relationships in Eq. (3.6.3) into the above equation:

$$\begin{Bmatrix} F_{y1} \\ -(M_{z1})_{RH} \\ F_{y2} \\ -(M_{z2})_{RH} \end{Bmatrix} = \frac{2EI_z}{l^3} \begin{bmatrix} 6 & -3l & -6 & -3l \\ -3l & 2l^2 & 3l & l^2 \\ -6 & 3l & 6 & 3l \\ -3l & l^2 & 3l & 2l^2 \end{bmatrix} \begin{Bmatrix} v_1 \\ -(\theta_{z1})_{RH} \\ v_2 \\ -(\theta_{z2})_{RH} \end{Bmatrix}. \tag{3.6.5}$$

Changing the signs of the second row and the fourth row and then the second column and the fourth column in the above equation, the equilibrium equation for the element shown in Figure 3.18 is

$$\begin{Bmatrix} F_{y1} \\ (M_{z1})_{RH} \\ F_{y2} \\ (M_{z2})_{RH} \end{Bmatrix} = \frac{2EI_z}{l^3} \begin{bmatrix} 6 & 3l & -6 & 3l \\ 3l & 2l^2 & -3l & l^2 \\ -6 & -3l & 6 & -3l \\ 3l & l^2 & -3l & 2l^2 \end{bmatrix} \begin{Bmatrix} v_1 \\ (\theta_{z1})_{RH} \\ v_2 \\ (\theta_{z2})_{RH} \end{Bmatrix}. \tag{3.6.6}$$

We also note that, following the procedure described in Section 3.3, transverse displacement $v$ is assumed to be cubic and can be expressed as

$$
\begin{aligned}
v &= N_1 v_1 + N_2 l(\theta_{z1})_{LH} + N_3 v_2 + N_4 (l\theta_{z2})_{LH} \\
&\rightarrow v = N_1 v_1 + N_2 l(-\theta_{z1})_{RH} + N_3 v_2 + N_4(-l\theta_{z2})_{RH},
\end{aligned}
\tag{3.6.7}
$$

where $N_1$, $N_2$, $N_3$, $N_4$ are given in Eq. (3.2.10). Following the FE formulation described in Section 3.2, the element load vector due to $p_y(x)$ is derived as

$$
\int_{x=x_1}^{x=x_2} \delta v p_y dx = \lfloor \delta v_1 \quad (\delta\theta_1)_{LH} \quad \delta v_2 \quad (\delta\theta_2)_{LH} \rfloor l \int_{s=0}^{s=1} \begin{Bmatrix} N_1 \\ N_2 l \\ N_3 \\ N_4 l \end{Bmatrix} p_y ds
\tag{3.6.8}
$$

or

$$
\int_{x=x_1}^{x=x_2} \delta v p_y dx = \lfloor \delta v_1 \quad (\delta\theta_1)_{RH} \quad \delta v_2 \quad (\delta\theta_2)_{RH} \rfloor l \int_{s=0}^{s=1} \begin{Bmatrix} N_1 \\ -N_2 l \\ N_3 \\ -N_4 l \end{Bmatrix} p_y ds = \delta \mathbf{d}^T \mathbf{Q}^e,
\tag{3.6.9}
$$

where

$$
\mathbf{Q}^e = l \int_{s=0}^{s=1} \begin{Bmatrix} N_1 \\ -N_2 l \\ N_3 \\ -N_4 l \end{Bmatrix} p_y ds : \text{element load vector due to } p_y
\tag{3.6.10}
$$

and

$$
\mathbf{d} = \begin{Bmatrix} v_1 \\ (\theta_{z1})_{RH} \\ v_2 \\ (\theta_{z2})_{RH} \end{Bmatrix} : \text{element DOF vector.}
\tag{3.6.11}
$$

## 3.7 Frame Structures

Frame structures, commonly used in the construction of buildings and bridges, can be modeled as an assemblage of slender bodies arbitrarily oriented in 3D space. In contrast to truss structures, it is assumed that the slender bodies are joined through rigid connections sharing common rotational DOF in addition to translational DOF. Accordingly, individual members in a frame structure can carry bending loads as well as torsional loads and axial forces. As the first step, we will consider planar frame structures deforming only in the $xz$-plane.

## 3.7.1 Planar Frame Structures

Figure 3.19 shows a planar frame element arbitrarily oriented in the *global* coordinate system in the $xz$-plane. The *local* coordinate system with $\tilde{x}$ and $\tilde{z}$-axes is defined such that the $\tilde{x}$-axis is along the element body axis. Figure 3.20 shows positive displacement, rotational angle, forces, and moments in the local $\tilde{x}$ and $\tilde{z}$ system.

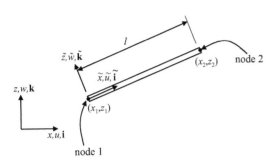

**Figure 3.19** Planar frame element arbitrarily oriented with respect to the global $x$ and $z$-axes (the $y$-axis and the $\tilde{y}$-axis are both perpendicular to the plane of the page and therefore parallel to one another; the local $\tilde{x}$ and $\tilde{z}$-axes are attached to the element)

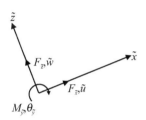

**Figure 3.20** Positive sign conventions for displacements, rotational angle, forces, and moment in the local coordinate system with the $y$-axis normal to the plane of the page

To derive the element stiffness matrix of a frame member or element, we first consider the equilibrium equation for the element in the local coordinate system, expressed as

$$\tilde{\mathbf{F}} = \tilde{\mathbf{k}}^e \tilde{\mathbf{d}}, \tag{3.7.1}$$

where $\tilde{\mathbf{F}}$ is the $6 \times 1$ column vector of nodal forces and moments such that

$$\tilde{\mathbf{F}}^T = \lfloor \tilde{F}_1 \quad \tilde{F}_2 \quad \tilde{F}_3 \quad \tilde{F}_4 \quad \tilde{F}_5 \quad \tilde{F}_6 \rfloor = \lfloor F_{\tilde{x}1} \quad F_{\tilde{z}1} \quad M_{\tilde{y}1} \quad F_{\tilde{x}2} \quad F_{\tilde{z}2} \quad M_{\tilde{y}2} \rfloor \tag{3.7.2}$$

and $\tilde{\mathbf{d}}$ is the column vector of nodal translational displacements and rotational angles such that

$$\tilde{\mathbf{d}}^T = \lfloor \tilde{d}_1 \quad \tilde{d}_2 \quad \tilde{d}_3 \quad \tilde{d}_4 \quad \tilde{d}_5 \quad \tilde{d}_6 \rfloor = \lfloor \tilde{u}_1 \quad \tilde{w}_1 \quad \theta_{\tilde{y}1} \quad \tilde{u}_2 \quad \tilde{w}_2 \quad \theta_{\tilde{y}2} \rfloor. \tag{3.7.3}$$

The column vectors $\tilde{\mathbf{F}}$ and $\tilde{\mathbf{d}}$ are connected through $\tilde{\mathbf{k}}^e$, which is the $6 \times 6$ element stiffness matrix. The element stiffness matrix can be constructed by assembling the contributions from the two-node uniaxial element and the two-node bending elements as follows.

For the two-node element under axial force described in Chapter 1, the equilibrium equation relating the nodal forces and the nodal displacements is

$$\begin{Bmatrix} F_{\tilde{x}1} = \tilde{F}_1 \\ F_{\tilde{x}2} = \tilde{F}_4 \end{Bmatrix} = \frac{EA}{l} \begin{matrix} (1) \\ (4) \end{matrix} \overset{(1)\ \ (4)}{\begin{bmatrix} 1 & -1 \\ -1 & 1 \end{bmatrix}} \begin{Bmatrix} \tilde{u}_1 = \tilde{d}_1 \\ \tilde{u}_2 = \tilde{d}_4 \end{Bmatrix},$$  (3.7.4)

using the notation in Figures 3.19 and 3.20. In the above equation, the numbers in parentheses indicate the row numbers and column numbers of each entry in the $6 \times 6$ element stiffness matrix.

For the two-node element undergoing pure bending in the $\tilde{x}\tilde{z}$-plane:

$$\begin{Bmatrix} F_{\tilde{z}1} = \tilde{F}_2 \\ M_{\tilde{y}1} = \tilde{F}_3 \\ F_{\tilde{z}2} = \tilde{F}_5 \\ M_{\tilde{y}2} = \tilde{F}_6 \end{Bmatrix} = \frac{2EI_y}{l^3} \begin{matrix} (2) \\ (3) \\ (5) \\ (6) \end{matrix} \overset{(2)\ \ (3)\ \ (5)\ \ (6)}{\begin{bmatrix} 6 & -3l & -6 & -3l \\ -3l & 2l^2 & 3l & l^2 \\ -6 & 3l & 6 & 3l \\ -3l & l^2 & 3l & 2l^2 \end{bmatrix}} \begin{Bmatrix} \tilde{w}_1 = \tilde{d}_2 \\ \theta_{\tilde{y}1} = \tilde{d}_3 \\ \tilde{w}_2 = \tilde{d}_5 \\ \theta_{\tilde{y}2} = \tilde{d}_6 \end{Bmatrix},$$  (3.7.5)

using the notation in Figures 3.19 and 3.20. The two equations can be combined into a single equation with the following element stiffness matrix in the local $\tilde{x}\tilde{z}$-plane:

$$\tilde{\mathbf{k}}^e = \begin{bmatrix} \dfrac{EA}{l} & 0 & 0 & -\dfrac{EA}{l} & 0 & 0 \\[2mm] 0 & \dfrac{12EI_y}{l^3} & -\dfrac{6EI_y}{l^2} & 0 & -\dfrac{12EI_y}{l^3} & -\dfrac{6EI_y}{l^2} \\[2mm] 0 & -\dfrac{6EI_y}{l^2} & \dfrac{4EI_y}{l} & 0 & \dfrac{6EI_y}{l^2} & \dfrac{2EI_y}{l} \\[2mm] -\dfrac{EA}{l} & 0 & 0 & \dfrac{EA}{l} & 0 & 0 \\[2mm] 0 & -\dfrac{12EI_y}{l^3} & \dfrac{6EI_y}{l^2} & 0 & \dfrac{12EI_y}{l^3} & \dfrac{6EI_y}{l^2} \\[2mm] 0 & -\dfrac{6EI_y}{l^2} & \dfrac{2EI_y}{l} & 0 & \dfrac{6EI_y}{l^2} & \dfrac{4EI_y}{l} \end{bmatrix}.$$  (3.7.6)

### Element Stiffness Matrix in the Global Coordinate System

For an element arbitrarily oriented as shown in Figure 3.19, positive displacements, rotational angle, forces, and moment in the global coordinate system are shown in Figure 3.21. The equilibrium equation in the global coordinate system can then be expressed as

$$\mathbf{F} = \mathbf{k}^e \mathbf{d},$$  (3.7.7)

where $\mathbf{F}$ is the $6 \times 1$ column vector of nodal forces and moments in the global coordinate system such that

$$\mathbf{F}^{\mathrm{T}} = \lfloor F_1 \ \ F_2 \ \ F_3 \ \ F_4 \ \ F_5 \ \ F_6 \rfloor = \lfloor F_{x1} \ \ F_{z1} \ \ M_{y1} \ \ F_{x2} \ \ F_{z2} \ \ M_{y2} \rfloor$$  (3.7.8)

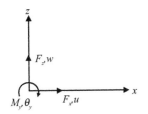

**Figure 3.21** Positive displacements, rotational angle, forces, and moment in the global coordinates system

and **d** is the 6 × 1 column vector of nodal translational displacements and rotational angles in the global coordinate system such that

$$\mathbf{d}^{\mathrm{T}} = \lfloor d_1 \quad d_2 \quad d_3 \quad d_4 \quad d_5 \quad d_6 \rfloor = \lfloor u_1 \quad w_1 \quad \theta_{y1} \quad u_2 \quad w_2 \quad \theta_{y2} \rfloor. \qquad (3.7.9)$$

The column vectors **F** and **d** are connected through $\mathbf{k}^e$, which is the 6 × 6 element stiffness matrix in the global coordinate system. The element stiffness matrix $\mathbf{k}^e$ defined with respect to the global $x$, $z$-coordinate system can be constructed from $\tilde{\mathbf{k}}^e$ once the direction cosines between the global $x$, $z$-coordinate system and the local $\tilde{x}$, $\tilde{z}$ system attached to the element are provided.

Note that the displacement vector **u** can be written in either coordinate system as

$$\mathbf{u} = u\mathbf{i} + w\mathbf{k} = \tilde{u}\tilde{\mathbf{i}} + \tilde{w}\tilde{\mathbf{k}}. \qquad (3.7.10)$$

Then

$$\tilde{u} = \tilde{\mathbf{i}} \cdot \mathbf{u} = \tilde{\mathbf{i}} \cdot (u\mathbf{i} + w\mathbf{k}) = (\tilde{\mathbf{i}} \cdot \mathbf{i})u + (\tilde{\mathbf{i}} \cdot \mathbf{k})w,$$
$$\tilde{w} = \tilde{\mathbf{k}} \cdot \mathbf{u} = \tilde{\mathbf{k}} \cdot (u\mathbf{i} + w\mathbf{k}) = (\tilde{\mathbf{k}} \cdot \mathbf{i})u + (\tilde{\mathbf{k}} \cdot \mathbf{k})w. \qquad (3.7.11)$$

From Figure 3.19:

$$l\tilde{\mathbf{i}} = (x_2 - x_1)\mathbf{i} + (z_2 - z_1)\mathbf{k} \rightarrow \tilde{\mathbf{i}} = \frac{(x_2 - x_1)}{l}\mathbf{i} + \frac{(z_2 - z_1)}{l}\mathbf{k},$$
$$\tilde{\mathbf{k}} = -\frac{(z_2 - z_1)}{l}\mathbf{i} + \frac{(x_2 - x_1)}{l}\mathbf{k}, \qquad (3.7.12)$$

where

$$l = \sqrt{(x_2 - x_1)^2 + (z_2 - z_1)^2}. \qquad (3.7.13)$$

The direction cosines are then

$$\tilde{\mathbf{i}} \cdot \mathbf{i} = \frac{(x_2 - x_1)}{l}, \tilde{\mathbf{i}} \cdot \mathbf{k} = \frac{(z_2 - z_1)}{l}, \tilde{\mathbf{k}} \cdot \mathbf{i} = -\frac{(z_2 - z_1)}{l}, \tilde{\mathbf{k}} \cdot \mathbf{k} = \frac{(x_2 - x_1)}{l}. \qquad (3.7.14)$$

For the planar frame, the $y$-axis is in parallel with the $\tilde{y}$-axis. Accordingly, for rotational angles:

$$\theta_y = \theta_{\tilde{y}}. \qquad (3.7.15)$$

Then, from Eqs (3.7.11) and (3.7.15):

$$\left\{ \begin{array}{c} \tilde{u} \\ \tilde{w} \\ \theta_{\tilde{y}} \end{array} \right\} = \begin{bmatrix} \tilde{\mathbf{i}} \cdot \mathbf{i} & \tilde{\mathbf{i}} \cdot \mathbf{k} & 0 \\ \tilde{\mathbf{k}} \cdot \mathbf{i} & \tilde{\mathbf{k}} \cdot \mathbf{k} & 0 \\ 0 & 0 & 1 \end{bmatrix} \left\{ \begin{array}{c} u \\ w \\ \theta_y \end{array} \right\}. \tag{3.7.16}$$

Accordingly, the DOF vector with respect to the local $\tilde{x}, \tilde{z}$ system can be expressed in terms of the DOF vector in the global $x, z$ system as

$$\tilde{\mathbf{d}} = \mathbf{T}\mathbf{d}, \tag{3.7.17}$$

where

$$\mathbf{T} = \begin{bmatrix} \tilde{\mathbf{i}} \cdot \mathbf{i} & \tilde{\mathbf{i}} \cdot \mathbf{k} & 0 & 0 & 0 & 0 \\ \tilde{\mathbf{k}} \cdot \mathbf{i} & \tilde{\mathbf{k}} \cdot \mathbf{k} & 0 & 0 & 0 & 0 \\ 0 & 0 & 1 & 0 & 0 & 0 \\ 0 & 0 & 0 & \tilde{\mathbf{i}} \cdot \mathbf{i} & \tilde{\mathbf{i}} \cdot \mathbf{k} & 0 \\ 0 & 0 & 0 & \tilde{\mathbf{k}} \cdot \mathbf{i} & \tilde{\mathbf{k}} \cdot \mathbf{k} & 0 \\ 0 & 0 & 0 & 0 & 0 & 1 \end{bmatrix} : 6 \times 6 \text{ transformation matrix.} \tag{3.7.18}$$

Let us now consider the force terms in Eqs (3.7.2) and (3.7.8). Following the discussion of the displacement vector in Eq. (3.7.10), a force vector can similarly be written in either coordinate system as

$$F_x \mathbf{i} + F_z \mathbf{k} = F_{\tilde{x}} \tilde{\mathbf{i}} + F_{\tilde{z}} \tilde{\mathbf{k}}. \tag{3.7.19}$$

Then

$$\left\{ \begin{array}{c} F_x \\ F_z \end{array} \right\} = \begin{bmatrix} \mathbf{i} \cdot \tilde{\mathbf{i}} & \mathbf{i} \cdot \tilde{\mathbf{k}} \\ \mathbf{k} \cdot \tilde{\mathbf{i}} & \mathbf{k} \cdot \tilde{\mathbf{k}} \end{bmatrix} \left\{ \begin{array}{c} F_{\tilde{x}} \\ F_{\tilde{z}} \end{array} \right\}. \tag{3.7.20}$$

Since the $y$-axis is parallel to the $\tilde{y}$-axis:

$$M_y = M_{\tilde{y}}. \tag{3.7.21}$$

Then

$$\left\{ \begin{array}{c} F_x \\ F_z \\ M_y \end{array} \right\} = \begin{bmatrix} \mathbf{i} \cdot \tilde{\mathbf{i}} & \mathbf{i} \cdot \tilde{\mathbf{k}} & 0 \\ \mathbf{k} \cdot \tilde{\mathbf{i}} & \mathbf{k} \cdot \tilde{\mathbf{k}} & 0 \\ 0 & 0 & 1 \end{bmatrix} \left\{ \begin{array}{c} F_{\tilde{x}} \\ F_{\tilde{z}} \\ M_{\tilde{y}} \end{array} \right\}. \tag{3.7.22}$$

Accordingly, it can be shown that

$$\mathbf{F} = \mathbf{T}^{\mathrm{T}} \tilde{\mathbf{F}}. \tag{3.7.23}$$

Introducing Eq. (3.7.1) and then Eq. (3.7.17) into the above equation:

$$\mathbf{F} = \mathbf{T}^{\mathrm{T}} \tilde{\mathbf{k}}^e \tilde{\mathbf{d}} = \mathbf{T}^{\mathrm{T}} \tilde{\mathbf{k}}^e \mathbf{T} \mathbf{d}. \tag{3.7.24}$$

Comparing the above equation with Eq. (3.7.7), the $6 \times 6$ element stiffness matrix in the global coordinate system is

$$\mathbf{k}^e = \mathbf{T}^{\mathrm{T}} \tilde{\mathbf{k}}^e \mathbf{T}. \tag{3.7.25}$$

## Example 3.6

For the planar frame structure shown in Figure 3.22, $E = 200$ GPa, $A = 0.05$ m$^2$, $I_y = 3.5 \times 10^{-4}$ m$^4$, $L = 10\sqrt{2}$ m. Determine the displacements and rotation at the node at which a downward force of $P$ is acting as follows:

(a) Construct the element stiffness matrices in the global coordinate system.
(b) Construct the equilibrium equation for the planar frame and determine the nodal DOF.
(c) Determine the force and moment distribution for each element.

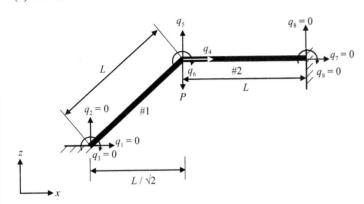

**Figure 3.22** A planar frame structure subjected to a downward force of $P$

## Solution:

(a) Element #1: Element node 1 is at the left bottom end. Then

$$x_2 - x_1 = 10,\ z_2 - z_1 = 10,\ l = 10\sqrt{2} \rightarrow \tilde{\mathbf{i}} = \frac{1}{\sqrt{2}}\mathbf{i} + \frac{1}{\sqrt{2}}\mathbf{k},\ \tilde{\mathbf{k}} = -\frac{1}{\sqrt{2}}\mathbf{i} + \frac{1}{\sqrt{2}}\mathbf{k}.$$

Accordingly, $\tilde{\mathbf{i}} \cdot \mathbf{i} = 1/\sqrt{2}$, $\tilde{\mathbf{i}} \cdot \mathbf{k} = 1/\sqrt{2}$, $\tilde{\mathbf{k}} \cdot \mathbf{i} = -1/\sqrt{2}$, $\tilde{\mathbf{k}} \cdot \mathbf{k} = 1/\sqrt{2}$. The transformation matrix is

$$\mathbf{T} = \begin{bmatrix} 1/\sqrt{2} & 1/\sqrt{2} & 0 & 0 & 0 & 0 \\ -1/\sqrt{2} & 1/\sqrt{2} & 0 & 0 & 0 & 0 \\ 0 & 0 & 1 & 0 & 0 & 0 \\ 0 & 0 & 0 & 1/\sqrt{2} & 1/\sqrt{2} & 0 \\ 0 & 0 & 0 & -1/\sqrt{2} & 1/\sqrt{2} & 0 \\ 0 & 0 & 0 & 0 & 0 & 1 \end{bmatrix}.$$

The element stiffness matrix in the global coordinate system is then

$$\mathbf{k}^1 = \mathbf{T}^T \tilde{\mathbf{k}}^1 \mathbf{T}.$$

Element #2: For this element, node 1 is located at the left end, and $\tilde{\mathbf{i}} = \mathbf{i}$, $\tilde{\mathbf{k}} = \mathbf{k}$, and $\mathbf{T}$ is an identity matrix. Accordingly, there is no need for transformation and the element stiffness matrix is given in Eq. (3.7.6).

(b) For the global load vector, $F_5 = -P$ for the applied load. After the global stiffness matrix is assembled from the element stiffness matrices and the geometric constraints are prescribed, the resulting equation can be solved for unknown DOF such that

$$q_4 = 1.4101 \times 10^{-9} P \, \text{m},$$
$$q_5 = -4.2326 \times 10^{-9} P \, \text{m},$$
$$q_6 = -1.2868 \times 10^{-11} P \, \text{rad}.$$

(c) For element #1, the two nodal displacements and rotational angle in the local coordinate system can be determined using Eq. (3.7.16) such that

$$\begin{Bmatrix} \tilde{u} \\ \tilde{w} \\ \theta_{\tilde{y}} \end{Bmatrix} = \begin{bmatrix} \tilde{i}\cdot i & \tilde{i}\cdot k & 0 \\ \tilde{k}\cdot i & \tilde{k}\cdot k & 0 \\ 0 & 0 & 1 \end{bmatrix} \begin{Bmatrix} u \\ w \\ \theta_y \end{Bmatrix} \rightarrow \begin{Bmatrix} \tilde{u}_2 \\ \tilde{w}_2 \\ \theta_{\tilde{y}2} \end{Bmatrix} = \begin{bmatrix} 1/\sqrt{2} & 1/\sqrt{2} & 0 \\ -1/\sqrt{2} & 1/\sqrt{2} & 0 \\ 0 & 0 & 1 \end{bmatrix} \begin{Bmatrix} q_4 \\ q_5 \\ q_6 \end{Bmatrix}.$$

We can then use Eq. (3.7.1) to determine nodal forces and moments of the element in the local coordinate system, which can be used to determine the force and moment distributions along element #1. For element #2, the transformation is not needed. A similar approach can be used to determine the force and moment distribution. The details are left as an exercise for the reader.

## 3.7.2 Frame Element in 3D Space

Consider a frame element oriented arbitrarily in the global coordinate system as shown in Figure 3.23. The frame element can carry axial, bending, and torsional loads. As shown in Figure 3.24, the rotational angles and moments follow the RH rule. To derive the element stiffness matrix for a frame member, we first consider the element with the local coordinate system aligned with the body axes:

$$\tilde{\mathbf{F}} = \tilde{\mathbf{k}}^e \tilde{\mathbf{d}}, \tag{3.7.26}$$

where $\tilde{\mathbf{F}}$ is the $12 \times 1$ column vector of nodal forces and moments such that

$$\tilde{\mathbf{F}}^T = \lfloor \tilde{F}_1 \ \ \tilde{F}_2 \ \ \tilde{F}_3 \ \ \tilde{F}_4 \ \ \tilde{F}_5 \ \ \tilde{F}_6 \ \ \tilde{F}_7 \ \ \tilde{F}_8 \ \ \tilde{F}_9 \ \ \tilde{F}_{10} \ \ \tilde{F}_{11} \ \ \tilde{F}_{12} \rfloor$$
$$= \lfloor F_{\tilde{x}1} \ \ F_{\tilde{y}1} \ \ F_{\tilde{z}1} \ \ M_{\tilde{x}1} \ \ M_{\tilde{y}1} \ \ M_{\tilde{z}1} \ \ F_{\tilde{x}2} \ \ F_{\tilde{y}2} \ \ F_{\tilde{z}2} \ \ M_{\tilde{x}2} \ \ M_{\tilde{y}2} \ \ M_{\tilde{z}2} \rfloor \tag{3.7.27}$$

and $\tilde{\mathbf{d}}$ is the $12 \times 1$ column vector of nodal translational displacements and rotational angles, such that

$$\tilde{\mathbf{d}}^T = \lfloor \tilde{d}_1 \ \ \tilde{d}_2 \ \ \tilde{d}_3 \ \ \tilde{d}_4 \ \ \tilde{d}_5 \ \ \tilde{d}_6 \ \ \tilde{d}_7 \ \ \tilde{d}_8 \ \ \tilde{d}_9 \ \ \tilde{d}_{10} \ \ \tilde{d}_{11} \ \ \tilde{d}_{12} \rfloor$$
$$= \lfloor \tilde{u}_1 \ \ \tilde{v}_1 \ \ \tilde{w}_1 \ \ \theta_{\tilde{x}1} \ \ \theta_{\tilde{y}1} \ \ \theta_{\tilde{z}1} \ \ \tilde{u}_2 \ \ \tilde{v}_2 \ \ \tilde{w}_2 \ \ \theta_{\tilde{x}2} \ \ \theta_{\tilde{y}2} \ \ \theta_{\tilde{z}2} \rfloor. \tag{3.7.28}$$

The column vectors $\tilde{\mathbf{F}}$ and $\tilde{\mathbf{d}}$ are connected through $\tilde{\mathbf{k}}^e$, which is a $12 \times 12$ element stiffness matrix. The element stiffness matrix can be constructed by assembling the contributions from the two-node uniaxial element, torsional element, and bending elements as follows.

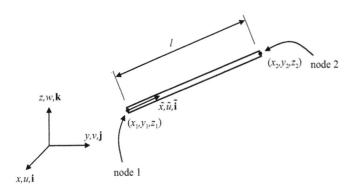

**Figure 3.23** Two-node frame element inclined with respect to the global coordinate system in 3D space

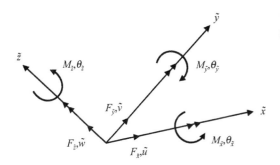

**Figure 3.24** Positive displacements, rotational angles, forces, and moments in the local coordinate system

## The Slender Body under Axial Force

For the two-node element under axial force described in Chapter 1, the equilibrium equation relating the nodal forces and the nodal displacements is

$$\begin{Bmatrix} F_{\tilde{x}1} = \tilde{F}_1 \\ F_{\tilde{x}2} = \tilde{F}_7 \end{Bmatrix} = \frac{EA}{l} \begin{matrix} (1) & (7) \\ \begin{bmatrix} 1 & -1 \\ -1 & 1 \end{bmatrix} \end{matrix} \begin{matrix} (1) \\ (7) \end{matrix} \begin{Bmatrix} \tilde{u}_1 = \tilde{d}_1 \\ \tilde{u}_2 = \tilde{d}_7 \end{Bmatrix}. \tag{3.7.29}$$

In the above equation, the numbers in parentheses indicate the row numbers and column numbers of each entry in the $12 \times 12$ element stiffness matrix.

For the same body under torsional moment, the equilibrium equation is

$$\begin{Bmatrix} M_{\tilde{x}1} = \tilde{F}_4 \\ M_{\tilde{x}2} = \tilde{F}_{10} \end{Bmatrix} = \frac{GJ}{l} \begin{matrix} (4) & (10) \\ \begin{bmatrix} 1 & -1 \\ -1 & 1 \end{bmatrix} \end{matrix} \begin{matrix} (4) \\ (10) \end{matrix} \begin{Bmatrix} \theta_{\tilde{x}1} = \tilde{d}_4 \\ \theta_{\tilde{x}2} = \tilde{d}_{10} \end{Bmatrix}. \tag{3.7.30}$$

For pure bending in the $\tilde{x}\tilde{z}$-plane:

$$\begin{Bmatrix} F_{\tilde{z}1} = \tilde{F}_3 \\ M_{\tilde{y}1} = \tilde{F}_5 \\ F_{\tilde{z}2} = \tilde{F}_9 \\ M_{\tilde{y}2} = \tilde{F}_{11} \end{Bmatrix} = \frac{2EI_y}{l^3} \begin{matrix} (3) \\ (5) \\ (9) \\ (11) \end{matrix} \begin{bmatrix} \overset{(3)\ \ (5)\ \ (9)\ \ (11)}{6} & -3l & -6 & -3l \\ -3l & 2l^2 & 3l & l^2 \\ -6 & 3l & 6 & 3l \\ -3l & l^2 & 3l & 2l^2 \end{bmatrix} \begin{Bmatrix} \tilde{w}_1 = \tilde{d}_3 \\ \theta_{\tilde{y}1} = \tilde{d}_5 \\ \tilde{w}_2 = \tilde{d}_9 \\ \theta_{\tilde{y}2} = \tilde{d}_{11} \end{Bmatrix}. \tag{3.7.31}$$

For pure bending in the $\tilde{x}\tilde{y}$-plane:

$$\begin{Bmatrix} F_{\tilde{y}1} = \tilde{F}_2 \\ M_{\tilde{z}1} = \tilde{F}_6 \\ F_{\tilde{y}2} = \tilde{F}_8 \\ M_{\tilde{z}2} = \tilde{F}_{12} \end{Bmatrix} = \frac{2EI_z}{l^3} \begin{matrix} (2) \\ (6) \\ (8) \\ (12) \end{matrix} \begin{bmatrix} \overset{(2)\ \ (6)\ \ (8)\ \ (12)}{6} & 3l & -6 & 3l \\ 3l & 2l^2 & -3l & l^2 \\ -6 & -3l & 6 & -3l \\ 3l & l^2 & -3l & 2l^2 \end{bmatrix} \begin{Bmatrix} \tilde{v}_1 = \tilde{d}_2 \\ \theta_{\tilde{z}1} = \tilde{d}_6 \\ \tilde{v}_2 = \tilde{d}_8 \\ \theta_{\tilde{z}2} = \tilde{d}_{12} \end{Bmatrix}. \tag{3.7.32}$$

## Element Stiffness Matrix for the Local Coordinate System

After accounting for all contributions, the stiffness matrix of the frame element in the local coordinate system can be expressed as shown in Eq. (3.7.33):

$$\tilde{k}^e = \begin{bmatrix}
\frac{EA}{l} & 0 & 0 & 0 & 0 & 0 & -\frac{EA}{l} & 0 & 0 & 0 & 0 & 0 \\
0 & \frac{12EI_z}{l^3} & 0 & 0 & 0 & \frac{6EI_z}{l^2} & 0 & -\frac{12EI_z}{l^3} & 0 & 0 & 0 & \frac{6EI_z}{l^2} \\
0 & 0 & \frac{12EI_y}{l^3} & 0 & -\frac{6EI_y}{l^2} & 0 & 0 & 0 & -\frac{12EI_y}{l^3} & 0 & -\frac{6EI_y}{l^2} & 0 \\
0 & 0 & 0 & \frac{GJ}{l} & 0 & 0 & 0 & 0 & 0 & -\frac{GJ}{l} & 0 & 0 \\
0 & 0 & -\frac{6EI_y}{l^2} & 0 & \frac{4EI_y}{l} & 0 & 0 & 0 & \frac{6EI_y}{l^2} & 0 & \frac{2EI_y}{l} & 0 \\
0 & \frac{6EI_z}{l^2} & 0 & 0 & 0 & \frac{4EI_z}{l} & 0 & -\frac{6EI_z}{l^2} & 0 & 0 & 0 & \frac{2EI_z}{l} \\
-\frac{EA}{l} & 0 & 0 & 0 & 0 & 0 & \frac{EA}{l} & 0 & 0 & 0 & 0 & 0 \\
0 & -\frac{12EI_z}{l^3} & 0 & 0 & 0 & -\frac{6EI_z}{l^2} & 0 & \frac{12EI_z}{l^3} & 0 & 0 & 0 & -\frac{6EI_z}{l^2} \\
0 & 0 & -\frac{12EI_y}{l^3} & 0 & \frac{6EI_y}{l^2} & 0 & 0 & 0 & \frac{12EI_y}{l^3} & 0 & \frac{6EI_y}{l^2} & 0 \\
0 & 0 & 0 & -\frac{GJ}{l} & 0 & 0 & 0 & 0 & 0 & \frac{GJ}{l} & 0 & 0 \\
0 & 0 & -\frac{6EI_y}{l^2} & 0 & \frac{2EI_y}{l} & 0 & 0 & 0 & \frac{6EI_y}{l^2} & 0 & \frac{4EI_y}{l} & 0 \\
0 & \frac{6EI_z}{l^2} & 0 & 0 & 0 & \frac{2EI_z}{l} & 0 & -\frac{6EI_z}{l^2} & 0 & 0 & 0 & \frac{4EI_z}{l}
\end{bmatrix} \tag{3.7.33}$$

### Element Stiffness Matrix for the Global Coordinate System

To construct the frame element stiffness matrix in the global coordinate system, a coordinate transformation must be performed. Referring to Figure 3.25, our goal is to determine the equilibrium equation in the global coordinate system as

$$\mathbf{F} = \mathbf{k}^e \mathbf{d}, \tag{3.7.34}$$

where $\mathbf{F}$ is the $12 \times 1$ column vector of nodal forces and moments in the global coordinate system such that

$$\mathbf{F}^T = \lfloor F_1 \quad F_2 \quad F_3 \quad F_4 \quad F_5 \quad F_6 \quad F_7 \quad F_8 \quad F_9 \quad F_{10} \quad F_{11} \quad F_{12} \rfloor$$
$$= \lfloor F_{x1} \quad F_{y1} \quad F_{z1} \quad M_{x1} \quad M_{y1} \quad M_{z1} \quad F_{x2} \quad F_{y2} \quad F_{z2} \quad M_{x2} \quad M_{y2} \quad M_{z2} \rfloor \tag{3.7.35}$$

and $\mathbf{d}$ is the $12 \times 1$-column vector of nodal translational displacements and rotational angles in the global coordinate system such that

$$\mathbf{d}^T = \lfloor d_1 \quad d_2 \quad d_3 \quad d_4 \quad d_5 \quad d_6 \quad d_7 \quad d_8 \quad d_9 \quad d_{10} \quad d_{11} \quad d_{12} \rfloor$$
$$= \lfloor u_1 \quad v_1 \quad w_1 \quad \theta_{x1} \quad \theta_{y1} \quad \theta_{z1} \quad u_2 \quad v_2 \quad w_2 \quad \theta_{x2} \quad \theta_{y2} \quad \theta_{z2} \rfloor. \tag{3.7.36}$$

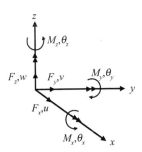

**Figure 3.25** Positive displacements, rotational angles, forces, and moments at each node in the global coordinates system

The element stiffness matrix $\mathbf{k}^e$ in Eq. (3.7.34), defined with respect to the global coordinate system, can be constructed from $\tilde{\mathbf{k}}^e$ using the coordinate transformation with the help of the direction cosines between the global $x$, $y$, $z$ coordinate system and the local $\tilde{x}, \tilde{y}, \tilde{z}$ system attached to the element.

The displacement vector $\mathbf{u}$ can be written in either coordinate system as

$$\mathbf{u} = u\mathbf{i} + v\mathbf{j} + w\mathbf{k} = \tilde{u}\tilde{\mathbf{i}} + \tilde{v}\tilde{\mathbf{j}} + \tilde{w}\tilde{\mathbf{k}}. \tag{3.7.37}$$

Then

$$\begin{Bmatrix} \tilde{u} \\ \tilde{v} \\ \tilde{w} \end{Bmatrix} = \begin{bmatrix} \tilde{\mathbf{i}} \cdot \mathbf{i} & \tilde{\mathbf{i}} \cdot \mathbf{j} & \tilde{\mathbf{i}} \cdot \mathbf{k} \\ \tilde{\mathbf{j}} \cdot \mathbf{i} & \tilde{\mathbf{j}} \cdot \mathbf{j} & \tilde{\mathbf{j}} \cdot \mathbf{k} \\ \tilde{\mathbf{k}} \cdot \mathbf{i} & \tilde{\mathbf{k}} \cdot \mathbf{j} & \tilde{\mathbf{k}} \cdot \mathbf{k} \end{bmatrix} \begin{Bmatrix} u \\ v \\ w \end{Bmatrix} = \Lambda \begin{Bmatrix} u \\ v \\ w \end{Bmatrix}, \tag{3.7.38}$$

where

$$\mathbf{\Lambda} = \begin{bmatrix} \tilde{\mathbf{i}} \cdot \mathbf{i} & \tilde{\mathbf{i}} \cdot \mathbf{j} & \tilde{\mathbf{i}} \cdot \mathbf{k} \\ \tilde{\mathbf{j}} \cdot \mathbf{i} & \tilde{\mathbf{j}} \cdot \mathbf{j} & \tilde{\mathbf{j}} \cdot \mathbf{k} \\ \tilde{\mathbf{k}} \cdot \mathbf{i} & \tilde{\mathbf{k}} \cdot \mathbf{j} & \tilde{\mathbf{k}} \cdot \mathbf{k} \end{bmatrix} : \text{matrix of direction cosines.} \tag{3.7.39}$$

For small rotational angles, one may define a rotation vector $\mathbf{\Theta}$ as

$$\mathbf{\Theta} = \theta_x \mathbf{i} + \theta_y \mathbf{j} + \theta_z \mathbf{k} = \theta_{\tilde{x}} \tilde{\mathbf{i}} + \theta_{\tilde{y}} \tilde{\mathbf{j}} + \theta_{\tilde{z}} \tilde{\mathbf{k}}. \tag{3.7.40}$$

The rotational angles in the two coordinate systems are then related as

$$\begin{Bmatrix} \theta_{\tilde{x}} \\ \theta_{\tilde{y}} \\ \theta_{\tilde{z}} \end{Bmatrix} = \mathbf{\Lambda} \begin{Bmatrix} \theta_x \\ \theta_y \\ \theta_z \end{Bmatrix}. \tag{3.7.41}$$

The DOF vector with respect to the local $\tilde{x}$, $\tilde{y}$, $\tilde{z}$ system can then be expressed in terms of the DOF vector in the global system as

$$\tilde{\mathbf{d}} = \mathbf{T}\mathbf{d}, \tag{3.7.42}$$

where

$$\mathbf{T} = \begin{bmatrix} \mathbf{\Lambda} & \mathbf{0} & \mathbf{0} & \mathbf{0} \\ \mathbf{0} & \mathbf{\Lambda} & \mathbf{0} & \mathbf{0} \\ \mathbf{0} & \mathbf{0} & \mathbf{\Lambda} & \mathbf{0} \\ \mathbf{0} & \mathbf{0} & \mathbf{0} & \mathbf{\Lambda} \end{bmatrix} : 12 \times 12 \text{ transformation matrix} \tag{3.7.43}$$

is constructed using the direction cosine matrix $\mathbf{\Lambda}$ defined in Eq. (3.7.39) and $\mathbf{0}$ which is a $3 \times 3$ null matrix.

A force vector can be written with respect to either coordinate system as

$$F_x \mathbf{i} + F_y \mathbf{j} + F_z \mathbf{k} = F_{\tilde{x}} \tilde{\mathbf{i}} + F_{\tilde{y}} \tilde{\mathbf{j}} + F_{\tilde{z}} \tilde{\mathbf{k}}. \tag{3.7.44}$$

Then it can be shown that

$$\begin{Bmatrix} F_x \\ F_y \\ F_z \end{Bmatrix} = \mathbf{\Lambda}^T \begin{Bmatrix} F_{\tilde{x}} \\ F_{\tilde{y}} \\ F_{\tilde{z}} \end{Bmatrix}. \tag{3.7.45}$$

The moment vector can similarly be written as

$$M_x \mathbf{i} + M_y \mathbf{j} + M_z \mathbf{k} = M_{\tilde{x}} \tilde{\mathbf{i}} + M_{\tilde{y}} \tilde{\mathbf{j}} + M_{\tilde{z}} \tilde{\mathbf{k}}. \tag{3.7.46}$$

It therefore follows that

$$\begin{Bmatrix} M_x \\ M_y \\ M_z \end{Bmatrix} = \mathbf{\Lambda}^T \begin{Bmatrix} M_{\tilde{x}} \\ M_{\tilde{y}} \\ M_{\tilde{z}} \end{Bmatrix}. \tag{3.7.47}$$

Accordingly:

$$\mathbf{F} = \mathbf{T}^\mathsf{T}\tilde{\mathbf{F}}. \tag{3.7.48}$$

Introducing Eq. (3.7.26) and then Eq. (3.7.42) into the above equation:

$$\mathbf{F} = \mathbf{T}^\mathsf{T}\tilde{\mathbf{k}}^e\tilde{\mathbf{d}} = \mathbf{T}^\mathsf{T}\tilde{\mathbf{k}}^e\mathbf{T}\mathbf{d}. \tag{3.7.49}$$

Comparing the above equation with Eq. (3.7.34), the $12 \times 12$ element stiffness matrix in the global coordinate system is

$$\mathbf{k}^e = \mathbf{T}^\mathsf{T}\tilde{\mathbf{k}}^e\mathbf{T}. \tag{3.7.50}$$

In order to appreciate the effectiveness of the frame element in 3D space, one can try a problem at the end of this chapter.

## PROBLEMS

Work with $\hat{\theta} = l\theta$ if possible. For problem sets 3.1–3.6, you may then find the solutions in non-dimensional form without using specific numerical values for geometry, Young's modulus, and loads. Alternatively, unless otherwise indicated, you may work with $E = 200$ GPa, $I_y = 3.5 \times 10^{-4}$ m$^4$, and $L = 5$ m.

**3.1** Consider a slender cantilevered body of constant $EI_y$ and length $L$ shown in Figure 3.26. The body is subjected to an applied force per unit length expressed as $p_z(x) = -p_o$, where $p_o$ is a given constant value. Use a model with two elements of equal length as shown in Figure 3.27.
   (a) Construct the global stiffness matrix.
   (b) Construct the global load vector in terms of $p_o$.
   (c) Apply the geometric boundary conditions and solve for the nodal DOF vector in terms of $p_o$.
   (d) Determine the reaction force and moment at the clamped end.
   (e) Compare with the exact solution given as follows:

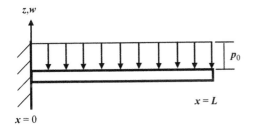

Figure 3.26  For Problem 3.1

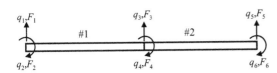

Figure 3.27  Two-element model for Problems 3.1, 3.2, and 3.4

$$V_z(x) = -p_0(L - x), \quad M_y(x) = \frac{1}{2}p_0(L - x)^2,$$

$$w = -\frac{p_0}{EI_y}\left(\frac{1}{24}x^4 - \frac{1}{6}Lx^3 + \frac{1}{4}L^2x^2\right) = -\frac{p_0L^4}{24EI_y}\left[\left(\frac{x}{L}\right)^4 - 4\left(\frac{x}{L}\right)^3 + 6\left(\frac{x}{L}\right)^2\right].$$

**3.2** Consider a slender body of constant $EI_y$ and length $L$ shown in Figure 3.28. The body is clamped at $x = 0$ and constrained against transverse displacement but free to rotate at $x = L/2$. The body is subjected to an applied force per unit length expressed as $p_z = -p_0$, where $p_o$ is a given constant value. Use a model with two elements of equal length to do as follows:

(a) Determine the unknown nodal DOF in terms of $p_o$.

(b) Find all reaction forces and moments that can be used to determine the shear force and moment distributions along the length.

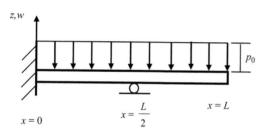

**Figure 3.28** For Problem 3.2

**3.3** Repeat Problem 3.2 using a model with four elements of equal length.

**3.4** A cantilevered slender body of constant $EI_y$ and length $L$ is clamped at $x = 0$ and subjected to an applied force per unit length expressed as follows:

$$p_z(x) = p_o\left[1 - \left(\frac{x}{L}\right)^2\right],$$

where $p_o$ is a given value. The structure is modeled using three elements of equal length as shown in Figure 3.29.

(a) Construct the element load vector for each element and assemble the global load vector.

(b) Construct the global stiffness matrix.

(c) Solve for the nodal DOF vector in terms of $p_o$.

(d) Determine the reaction force and moment at the clamped end.

**Figure 3.29** For Problem 3.4

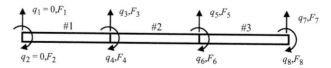

(e) Compare the nodal displacements and rotations with those from the exact solution, given as

$$w(x) = \frac{p_0 L^4}{360 E I_y}\left[-\left(\frac{x}{L}\right)^6 + 15\left(\frac{x}{L}\right)^4 - 40\left(\frac{x}{L}\right)^3 + 45\left(\frac{x}{L}\right)^2\right].$$

**3.5** Consider a cantilevered beam under tip load described in Example 3.5. The beam is placed on an elastic foundation. Determine the following for $\bar{k} = 3.0$ using a model with one element:

(a) Unknown nodal DOF.
(b) Reactions at the clamped end.
(c) Shear and moment distributions.

**3.6** Try Problem 3.5 with a model with two elements of equal length.

**3.7** A clamped beam of non-uniform cross-section is subjected to a tip force as shown in Figure 3.30. The area moment of inertia of the cross-section can be expressed as

$$I_y = I_0\left(1 + \alpha\frac{x}{L}\right),$$

where $I_0$ and $\alpha$ are given constant values. Do the following for $E I_0 = 2 \times 10^5$ N m$^2$, $L = 3.5$ m, and $\alpha = -0.5$.

(a) Use a model with one element to determine the unknown nodal DOF.
(b) Use a model with two elements of equal length to determine the unknown nodal DOF.
(c) Use a model with three elements of equal length to determine the unknown nodal DOF.

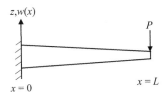

$z,w(x)$

$P$

$x = L$

$x = 0$

**Figure 3.30** For Problem 3.7

**3.8** The slender body described in Problem 3.7 is now subjected to an applied force per unit length expressed as

$$p_z(x) = p_o\left(1 - \frac{x}{L}\right),$$

where $p_o$ is a given constant value.

(a) Use a model with two elements of equal length to determine the unknown nodal DOF.
(b) Use a model with four elements of equal length to determine the unknown nodal DOF.

**3.9** For the planar frame structure described in Example 3.6 and Figure 3.22, the applied load is $F_4 = F_5 = 0$, $F_6 = Q$ N m.

(a) Determine the nodal DOF in terms of the applied moment $Q$.

(b) Determine the force and moment distributions in each element.

**3.10** The frame structure shown in Figure 3.22 is fixed at two ends and subjected to a force of $P$ in the direction normal to the plane such that $F_z = -P$ at the junction where the two elements are joined. For the frame structure, $E = 200$ GPa, $A = 0.05$ m$^2$, $I_y = I_z = 3.5 \times 10^{-4}$ m$^4$, $L = 10\sqrt{2}$ m.

(a) Construct the element stiffness matrices in the global coordinate system.

(b) Construct the equilibrium equation for the frame and determine the nodal DOF.

(c) Determine force and moment distributions for each element.

**3.11** The frame structure shown in Figure 3.31 is fixed at the two ends. The cross-section is square with side length of 0.2 m, length $L = 4$ m, and $E = 200$ GPa. Find the unknown nodal DOF for the following three cases.

Case 1: A force of 1 N applied at the midpoint on the horizontal part, in the in-plane downward direction.

Case 2: A force of 1 N applied at the junction where the two parts are joined, in the direction normal to the plane.

Case 3: A force of 1 N applied at the midpoint on the horizontal part, in the direction normal to the plane.

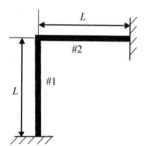

**Figure 3.31** For Problem 3.11

# 4  Structural Dynamics

In earlier chapters, we considered the problems to be static (i.e. independent of time). In this chapter, we consider time-dependent problems of discrete systems with $N$ degrees of freedom. For solids and structures, the equation of motion can be expressed in the following general form:

$$\mathbf{M\ddot{q} + C\dot{q} + Kq = F}$$

where $\mathbf{M}$ is an $N \times N$ global mass matrix, $\mathbf{C}$ is an $N \times N$ global damping matrix, $\mathbf{K}$ is an $N \times N$ global stiffness matrix, and $\mathbf{F}$ is an $N \times 1$ global load vector. In the above equation, a dot indicates a time derivative. Thus, $\mathbf{q}$ is a DOF vector, $\dot{\mathbf{q}}$ is a velocity vector and $\ddot{\mathbf{q}}$ is an acceleration vector. We will show how the FE formulation is used to construct the element mass matrices, which are assembled into the global mass matrix. A comprehensive formulation on the damping matrix is beyond the scope of this book. Accordingly, only a brief discussion will be given at the end of this chapter.

For construction of mass matrices, we consider slender bodies undergoing uniaxial, torsional, and bending motion as examples. For uniaxial or longitudinal vibration, we will incorporate dynamic effect by treating the inertia force as an "applied force." This results in the construction of element mass matrices which are assembled into the global mass matrix, in addition to the global stiffness matrix and the global load vector. Subsequently, we can observe that mathematical equivalence exists between torsional vibration and uniaxial vibration. For bending vibration, once again the dynamic effect is incorporated by treating the inertia force as an applied force to generate element mass matrices and then constructing the global mass matrix.

We will then consider free vibration to determine natural frequencies and natural modes of an FE model through eigenvalue analysis. We introduce numerical methods for integrating the equation of motion in time, which can be used to determine dynamic response under applied loads and given initial conditions.

Subsequently, we will introduce the Lagrange equation to demonstrate how it can be applied to construct equations of motion. Once again we will consider slender bodies undergoing uniaxial vibration, torsional vibration, and bending vibration. A formal derivation of the Lagrange equation will be considered in Chapter 6.

## 4.1  Mass Matrix and Equation of Motion

For dynamic problems, we may treat the inertia force as an applied force to generate element mass matrices which can be assembled into the global mass matrix for construction of the equation of motion. As examples, we will consider a slender body undergoing uniaxial or longitudinal vibration, torsional vibration, and bending vibration, while neglecting the effect of damping.

## 4.1.1   Uniaxial Vibration of a Slender Body

Consider a slender body subject to uniaxial forces, as shown in Figure 4.1, in which

$u(x, t)$: axial displacement
$f(x, t)$: applied force per unit length
$P(t)$: applied force at the tip $(x = L)$.

**Figure 4.1** Slender body undergoing uniaxial or longitudinal vibration

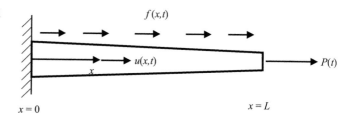

Recall that for a body in equilibrium:

$$\delta U - \delta W = 0. \tag{4.1.1}$$

Treating the reaction force $R_0$ at $x = 0$ as an "external" load allows the incremental work term to be written as

$$\delta W = \int_{x=0}^{x=L} \delta u f dx + (P \delta u)_{x=L} + (R_0 \delta u)_{x=0}. \tag{4.1.2}$$

The dynamic effect can be incorporated via treating the inertia force as an applied force per unit length. The inertial force associated with accelerations experienced by a segment of length $dx$, as shown in Figure 4.2, is

$$f dx = -(m dx)\ddot{u}, \tag{4.1.3}$$

where $m$ is the mass per unit length and $\ddot{u}$ is the acceleration of the segment. Then, for the entire body:

$$-\int_{x=0}^{x=L} \delta u f dx = \int_{x=0}^{x=L} \delta u \ddot{u} m dx. \tag{4.1.4}$$

**Figure 4.2** Segment of infinitesimal length $dx$

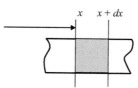

**Element Mass Matrix**

Recall that the assumed displacement for the two-node element is expressed as

$$u = (1 - s)u_1 + su_2 = N_1 u_1 + N_2 u_2, \qquad (4.1.5)$$

where

$$N_1 = 1 - s, \quad N_2 = s. \qquad (4.1.6)$$

Accordingly:

$$\ddot{u} = N_1 \ddot{u}_1 + N_2 \ddot{u}_2 = \lfloor N_1 \ N_2 \rfloor \begin{Bmatrix} \ddot{u}_1 \\ \ddot{u}_2 \end{Bmatrix}, \qquad (4.1.7)$$

$$\delta u = N_1 \delta u_1 + N_2 \delta u_2 = \lfloor \delta u_1 \quad \delta u_2 \rfloor \begin{Bmatrix} N_1 \\ N_2 \end{Bmatrix}. \qquad (4.1.8)$$

Substituting Eqs (4.1.7) and (4.1.8) into Eq. (4.1.4) for an element:

$$\int_{x=x_1}^{x=x_2} \delta u \ddot{u} m \, dx = \int_{s=0}^{s=1} \lfloor \delta u_1 \quad \delta u_2 \rfloor \begin{Bmatrix} N_1 \\ N_2 \end{Bmatrix} \lfloor N_1 \ N_2 \rfloor \begin{Bmatrix} \ddot{u}_1 \\ \ddot{u}_2 \end{Bmatrix} m l \, ds$$

$$= \lfloor \delta u_1 \quad \delta u_2 \rfloor \left( l \int_{s=0}^{s=1} \begin{bmatrix} N_1^2 & N_1 N_2 \\ N_2 N_1 & N_2^2 \end{bmatrix} m \, ds \right) \begin{Bmatrix} \ddot{u}_1 \\ \ddot{u}_2 \end{Bmatrix} \qquad (4.1.9)$$

$$= \lfloor \delta u_1 \quad \delta u_2 \rfloor \begin{bmatrix} m_{11}^e & m_{12}^e \\ m_{21}^e & m_{22}^e \end{bmatrix} \begin{Bmatrix} \ddot{u}_1 \\ \ddot{u}_2 \end{Bmatrix} = \delta \mathbf{d}^T \mathbf{m}^e \ddot{\mathbf{d}},$$

where

$$\mathbf{m}^e = \begin{bmatrix} m_{11}^e & m_{12}^e \\ m_{21}^e & m_{22}^e \end{bmatrix} = l \int_{s=0}^{s=1} \begin{bmatrix} N_1^2 & N_1 N_2 \\ N_2 N_1 & N_2^2 \end{bmatrix} m \, ds : \text{element mass matrix}, \qquad (4.1.10)$$

$$\ddot{\mathbf{d}} = \begin{Bmatrix} \ddot{u}_1 \\ \ddot{u}_2 \end{Bmatrix} : \text{element nodal acceleration vector}, \qquad (4.1.11)$$

$$\delta \mathbf{d} = \begin{Bmatrix} \delta u_1 \\ \delta u_2 \end{Bmatrix} : \text{element incremental DOF vector}. \qquad (4.1.12)$$

One can show that for the two-node uniaxial element with constant $m$, mass per length:

$$\mathbf{m}^e = \frac{ml}{6} \begin{bmatrix} 2 & 1 \\ 1 & 2 \end{bmatrix}. \qquad (4.1.13)$$

### Equation of Motion

For the entire structure:

$$-\int_{x=0}^{x=L} \delta u f dx = \int_{x=0}^{x=L} \delta u \ddot{u} m dx = \delta \mathbf{q}^T \mathbf{M} \ddot{\mathbf{q}}, \tag{4.1.14}$$

where

$\mathbf{M}$: global mass matrix
$\ddot{\mathbf{q}}$: global acceleration vector.

Note that the global mass matrix is assembled in the same manner as the global stiffness matrix. Then

$$\delta U - \delta W = \delta \mathbf{q}^T (\mathbf{K}\mathbf{q} + \mathbf{M}\ddot{\mathbf{q}} - \mathbf{F}) = \mathbf{0}. \tag{4.1.15}$$

Accordingly:

$$\mathbf{K}\mathbf{q} + \mathbf{M}\ddot{\mathbf{q}} - \mathbf{F} = \mathbf{0}, \tag{4.1.16}$$

or

$$\mathbf{M}\ddot{\mathbf{q}} + \mathbf{K}\mathbf{q} = \mathbf{F}. \tag{4.1.17}$$

Initial conditions are given at $t = 0$ as

$$\mathbf{q}(0) = \mathbf{q}_0, \quad \dot{\mathbf{q}}(0) = \dot{\mathbf{q}}_0, \tag{4.1.18}$$

where $\mathbf{q}_0$ and $\dot{\mathbf{q}}_0$ are vectors of prescribed values.

## 4.1.2   Torsional Vibration of a Slender Body

Consider a slender body under applied torques as shown in Figure 4.3 in which

$\phi(x, t)$: twist angle
$f_T(x, t)$: applied torque per unit length
$\bar{T}(t)$: torque applied at the tip $(x = L)$.

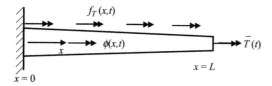

**Figure 4.3** Slender body undergoing torsional vibration

Treating inertia torque as an applied torque per length:

$$f_T dx = -(I dx)\ddot{\phi}, \tag{4.1.19}$$

where
$I(x)$: mass moment of inertia per unit length with respect to the axis of rotation
$\ddot{\phi}(x, t)$: rotational acceleration of segment $dx$.

Then, for the entire body:

$$-\int_{x=0}^{x=L} \delta\phi f_T dx = \int_{x=0}^{x=L} \delta\phi\ddot{\phi}I dx. \qquad (4.1.20)$$

Comparing Eq. (4.1.20) with Eq. (4.1.4), we observe that a mathematical equivalence exists between the uniaxial vibration and the torsional vibration as follows:

$$u \quad \leftrightarrow \quad \phi,$$
$$m \quad \leftrightarrow \quad I.$$

For a two-node element for torsion, the nodal DOF are $\phi_1$ and $\phi_2$. Utilizing the above equivalence and Eq. (4.1.10), the element mass matrix is then

$$\mathbf{m}^e = \begin{bmatrix} m^e_{11} & m^e_{12} \\ m^e_{21} & m^e_{22} \end{bmatrix} = l\int_{s=0}^{s=1} \begin{bmatrix} N_1^2 & N_1N_2 \\ N_2N_1 & N_2^2 \end{bmatrix} I ds. \qquad (4.1.21)$$

For constant $I$, the element mass matrix is

$$\mathbf{m}^e = \frac{Il}{6}\begin{bmatrix} 2 & 1 \\ 1 & 2 \end{bmatrix}. \qquad (4.1.22)$$

### 4.1.3  Bending Vibration of a Slender Body

Consider a cantilevered beam under applied loads as shown in Figure 4.4 in which

$p_z(x, t)$: applied force per unit length
$P_L(t)$: force applied at $x = L$
$M_L(t)$: moment applied at $x = L$
$w(x, t)$: transverse displacement.

For the applied loads shown in the figure, the incremental work is

$$\delta W = \int_{x=0}^{x=L} \delta w p_z dx + (M_L\delta\theta)_{x=L} + (P_L\delta w)_{x=L}. \qquad (4.1.23)$$

Neglecting the effect of cross-sectional rotation, the dynamic effect associated with the transverse motion can be incorporated via treating the inertia force as the applied force per unit length. For the segment of length $dx$:

**Figure 4.4** Cantilevered beam undergoing bending vibration

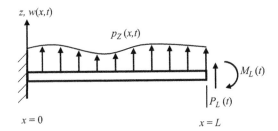

$$p_z dx = -(m dx)\ddot{w},\tag{4.1.24}$$

where $m$ is the mass per unit length. For the entire body:

$$-\int_{x=0}^{x=L} \delta w p_z dx = \int_{x=0}^{x=L} \delta w \ddot{w} m dx.\tag{4.1.25}$$

**Element Mass Matrix**

For the two-node bending element, recall that

$$w = \mathbf{N}\mathbf{d}, \quad \ddot{w} = \mathbf{N}\ddot{\mathbf{d}}, \quad \delta w = \mathbf{N}\delta\mathbf{d} = \delta\mathbf{d}^T\mathbf{N}^T,\tag{4.1.26}$$

where

$$\mathbf{N} = \lfloor N_1 \quad N_2 l \quad N_3 \quad N_4 l \rfloor,\tag{4.1.27}$$

with

$$\begin{aligned} N_1 &= 1 - 3s^2 + 2s^3, & N_2 &= -s + 2s^2 - s^3 \\ N_3 &= 3s^2 - 2s^3, & N_4 &= s^2 - s^3 \end{aligned}\tag{4.1.28}$$

and

$$\mathbf{d} = \begin{Bmatrix} w_1 \\ \theta_1 \\ w_2 \\ \theta_2 \end{Bmatrix}: \text{element DOF vector.}\tag{4.1.29}$$

Accordingly:

$$-\int_{x=x_1}^{x=x_2} \delta w p_z dx = \int_{x=x_1}^{x=x_2} \delta w \ddot{w} m dx = \int_{s=0}^{s=1} \delta\mathbf{d}^T\mathbf{N}^T\mathbf{N}\ddot{\mathbf{d}} m l ds$$
$$= \delta\mathbf{d}^T\left(l\int_{s=0}^{s=1}\mathbf{N}^T\mathbf{N}m ds\right)\ddot{\mathbf{d}} = \delta\mathbf{d}^T\mathbf{m}^e\ddot{\mathbf{d}},\tag{4.1.30}$$

where

$$\mathbf{m}^e = l\int_{s=0}^{s=1}\mathbf{N}^T\mathbf{N}m ds\tag{4.1.31}$$

is the element mass matrix. For constant $m$, it can be shown that

$$\mathbf{m}^e = \frac{ml}{420} \begin{bmatrix} 156 & -22l & 54 & 13l \\ -22l & 4l^2 & -13l & -3l^2 \\ 54 & -13l & 156 & 22l \\ 13l & -3l^2 & 22l & 4l^2 \end{bmatrix}. \tag{4.1.32}$$

If $\hat{\theta} = l\theta$ is introduced instead of $\theta$:

$$\mathbf{m}^e = \frac{ml}{420} \begin{bmatrix} 156 & -22 & 54 & 13 \\ -22 & 4 & -13 & -3 \\ 54 & -13 & 156 & 22 \\ 13 & -3 & 22 & 4 \end{bmatrix}. \tag{4.1.33}$$

**Equation of Motion**

For the entire structure:

$$-\int_{x=0}^{x=L} \delta w p_z dx = \int_{x=0}^{x=L} \delta w \ddot{w} m dx = \delta \mathbf{q}^{\mathrm{T}} \mathbf{M} \ddot{\mathbf{q}}, \tag{4.1.34}$$

where

$\mathbf{M}$: global mass matrix

$\ddot{\mathbf{q}}$: global acceleration vector.

The global mass matrix is assembled in the same manner as the global stiffness matrix. Then

$$\delta U - \delta W = \delta \mathbf{q}^{\mathrm{T}} (\mathbf{K}\mathbf{q} + \mathbf{M}\ddot{\mathbf{q}} - \mathbf{F}) = \mathbf{0} \rightarrow \mathbf{K}\mathbf{q} + \mathbf{M}\ddot{\mathbf{q}} - \mathbf{F} = \mathbf{0} \tag{4.1.35}$$

or

$$\mathbf{M}\ddot{\mathbf{q}} + \mathbf{K}\mathbf{q} = \mathbf{F}. \tag{4.1.36}$$

Initial conditions are given at $t = 0$ as

$$\mathbf{q}(0) = \mathbf{q}_0, \quad \dot{\mathbf{q}}(0) = \dot{\mathbf{q}}_0, \tag{4.1.37}$$

where $\mathbf{q}_0$ and $\dot{\mathbf{q}}_0$ are prescribed values.

## 4.2    Free Vibration Analysis

For free vibration ($\mathbf{F} = \mathbf{0}$) of a system with $N$ DOF:

$$\mathbf{M}\ddot{\mathbf{q}} + \mathbf{K}\mathbf{q} = \mathbf{0}, \tag{4.2.1}$$

where $\mathbf{q}$ is an $N \times 1$ DOF vector. The above equation implies that it is in the reduced form after geometric boundary conditions are applied and the equations with unknown reactions are deleted. Equation (4.2.1) is homogeneous and admits a solution of the following form:

$$\mathbf{q} = \boldsymbol{\varphi} e^{pt}, \tag{4.2.2}$$

where $p = \mathrm{Re}(p) + \mathrm{Im}(p) = \sigma \pm i\omega$. For oscillatory response with no energy loss (i.e. no damping), $\sigma = 0$ and $p = \pm i\omega$. Thus

$$\mathbf{q} = \boldsymbol{\varphi} e^{\pm i\omega t}. \tag{4.2.3}$$

Placing Eq. (4.2.3) into Eq. (4.2.1):

$$\left(\mathbf{K} - \omega^2 \mathbf{M}\right)\boldsymbol{\varphi} e^{\pm i\omega t} = \mathbf{0}. \tag{4.2.4}$$

Since $e^{\pm i\omega t}$ cannot be zero for all time $t$:

$$\left(\mathbf{K} - \omega^2 \mathbf{M}\right)\boldsymbol{\varphi} = \mathbf{0}. \tag{4.2.5}$$

Equation (4.2.5) is a homogeneous algebraic equation. For nontrivial $\boldsymbol{\varphi}$, the following equation must be satisfied:

$$\det\left(\mathbf{K} - \omega^2 \mathbf{M}\right) = 0, \tag{4.2.6}$$

where "det" stands for "determinant." Specific values of $\omega^2$ that satisfy Eq. (4.2.6) are called eigenvalues. For a system with $N$ DOF, there are $N$ eigenvalues counting multiple roots separately. For each $\omega = \omega_i$, Eq. (4.2.5) is expressed as

$$\left(\mathbf{K} - \omega_i^2 \mathbf{M}\right)\boldsymbol{\varphi}_i = \mathbf{0}, \tag{4.2.7}$$

where $\boldsymbol{\varphi}_i$ is the eigenvector corresponding to $\omega_i$, called natural frequency (in radians per second) of mode number $i$. The mathematical exercise of determining eigenvalues and eigenvectors is called eigenvalue analysis. Instead of using Eq. (4.2.6), one may write Eq. (4.2.5) in a standard form for eigenvalue analysis as follows:

$$\mathbf{K}\boldsymbol{\varphi} = \lambda \mathbf{M}\boldsymbol{\varphi}, \tag{4.2.8}$$

where

$$\lambda = \omega^2. \tag{4.2.9}$$

For the $i$th mode:

$$\mathbf{K}\boldsymbol{\varphi}_i = \lambda_i \mathbf{M}\boldsymbol{\varphi}_i \rightarrow \mathbf{K}\boldsymbol{\varphi}_i = \omega_i^2 \mathbf{M}\boldsymbol{\varphi}_i. \tag{4.2.10}$$

The natural mode corresponding to a natural frequency can be constructed using eigenvector $\boldsymbol{\varphi}_i$ and the assumed displacement. From Eq. (4.2.10), we observe that any constant multiple of an eigenvector is also an eigenvector. That is, an eigenvector (or a natural mode of free vibration) is determined within a constant multiple. For a system with $N$ DOF, there are $N$ natural frequencies and $N$ natural modes, counting multiple roots separately. For convenience, $\omega_i$ are arranged in increasing order of magnitude, starting from $\omega_1$ as the lowest frequency.

**Example 4.1**

Consider a slender body of length $L$ moving with a constant velocity on a track as shown in Figure 4.5. The body is assumed flexible in bending, but rigid axially. For free vibration analysis the body is modeled with two beam bending elements of equal length, as shown in Figure 4.6. For simplicity, assume $m$, $EI_y$ are constants.

*Note:* Introduce a new variable $\hat{\theta} = l\theta$ and $\bar{\lambda} = \omega^2 \frac{mL^4}{EI_y}$.

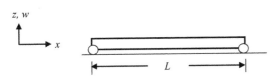

**Figure 4.5** Slender body moving with a constant velocity on a track

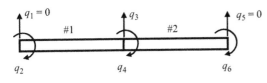

**Figure 4.6** Two-element model for bending vibration

(a) Construct the $6 \times 6$ global mass matrix.
(b) Construct the $6 \times 6$ global stiffness matrix.
(c) Apply the geometric constraints and set up the equation of motion.
(d) Compute eigenvalues $(\bar{\lambda}_1, \bar{\lambda}_2, \bar{\lambda}_3, \bar{\lambda}_4)$ and eigenvectors for free vibration. Scale eigenvectors such that the entry of largest magnitude in each eigenvector is equal to unity.
(e) Determine the natural frequencies and sketch the natural modes. Scale each mode such that maximum displacement is equal to unity. Compare with the exact solutions, given as follows:

$$\omega_n = n^2\pi^2\sqrt{\frac{EI_y}{mL^4}} : \text{natural frequency,} \quad \phi_n = \sin\frac{n\pi x}{L} : \text{natural mode}$$

where integer $n = 1, 2, \ldots$ is the mode number.

## Solution:

(a) According to Eq. (4.1.32), the element mass matrix is

$$\mathbf{m}^e = \frac{ml}{420}\begin{bmatrix} 156 & -22 & 54 & 13 \\ -22 & 4 & -13 & -3 \\ 54 & -13 & 156 & 22 \\ 13 & -3 & 22 & 4 \end{bmatrix} = \frac{mL}{840}\begin{bmatrix} 156 & -22 & 54 & 13 \\ -22 & 4 & -13 & -3 \\ 54 & -13 & 156 & 22 \\ 13 & -3 & 22 & 4 \end{bmatrix}.$$

Using the connectivity between the element DOF and the global DOF, the global mass matrix is assembled such that

$$\mathbf{M} = \frac{mL}{840} \begin{bmatrix} 156 & -22 & 54 & 13 & 0 & 0 \\ -22 & 4 & -13 & -3 & 0 & 0 \\ 54 & -13 & 312 & 0 & 54 & 13 \\ 13 & -3 & 0 & 8 & -13 & -3 \\ 0 & 0 & 54 & -13 & 156 & 22 \\ 0 & 0 & 13 & -3 & 22 & 4 \end{bmatrix}.$$

(b) From Chapter 3, the element stiffness matrix is

$$\mathbf{k}^e = \frac{2EI_y}{l^3} \begin{bmatrix} 6 & -3 & -6 & -3 \\ -3 & 2 & 3 & 1 \\ -6 & 3 & 6 & 3 \\ -3 & 1 & 3 & 2 \end{bmatrix} = \frac{16EI_y}{L^3} \begin{bmatrix} 6 & -3 & -6 & -3 \\ -3 & 2 & 3 & 1 \\ -6 & 3 & 6 & 3 \\ -3 & 1 & 3 & 2 \end{bmatrix}.$$

The global stiffness matrix is then assembled as

$$\mathbf{K} = \frac{16EI_y}{L^3} \begin{bmatrix} 6 & -3 & -6 & -3 & 0 & 0 \\ -3 & 2 & 3 & 1 & 0 & 0 \\ -6 & 3 & 12 & 0 & -6 & -3 \\ -3 & 1 & 0 & 4 & 3 & 1 \\ 0 & 0 & -6 & 3 & 6 & 3 \\ 0 & 0 & -3 & 1 & 3 & 2 \end{bmatrix}.$$

(c) Equation of motion: $\mathbf{M\ddot{q}} + \mathbf{Kq} = \mathbf{F}$, where $F_1 = R_1$, $F_2 = 0$, $F_3 = 0$, $F_4 = 0$, $F_5 = R_5$, $F_6 = 0$ and $q_1 = 0$, $\ddot{q}_1 = 0$, $q_5 = 0$, $\ddot{q}_5 = 0$. Applying the geometric constraints or boundary conditions and deleting equations with reaction loads $R_1$ and $R_5$, the global mass and stiffness matrices are reduced as follows:

$$\mathbf{M} = \frac{mL}{840} \begin{bmatrix} 4 & -13 & -3 & 0 \\ -13 & 312 & 0 & 13 \\ -3 & 0 & 8 & -3 \\ 0 & 13 & -3 & 4 \end{bmatrix}, \quad \mathbf{K} = \frac{16EI_y}{L^3} \begin{bmatrix} 2 & 3 & 1 & 0 \\ 3 & 12 & 0 & -3 \\ 1 & 0 & 4 & 1 \\ 0 & -3 & 1 & 2 \end{bmatrix}.$$

The equation of motion is then $\mathbf{M\ddot{q}} + \mathbf{Kq} = \mathbf{0}$, where

$$\mathbf{q} = \begin{Bmatrix} q_2 \\ q_3 \\ q_4 \\ q_6 \end{Bmatrix} = \begin{Bmatrix} \phi_2 \\ \phi_3 \\ \phi_4 \\ \phi_6 \end{Bmatrix} e^{\pm i\omega t}.$$

(d) For free vibration, $\mathbf{K}\boldsymbol{\varphi} = \omega^2 \mathbf{M}\boldsymbol{\varphi}$:

$$\frac{16EI_y}{L^3}\begin{bmatrix} 2 & 3 & 1 & 0 \\ 3 & 12 & 0 & -3 \\ 1 & 0 & 4 & 1 \\ 0 & -3 & 1 & 2 \end{bmatrix}\begin{Bmatrix} \phi_2 \\ \phi_3 \\ \phi_4 \\ \phi_6 \end{Bmatrix} = \omega^2 \frac{mL}{840}\begin{bmatrix} 4 & -13 & -3 & 0 \\ -13 & 312 & 0 & 13 \\ -3 & 0 & 8 & -3 \\ 0 & 13 & -3 & 4 \end{bmatrix}\begin{Bmatrix} \phi_2 \\ \phi_3 \\ \phi_4 \\ \phi_6 \end{Bmatrix},$$

$$\begin{bmatrix} 2 & 3 & 1 & 0 \\ 3 & 12 & 0 & -3 \\ 1 & 0 & 4 & 1 \\ 0 & -3 & 1 & 2 \end{bmatrix}\begin{Bmatrix} \phi_2 \\ \phi_3 \\ \phi_4 \\ \phi_6 \end{Bmatrix} = \bar{\lambda}\frac{1}{13{,}440}\begin{bmatrix} 4 & -13 & -3 & 0 \\ -13 & 312 & 0 & 13 \\ -3 & 0 & 8 & -3 \\ 0 & 13 & -3 & 4 \end{bmatrix}\begin{Bmatrix} \phi_2 \\ \phi_3 \\ \phi_4 \\ \phi_6 \end{Bmatrix}.$$

For eigenvalue analysis on Matlab, the above equation can be expressed as $\mathbf{Ax} = \bar{\lambda}\mathbf{Bx}$ where

$$\mathbf{A} = \begin{bmatrix} 2 & 3 & 1 & 0 \\ 3 & 12 & 0 & -3 \\ 1 & 0 & 4 & 1 \\ 0 & -3 & 1 & 2 \end{bmatrix}, \quad \mathbf{x} = \begin{Bmatrix} \phi_2 \\ \phi_3 \\ \phi_4 \\ \phi_6 \end{Bmatrix}, \quad \mathbf{B} = \frac{1}{13{,}440}\begin{bmatrix} 4 & -13 & -3 & 0 \\ -13 & 312 & 0 & 13 \\ -3 & 0 & 8 & -3 \\ 0 & 13 & -3 & 4 \end{bmatrix}.$$

One may then use the command $[\mathtt{X}, \mathtt{D}] = \mathtt{eig}\,(\mathtt{A}, \mathtt{B})$ in Matlab to determine eigenvalues and eigenvectors. The output is as follows:

$\mathtt{D}$: diagonal matrix of : diagonal matrix of eigenvalues
$\mathtt{X}$: full matrix in which the eigenvectors corresponding to the eigenvalues are stored columnwise.

Carrying out the eigenvalue analysis, we find that

$$\bar{\lambda}_1 = 98.180, \quad \bar{\lambda}_2 = 1920.0, \quad \bar{\lambda}_3 = 12{,}131, \quad \bar{\lambda}_4 = 40{,}320.$$

Also, after scaling, the eigenvectors can be written as

$$\boldsymbol{\varphi}_1 = \begin{Bmatrix} -1 \\ 0.63676 \\ 0 \\ 1 \end{Bmatrix}, \quad \boldsymbol{\varphi}_2 = \begin{Bmatrix} 1 \\ 0 \\ -1 \\ 1 \end{Bmatrix}, \quad \boldsymbol{\varphi}_3 = \begin{Bmatrix} 1 \\ 0.109298 \\ 0 \\ -1 \end{Bmatrix}, \quad \boldsymbol{\varphi}_4 = \begin{Bmatrix} 1 \\ 0 \\ 1 \\ 1 \end{Bmatrix}.$$

Expanding to include all six DOF:

$$\boldsymbol{\varphi}_1 = \begin{Bmatrix} 0 \\ -1 \\ 0.63676 \\ 0 \\ 0 \\ 1 \end{Bmatrix}, \quad \boldsymbol{\varphi}_2 = \begin{Bmatrix} 0 \\ 1 \\ 0 \\ -1 \\ 0 \\ 1 \end{Bmatrix}, \quad \boldsymbol{\varphi}_3 = \begin{Bmatrix} 0 \\ 1 \\ 0.109298 \\ 0 \\ 0 \\ -1 \end{Bmatrix}, \quad \boldsymbol{\varphi}_4 = \begin{Bmatrix} 0 \\ 1 \\ 0 \\ 1 \\ 0 \\ 1 \end{Bmatrix}.$$

(e) Natural frequencies are as follows:

$$\omega_1 = \sqrt{\bar{\lambda}_1 \frac{EI_y}{mL^4}} = 9.9086\sqrt{\frac{EI_y}{mL^4}}, \quad \text{Exact } \omega_1 = 9.8696\sqrt{\frac{EI_y}{mL^4}}.$$

$$\omega_2 = \sqrt{\bar{\lambda}_2 \frac{EI_y}{mL^4}} = 43.818\sqrt{\frac{EI_y}{mL^4}}, \quad \text{Exact } \omega_2 = 39.4784\sqrt{\frac{EI_y}{mL^4}}.$$

$$\omega_3 = \sqrt{\bar{\lambda}_3 \frac{EI_y}{mL^4}} = 110.14\sqrt{\frac{EI_y}{mL^4}}, \quad \text{Exact } \omega_3 = 88.8264\sqrt{\frac{EI_y}{mL^4}}.$$

$$\omega_4 = \sqrt{\bar{\lambda}_4 \frac{EI_y}{mL^4}} = 200.80\sqrt{\frac{EI_y}{mL^4}}, \quad \text{Exact } \omega_4 = 157.914\sqrt{\frac{EI_y}{mL^4}}.$$

We note that as the mode number increases, the solution accuracy decreases, indicating that a model with more elements is needed to improve the accuracy of FE solutions. Noting that the assumed $w$ is cubic in an element, one can plot the mode shapes using each eigenvector as follows.

$$\text{For element } \#k : w^k = \mathbf{N}\hat{\mathbf{d}}^k = \hat{\mathbf{N}}\boldsymbol{\varphi}^k e^{\pm i\omega t} = \phi^k(x)e^{\pm i\omega t},$$

where $\phi^k(x) = \hat{\mathbf{N}}\boldsymbol{\varphi}^k$: mode shape for the element.

For the first mode: $\phi_1^k(x) = \hat{\mathbf{N}}\boldsymbol{\varphi}_1^k$

Then for element #1: $x = x_1 + ls = sL/2 \rightarrow s = 2x/L$

$$\phi_1^1(x) = \hat{\mathbf{N}}\boldsymbol{\varphi}_1^1 = \lfloor N_1 \quad N_2 \quad N_3 \quad N_4 \rfloor \begin{Bmatrix} 0 \\ -1 \\ 0.63676 \\ 0 \end{Bmatrix};$$

and for element #2: $x = x_1 + ls = L/2(1 + s) \rightarrow s = 2x/L - 1$

$$\phi_1^2(x) = \hat{\mathbf{N}}\boldsymbol{\varphi}_1^2 = \lfloor N_1 \quad N_2 \quad N_3 \quad N_4 \rfloor \begin{Bmatrix} 0.63676 \\ 0 \\ 0 \\ 1 \end{Bmatrix}.$$

Mode shapes for modes 2, 3, and 4 can be similarly determined and plotted. In Figures 4.7–4.10 the mode shapes are scaled such that the maximum value is equal to unity. We observe that the discrepancy between the exact mode shape and the mode shape obtained by the two-element model increases as the mode number increases when the mode shape becomes more complicated.

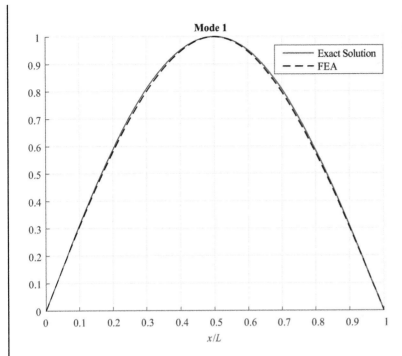

**Figure 4.7** Shape of natural mode for mode 1

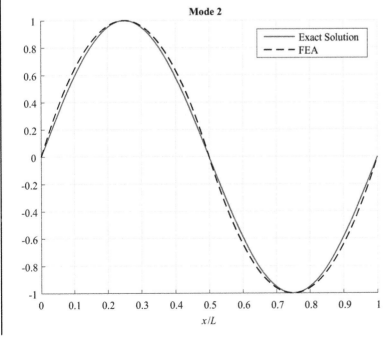

**Figure 4.8** Shape of natural mode for mode 2

**Figure 4.9** Shape of natural mode for mode 3

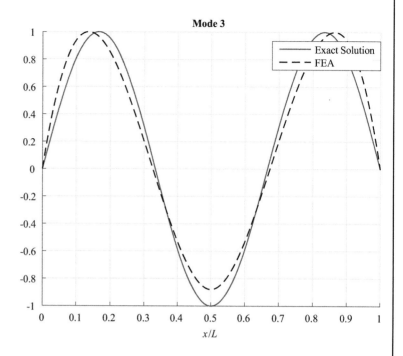

**Figure 4.10** Shape of natural mode for mode 4

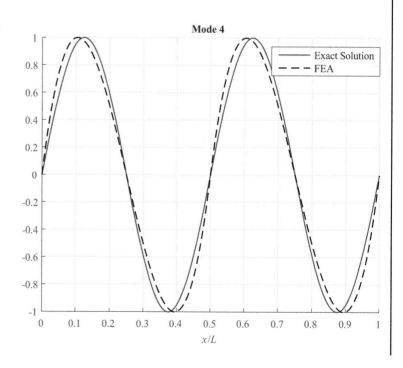

## 4.3    Numerical Integration in Time

Consider an equation of motion, including now the damping effect for generality:

$$\mathbf{M}\ddot{\mathbf{q}} + \mathbf{C}\dot{\mathbf{q}} + \mathbf{K}\mathbf{q} = \mathbf{F}, \tag{4.3.1}$$

where $\mathbf{C}$ is a damping matrix. A simple example of a damping matrix is given in Section 4.5 of this chapter. In addition to boundary conditions, initial conditions are given at $t = 0$ as

$$\mathbf{q}(0) = \mathbf{q}_0, \quad \dot{\mathbf{q}}(0) = \dot{\mathbf{q}}_0. \tag{4.3.2}$$

There exist various methods for numerical integration in time, which can be found in the open literature. Among these, we will consider two schemes as follows:

(a) trapezoidal rule
(b) central difference scheme.

For numerical integration in time, we divide the time span of interest into many time increments, each of size $\Delta t$, as shown in Figure 4.11. The integers in the figure indicate the instances in time, or *time steps*. Then

$$t_{n+1} = t_n + \Delta t, \quad \mathbf{q}_n = \mathbf{q}(t_n), \quad \mathbf{q}_{n+1} = \mathbf{q}(t_{n+1}). \tag{4.3.3}$$

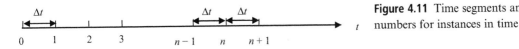

**Figure 4.11** Time segments and integer numbers for instances in time

### 4.3.1    Trapezoidal Rule

For the trapezoidal rule, we assume that

$$\mathbf{q}_{n+1} = \mathbf{q}_n + \frac{1}{2}\left(\dot{\mathbf{q}}_n + \dot{\mathbf{q}}_{n+1}\right)\Delta t, \tag{4.3.4}$$

$$\dot{\mathbf{q}}_{n+1} = \dot{\mathbf{q}}_n + \frac{1}{2}\left(\ddot{\mathbf{q}}_n + \ddot{\mathbf{q}}_{n+1}\right)\Delta t. \tag{4.3.5}$$

Note that over the time segment between time $t_n$ and $t_{n+1}$:

$$\frac{1}{2}\left(\dot{\mathbf{q}}_n + \dot{\mathbf{q}}_{n+1}\right): \text{average velocity}$$

$$\frac{1}{2}\left(\ddot{\mathbf{q}}_n + \ddot{\mathbf{q}}_{n+1}\right): \text{average acceleration.}$$

From Eq. (4.3.4):

$$\dot{\mathbf{q}}_{n+1} = \frac{2}{\Delta t}\left(\mathbf{q}_{n+1} - \mathbf{q}_n\right) - \dot{\mathbf{q}}_n. \qquad (4.3.6)$$

Substituting Eq. (4.3.5) into Eq. (4.3.4):

$$\mathbf{q}_{n+1} = \mathbf{q}_n + \dot{\mathbf{q}}_n \Delta t + \left(\frac{\Delta t}{2}\right)^2 \left(\ddot{\mathbf{q}}_n + \ddot{\mathbf{q}}_{n+1}\right). \qquad (4.3.7)$$

Solving Eq. (4.3.7) for $\ddot{\mathbf{q}}_{n+1}$:

$$\ddot{\mathbf{q}}_{n+1} = \left(\frac{2}{\Delta t}\right)^2 \left(\mathbf{q}_{n+1} - \mathbf{q}_n\right) - \left(\frac{4}{\Delta t}\right)\dot{\mathbf{q}}_n - \ddot{\mathbf{q}}_n. \qquad (4.3.8)$$

At time $t = t_{n+1}$, the equation of motion is

$$\mathbf{M}\ddot{\mathbf{q}}_{n+1} + \mathbf{C}\dot{\mathbf{q}}_{n+1} + \mathbf{K}\mathbf{q}_{n+1} = \mathbf{F}_{n+1}. \qquad (4.3.9)$$

Substituting Eqs (4.3.6) and (4.3.8) into Eq. (4.3.9):

$$\mathbf{M}\left[\left(\frac{2}{\Delta t}\right)^2 \left(\mathbf{q}_{n+1} - \mathbf{q}_n\right) - \left(\frac{4}{\Delta t}\right)\dot{\mathbf{q}}_n - \ddot{\mathbf{q}}_n\right] + \mathbf{C}\left[\frac{2}{\Delta t}\left(\mathbf{q}_{n+1} - \mathbf{q}_n\right) - \dot{\mathbf{q}}_n\right] + \mathbf{K}\mathbf{q}_{n+1} = \mathbf{F}_{n+1}.$$

$$(4.3.10)$$

Rearranging:

$$\left[\mathbf{K} + \frac{2}{\Delta t}\mathbf{C} + \left(\frac{2}{\Delta t}\right)^2 \mathbf{M}\right]\mathbf{q}_{n+1} = \mathbf{F}_{n+1} + \mathbf{C}\left[\frac{2}{\Delta t}\mathbf{q}_n + \dot{\mathbf{q}}_n\right] + \mathbf{M}\left[\left(\frac{2}{\Delta t}\right)^2 \mathbf{q}_n + \left(\frac{4}{\Delta t}\right)\dot{\mathbf{q}}_n + \ddot{\mathbf{q}}_n\right].$$

$$(4.3.11)$$

We can solve the above equation for $\mathbf{q}_{n+1}$ once the right-hand side is known. We may then use Eqs (4.3.6) and (4.3.8) to determine $\dot{\mathbf{q}}_{n+1}$ and $\ddot{\mathbf{q}}_{n+1}$. Note that the trapezoidal rule involves two steps in time.

To start the solution process, we may choose a time increment $\Delta t$ and set $n = 0$ in Eq. (4.3.11). Then

$$\left[\mathbf{K} + \frac{2}{\Delta t}\mathbf{C} + \left(\frac{2}{\Delta t}\right)^2 \mathbf{M}\right]\mathbf{q}_1 = \mathbf{F}_1 + \mathbf{C}\left[\frac{2}{\Delta t}\mathbf{q}_0 + \dot{\mathbf{q}}_0\right] + \mathbf{M}\left[\left(\frac{2}{\Delta t}\right)^2 \mathbf{q}_0 + \left(\frac{4}{\Delta t}\right)\dot{\mathbf{q}}_0 + \ddot{\mathbf{q}}_0\right], \quad (4.3.12)$$

where $\mathbf{q}_0$ and $\dot{\mathbf{q}}_0$ are the initial displacement vector and the initial velocity vector, respectively. The initial acceleration vector $\ddot{\mathbf{q}}_0$ is obtained from the equation of motion at $t = 0$ as follows:

$$\mathbf{M}\ddot{\mathbf{q}}_0 = \mathbf{F}_0 - \mathbf{C}\dot{\mathbf{q}}_0 - \mathbf{K}\mathbf{q}_0. \qquad (4.3.13)$$

Equation (4.3.12) can be used to determine the $\mathbf{q}_1$ vector at time $t_1 = \Delta t$. We can then use Eqs (4.3.6) and (4.3.8) to determine the velocity and acceleration at $t_1 = \Delta t$ as follows:

$$\dot{\mathbf{q}}_1 = \frac{2}{\Delta t}(\mathbf{q}_1 - \mathbf{q}_0) - \dot{\mathbf{q}}_0, \quad \ddot{\mathbf{q}}_1 = \left(\frac{2}{\Delta t}\right)^2 (\mathbf{q}_1 - \mathbf{q}_0) - \left(\frac{4}{\Delta t}\right)\dot{\mathbf{q}}_0 - \ddot{\mathbf{q}}_0. \tag{4.3.14}$$

To march further in time, we set $n = 1$, and then from Eq. (4.3.11):

$$\left[\mathbf{K} + \frac{2}{\Delta t}\mathbf{C} + \left(\frac{2}{\Delta t}\right)^2 \mathbf{M}\right]\mathbf{q}_2 = \mathbf{F}_2 + \mathbf{C}\left[\frac{2}{\Delta t}\mathbf{q}_1 + \dot{\mathbf{q}}_1\right] + \mathbf{M}\left[\left(\frac{2}{\Delta t}\right)^2 \mathbf{q}_1 + \left(\frac{4}{\Delta t}\right)\dot{\mathbf{q}}_1 + \ddot{\mathbf{q}}_1\right], \tag{4.3.15}$$

In the above equation, the right-hand side is known. Accordingly, it can be solved for $\mathbf{q}_2$ at $t_2 = 2\Delta t$. Once $\mathbf{q}_2$ is determined, we again use Eqs (4.3.6) and (4.3.8) to determine the velocity and acceleration at time $t_2 = 2\Delta t$ as follows:

$$\dot{\mathbf{q}}_2 = \frac{2}{\Delta t}(\mathbf{q}_2 - \mathbf{q}_1) - \dot{\mathbf{q}}_1, \quad \ddot{\mathbf{q}}_2 = \left(\frac{2}{\Delta t}\right)^2 (\mathbf{q}_2 - \mathbf{q}_1) - \left(\frac{4}{\Delta t}\right)\dot{\mathbf{q}}_1 - \ddot{\mathbf{q}}_1. \tag{4.3.16}$$

To march further in time, we set $n = 2$ and then use Eq. (4.3.11) to repeat the process of finding the DOF vector, velocity vector, and acceleration vector. This process can be continued in a recursive manner until the time span of our interest is covered.

The trapezoidal rule is called an implicit method in that the assumptions in Eqs (4.3.4) and (4.3.5) involve unknown velocity and acceleration at time $t_{n+1}$ on the right-hand side. It is known that the trapezoidal rule is numerically stable for any $\Delta t$ in that the solution does not blow up, but $\Delta t$ must be small to ensure accuracy of the solution. A good scheme for numerical integration in time must conserve the total energy when it is applied to an undamped free vibration problem. One may try the problems at the end of this chapter to appreciate the effectiveness of the trapezoidal rule.

## Example 4.2

Consider a single DOF system with the following equation of motion and the initial condition as

$$m\ddot{q} + kq = F,$$
$$q(0) = 0, \quad \dot{q}(0) = 1.0 \text{ m/s},$$

where $m = 1.0$ kg, $k = 4\pi^2 \text{N/m}$, $F = 0$. Use the trapezoidal rule with $\Delta t = 0.01$s to integrate up to $t = 8$ s. Plot $q$ vs. time $t$ . Compare with the exact solution given as

$$q = \frac{1}{2\pi}\sin 2\pi t.$$

## Solution:

Figure 4.12 shows a good agreement between the solution by the trapezoidal rule and the exact solution.

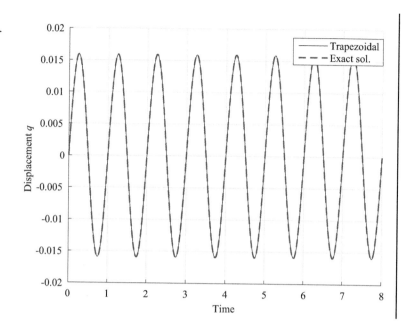

**Figure 4.12** Displacement vs. time (trapezoidal rule with $\Delta t = 0.01$ s)

## 4.3.2  Central Difference Scheme

Based on the Taylor series expansion of $\mathbf{q}$ at $t_{n+1}$ and $t_{n-1}$, the central difference scheme (CDS) assumes that

$$\mathbf{q}_{n+1} = \mathbf{q}_n + \dot{\mathbf{q}}_n \Delta t + \frac{1}{2} \ddot{\mathbf{q}}_n (\Delta t)^2, \tag{4.3.17}$$

$$\mathbf{q}_{n-1} = \mathbf{q}_n - \dot{\mathbf{q}}_n \Delta t + \frac{1}{2} \ddot{\mathbf{q}}_n (\Delta t)^2, \tag{4.3.18}$$

where the cubic and higher-order terms have been neglected. Summing Eqs (4.3.17) and (4.3.18):

$$\mathbf{q}_{n+1} + \mathbf{q}_{n-1} = 2\mathbf{q}_n + \ddot{\mathbf{q}}_n (\Delta t)^2. \tag{4.3.19}$$

Solving the above equation for acceleration yields

$$\ddot{\mathbf{q}}_n = \frac{1}{(\Delta t)^2} \left( \mathbf{q}_{n+1} - 2\mathbf{q}_n + \mathbf{q}_{n-1} \right). \tag{4.3.20}$$

Subtracting Eq. (4.3.18) from Eq. (4.3.17):

$$\mathbf{q}_{n+1} - \mathbf{q}_{n-1} = 2\dot{\mathbf{q}}_n \Delta t. \tag{4.3.21}$$

Solving the above equation for velocity yields

$$\dot{\mathbf{q}}_n = \frac{1}{2\Delta t}\left(\mathbf{q}_{n+1} - \mathbf{q}_{n-1}\right). \tag{4.3.22}$$

At time $t = t_n$, the equation of motion is

$$\mathbf{M}\ddot{\mathbf{q}}_n + \mathbf{C}\dot{\mathbf{q}}_n + \mathbf{K}\mathbf{q}_n = \mathbf{F}_n. \tag{4.3.23}$$

Inserting Eqs (4.3.20) and (4.3.22) into Eq. (4.3.23):

$$\mathbf{M}\frac{1}{(\Delta t)^2}\left(\mathbf{q}_{n+1} - 2\mathbf{q}_n + \mathbf{q}_{n-1}\right) + \mathbf{C}\frac{1}{2\Delta t}\left(\mathbf{q}_{n+1} - \mathbf{q}_{n-1}\right) + \mathbf{K}\mathbf{q}_n = \mathbf{F}_n. \tag{4.3.24}$$

Multiplying the above equation with $(\Delta t)^2$ and rearranging:

$$\left(\mathbf{M} + \frac{\Delta t}{2}\mathbf{C}\right)\mathbf{q}_{n+1} = \mathbf{M}(2\mathbf{q}_n - \mathbf{q}_{n-1}) + \frac{\Delta t}{2}\mathbf{C}\mathbf{q}_{n-1} + (\Delta t)^2(\mathbf{F}_n - \mathbf{K}\mathbf{q}_n). \tag{4.3.25}$$

The above equation can be solved for $\mathbf{q}_{n+1}$ once the right-hand side is known. Equations (4.3.20) and (4.3.22) can then be used to solve for acceleration $\ddot{\mathbf{q}}_n$ and velocity $\dot{\mathbf{q}}_n$. Note that the CDS involves three steps in time, in contrast to the trapezoidal rule which involves only two steps.

To start the solution process, we may choose a time increment $\Delta t$ and set $n = 0$ in Eq. (4.3.25). Then

$$\left(\mathbf{M} + \frac{\Delta t}{2}\mathbf{C}\right)\mathbf{q}_1 = \mathbf{M}(2\mathbf{q}_0 - \mathbf{q}_{-1}) + \frac{\Delta t}{2}\mathbf{C}\mathbf{q}_{-1} + (\Delta t)^2(\mathbf{F}_0 - \mathbf{K}\mathbf{q}_0). \tag{4.3.26}$$

Observe that in the above equation there exists $\mathbf{q}_{-1}$ which has no physical meaning. How do we find $\mathbf{q}_{-1}$? We may find it from Eq. (4.3.18) with $n = 0$ such that

$$\mathbf{q}_{-1} = \mathbf{q}_0 - \dot{\mathbf{q}}_0\Delta t + \frac{1}{2}\ddot{\mathbf{q}}_0(\Delta t)^2. \tag{4.3.27}$$

Note that we accept the above equation based on the Taylor expansion. The initial acceleration vector $\ddot{\mathbf{q}}_0$ is obtained from the equation of motion at $t = 0$ as follows:

$$\mathbf{M}\ddot{\mathbf{q}}_0 = \mathbf{F}_0 - \mathbf{C}\dot{\mathbf{q}}_0 - \mathbf{K}\mathbf{q}_0. \tag{4.3.28}$$

The right-hand side of Eq. (4.3.26) is now known, and we can then solve the equation to determine the $\mathbf{q}_1$ vector at time $t_1 = \Delta t$.

To march further in time, we set $n = 1$ and use Eq. (4.3.25) such that

$$\left(\mathbf{M} + \frac{\Delta t}{2}\mathbf{C}\right)\mathbf{q}_2 = \mathbf{M}(2\mathbf{q}_1 - \mathbf{q}_0) + \frac{\Delta t}{2}\mathbf{C}\mathbf{q}_0 + (\Delta t)^2(\mathbf{F}_1 - \mathbf{K}\mathbf{q}_1). \tag{4.3.29}$$

The right-hand side of the above equation is known. Accordingly, it can be used to solve the $\mathbf{q}_2$ vector at time $t_2 = 2\Delta t$. We can then use Eqs (4.3.20) and (4.3.22) to determine the velocity and acceleration at time $t_1 = \Delta t$ as

$$\dot{\mathbf{q}}_1 = \frac{1}{2\Delta t}(\mathbf{q}_2 - \mathbf{q}_0), \quad \ddot{\mathbf{q}}_1 = \frac{1}{(\Delta t)^2}(\mathbf{q}_2 - 2\mathbf{q}_1 + \mathbf{q}_0). \tag{4.3.30}$$

To march to the next time step, we set $n = 2$ and use Eq. (4.3.25) such that

$$\left(\mathbf{M} + \frac{\Delta t}{2}\mathbf{C}\right)\mathbf{q}_3 = \mathbf{M}(2\mathbf{q}_2 - \mathbf{q}_1) + \frac{\Delta t}{2}\mathbf{C}\mathbf{q}_1 + (\Delta t)^2(\mathbf{F}_2 - \mathbf{K}\mathbf{q}_2). \tag{4.3.31}$$

The right-hand side of the above equation is now known. Accordingly, it can be solved for the $\mathbf{q}_3$ vector at time $t_3 = 3\Delta t$. We can then use Eqs (4.3.20) and (4.3.22) to determine the velocity and acceleration at time $t_2 = 2\Delta t$ as

$$\dot{\mathbf{q}}_2 = \frac{1}{2\Delta t}(\mathbf{q}_3 - \mathbf{q}_1), \quad \ddot{\mathbf{q}}_2 = \frac{1}{(\Delta t)^2}(\mathbf{q}_3 - 2\mathbf{q}_2 + \mathbf{q}_1). \tag{4.3.32}$$

To march further in time, we set $n = 3$ and use Eq. (4.3.25) to repeat the process of finding the DOF vector at time $t_4 = 4\Delta t$. We then use Eqs (4.3.20) and (4.3.22) to determine the acceleration vector and velocity vector at time $t_3 = 3\Delta t$. We can repeat this process until the time span of our interest is covered.

Note that there is no need to assemble the global stiffness matrix as we can evaluate the last term in Eq. (4.3.25) at the element level and assemble the resulting column vectors. It is known that for the CDS, the following condition must be satisfied for numerical stability:

$$\Delta t < \frac{2}{\omega_{max}}, \tag{4.3.33}$$

where $\omega_{max}$ is the maximum natural frequency of a model. Otherwise, the numerical solution will blow up. As the total number of DOF increases with mesh refinement, $\omega_{max}$ of the model increases and the required time increment $\Delta t$ gets smaller. Once again, a good scheme for numerical integration in time must conserve the total energy when it is applied to an undamped free vibration problem. The reader may try the problems at the end of this chapter to examine the effectiveness of the CDS.

## 4.4    Equation of Motion via the Lagrange Equation

For a system with $N$ DOF $(q_1, q_2, \dots, q_N)$ subjected to applied forces or moments $F_1, F_2, \dots, F_N$, the kinetic energy $T$ and potential energy $V$ of the system can be expressed as

$$\begin{aligned} T &= T(\dot{q}_1, \dot{q}_2, \dots, \dot{q}_N, q_1, q_2, \dots, q_N), \\ V &= V(q_1, q_2, \dots, q_N). \end{aligned} \tag{4.4.1}$$

Note that kinetic energy can be a function of $q_i$ as well as $\dot{q}_i$ when a system has angular displacements as degrees of freedom. An example of potential energy is strain energy. It can then be shown that the dynamic equation of motion for such a system can be obtained from the following equation, called the Lagrange equation:

$$\frac{d}{dt}\left(\frac{\partial T}{\partial \dot{q}_i}\right) - \frac{\partial T}{\partial q_i} + \frac{\partial V}{\partial q_i} = F_i \quad (i = 1, 2, \dots, N). \tag{4.4.2}$$

A formal derivation of the above equation is in Section 6.4 of Chapter 6. Setting $V = U$, the Lagrange equation can be written in matrix form as

$$\frac{d}{dt}\left(\frac{\partial T}{\partial \dot{\mathbf{q}}}\right) - \frac{\partial T}{\partial \mathbf{q}} + \frac{\partial U}{\partial \mathbf{q}} = \mathbf{F}. \tag{4.4.3}$$

where

$$\frac{\partial U}{\partial \mathbf{q}} = \left\{ \begin{array}{c} \dfrac{\partial U}{\partial q_1} \\ .. \\ .. \\ \dfrac{\partial U}{\partial q_N} \end{array} \right\}. \tag{4.4.4}$$

Similar expressions hold for other terms with derivatives.

### Slender Body Undergoing Uniaxial or Longitudinal Motion

The kinetic energy $dT$ for a segment of length $dx$ as shown in Figure 4.2 is

$$dT = \frac{1}{2}(mdx)\dot{u}^2, \tag{4.4.5}$$

where $m$ is mass per unit length and $\dot{u}$ is axial velocity. Then, for the entire body:

$$T = \int dT = \frac{1}{2}\int_{x=0}^{x=L} \dot{u}^2 m\, dx. \tag{4.4.6}$$

For the two-node element:

$$\dot{u} = N_1 \dot{u}_1 + N_2 \dot{u}_2 = \lfloor N_1 \quad N_2 \rfloor \left\{ \begin{array}{c} \dot{u}_1 \\ \dot{u}_2 \end{array} \right\} \tag{4.4.7}$$

or

$$\dot{u} = \lfloor \dot{u}_1 \quad \dot{u}_2 \rfloor \left\{ \begin{array}{c} N_1 \\ N_2 \end{array} \right\}. \tag{4.4.8}$$

Accordingly, for the element:

$$\begin{aligned} T_e &= \frac{1}{2}\int_{x=x_1}^{x=x_2} \dot{u}^2 m\, dx = \frac{1}{2}\int_{s=0}^{s=1} \lfloor \dot{u}_1 \quad \dot{u}_2 \rfloor \left\{ \begin{array}{c} N_1 \\ N_2 \end{array} \right\} \lfloor N_1 \quad N_2 \rfloor \left\{ \begin{array}{c} \dot{u}_1 \\ \dot{u}_2 \end{array} \right\} m l\, ds \\[2mm] &= \frac{1}{2}\lfloor \dot{u}_1 \quad \dot{u}_2 \rfloor \left( l \int_{s=0}^{s=1} \left[ \begin{array}{cc} N_1^2 & N_1 N_2 \\ N_2 N_1 & N_2^2 \end{array} \right] m\, ds \right) \left\{ \begin{array}{c} \dot{u}_1 \\ \dot{u}_2 \end{array} \right\} \\[2mm] &= \frac{1}{2}\lfloor \dot{u}_1 \quad \dot{u}_2 \rfloor \left[ \begin{array}{cc} m_{11}^e & m_{12}^e \\ m_{21}^e & m_{22}^e \end{array} \right] \left\{ \begin{array}{c} \dot{u}_1 \\ \dot{u}_2 \end{array} \right\} = \frac{1}{2}\dot{\mathbf{d}}^T \mathbf{m}^e \dot{\mathbf{d}}, \end{aligned} \tag{4.4.9}$$

where $\mathbf{m}^e$ is the element mass matrix which is identical to that in Eq. (4.1.10). For the entire body:

$$T = \frac{1}{2}\dot{\mathbf{q}}^{\mathrm{T}}\mathbf{M}\dot{\mathbf{q}}, \tag{4.4.10}$$

where $\mathbf{M}$ is the global mass matrix. Then

$$\frac{d}{dt}\left(\frac{\partial T}{\partial \dot{\mathbf{q}}}\right) = \frac{d}{dt}(\mathbf{M}\dot{\mathbf{q}}) = \mathbf{M}\ddot{\mathbf{q}}, \quad \frac{\partial T}{\partial \mathbf{q}} = \mathbf{0}. \tag{4.4.11}$$

Recall that

$$U = \frac{1}{2}\mathbf{q}^T\mathbf{Kq} \rightarrow \delta U = \delta\mathbf{q}^T\mathbf{Kq}. \tag{4.4.12}$$

Also

$$\delta U = \frac{\partial U}{\partial q_1}\delta q_1 + \cdots + \frac{\partial U}{\partial q_N}\delta q_N = \lfloor \delta q_1 \quad \cdots \quad \cdots \quad \delta q_N \rfloor \begin{Bmatrix} \dfrac{\partial U}{\partial q_1} \\ .. \\ .. \\ \dfrac{\partial U}{\partial q_N} \end{Bmatrix} \tag{4.4.13}$$

or

$$\delta U = \delta\mathbf{q}^{\mathrm{T}}\frac{\partial U}{\partial \mathbf{q}}. \tag{4.4.14}$$

Comparing Eq. (4.4.14) with Eq. (4.4.12):

$$\frac{\partial U}{\partial \mathbf{q}} = \mathbf{Kq}. \tag{4.4.15}$$

Plugging Eqs (4.4.11) and (4.4.15) into Eq. (4.4.3), we then obtain the equation of motion as

$$\mathbf{M}\ddot{\mathbf{q}} + \mathbf{Kq} = \mathbf{F}. \tag{4.4.16}$$

### Property of Element Mass Matrix

According to Eq. (4.4.9):

$$T_e = \frac{1}{2}\int\limits_{x=x_1}^{x=x_2} \dot{u}^2 m\,dx = \frac{1}{2}\lfloor \dot{u}_1 \quad \dot{u}_2 \rfloor \begin{bmatrix} m_{11}^e & m_{12}^e \\ m_{21}^e & m_{22}^e \end{bmatrix}\begin{Bmatrix} \dot{u}_1 \\ \dot{u}_2 \end{Bmatrix}. \tag{4.4.17}$$

Consider now a case in which the element is moving with uniform velocity such that

$$\dot{u} = c_1. \tag{4.4.18}$$

Placing Eq. (4.4.18) into Eq. (4.4.17):

$$\frac{1}{2}c_1^2 \int\limits_{x=x_1}^{x=x_2} mdx = \frac{1}{2}\lfloor c_1 \quad c_1 \rfloor \begin{bmatrix} m_{11}^e & m_{12}^e \\ m_{21}^e & m_{22}^e \end{bmatrix} \begin{Bmatrix} c_1 \\ c_1 \end{Bmatrix} = \frac{1}{2}c_1^2\lfloor 1 \quad 1 \rfloor \begin{bmatrix} m_{11}^e & m_{12}^e \\ m_{21}^e & m_{22}^e \end{bmatrix} \begin{Bmatrix} 1 \\ 1 \end{Bmatrix}$$
(4.4.19)
$$= \frac{1}{2}c_1^2 \lfloor m_{11}^e + m_{21}^e \quad m_{12}^e + m_{22}^e \rfloor \begin{Bmatrix} 1 \\ 1 \end{Bmatrix} = \frac{1}{2}c_1^2 (m_{11}^e + m_{21}^e + m_{12}^e + m_{22}^e).$$

Thus

$$\int\limits_{x=x_1}^{x=x_2} mdx = m_{11}^e + m_{21}^e + m_{12}^e + m_{22}^e.$$
(4.4.20)

Equation (4.4.20) shows that the sum of all entries in the element mass matrix is equal to the total mass of the element.

### Slender Body Undergoing Torsional Vibration

Kinetic energy $dT$ for a segment of length $dx$ is

$$dT = \frac{1}{2}(Idx)\dot{\phi}^2.$$
(4.4.21)

For the entire body, the kinetic energy is expressed as

$$T = \int dT = \frac{1}{2} \int\limits_{x=0}^{x=L} \dot{\phi}^2 Idx.$$
(4.4.22)

Comparing the above expression with that for a slender body undergoing uniaxial motion, we once again observe mathematical equivalence between the uniaxial vibration and torsional vibration as follows:

$$\begin{aligned} u &\leftrightarrow \phi, \\ m &\leftrightarrow I. \end{aligned}$$

### Slender Body Undergoing Bending Vibration

Kinetic energy $dT$ of a segment of length $dx$ is

$$dT = \frac{1}{2}(mdx)\dot{w}^2,$$
(4.4.23)

neglecting the contribution from rotation associated with bending deformation. Kinetic energy for the entire body is

$$T = \int dT = \frac{1}{2} \int\limits_{x=0}^{x=L} \dot{w}^2 m dx. \tag{4.4.24}$$

For an element, recall that

$$\dot{w} = \mathbf{N}\dot{\mathbf{d}} = \dot{\mathbf{d}}^{\mathrm{T}}\mathbf{N}^{\mathrm{T}}. \tag{4.4.25}$$

Placing Eq. (4.4.25) into the kinetic energy expression for the element:

$$T_e = \frac{1}{2} \int\limits_{x=x_1}^{x=x_2} \dot{w}^2 m dx = \frac{1}{2} \int\limits_{s=0}^{s=1} \dot{\mathbf{d}}^{\mathrm{T}}\mathbf{N}^{\mathrm{T}}\mathbf{N}\dot{\mathbf{d}} m l ds$$

$$= \frac{1}{2}\dot{\mathbf{d}}^{\mathrm{T}} \left( l \int\limits_{s=0}^{s=1} m\mathbf{N}^{\mathrm{T}}\mathbf{N}ds \right)\dot{\mathbf{d}} = \frac{1}{2}\dot{\mathbf{d}}^{\mathrm{T}}\mathbf{m}^e\dot{\mathbf{d}}, \tag{4.4.26}$$

where

$$\mathbf{m}^e = l \int\limits_{s=0}^{s=1} \mathbf{N}^{\mathrm{T}}\mathbf{N}m ds : \text{element mass matrix} \tag{4.4.27}$$

is identical to that in Eq. (4.1.31). For the entire body:

$$T = \frac{1}{2}\dot{\mathbf{q}}^{\mathrm{T}}\mathbf{M}\dot{\mathbf{q}}, \tag{4.4.28}$$

where **M** is the global mass matrix.

## 4.5  Damping Matrix

A comprehensive formulation of the damping matrix is beyond the scope of this book. However, as a simple example, we will consider viscous damping associated with bending of a slender body. In this case, the damping force can be treated as a distributed transverse force per unit length as

$$(p_z)_{damping} dx = -(cdx)\dot{w}, \tag{4.5.1}$$

where $c$ is the damping constant per unit length. The first term on the right-hand side of Eq. (4.1.23) can then be expressed as

$$\int\limits_{x=0}^{x=L} \delta w (p_z)_{damping} dx = -\int\limits_{x=0}^{x=L} \delta w \dot{w} c dx. \tag{4.5.2}$$

Introducing $w$ and $\delta w$ in Eq. (4.1.26) into an element:

$$\int_{x=x_1}^{x=x_2} \delta w (p_z)_{damping} dx = -\int_{x=x_1}^{x=x_2} \delta w \dot{w} c\, dx = -\int_{s=0}^{s=1} \delta \mathbf{d}^{\mathsf{T}} \mathbf{N}^{\mathsf{T}} \mathbf{N} \dot{\mathbf{d}} c l\, ds$$

$$= -\delta \mathbf{d}^{\mathsf{T}} \left( l \int_{s=0}^{s=1} \mathbf{N}^{\mathsf{T}} \mathbf{N} c\, ds \right) \dot{\mathbf{d}} = -\delta \mathbf{d}^{\mathsf{T}} \mathbf{c}^e \dot{\mathbf{d}},$$

(4.5.3)

where

$$\mathbf{c}^e = l \int_{s=0}^{s=1} \mathbf{N}^{\mathsf{T}} \mathbf{N} c\, ds : \text{element damping matrix.}$$

(4.5.4)

For constant $c$, it can be shown that

$$\mathbf{c}^e = \frac{cl}{420} \begin{bmatrix} 156 & -22l & 54 & 13l \\ -22l & 4l^2 & -13l & -3l^2 \\ 54 & -13l & 156 & 22l \\ 13l & -3l^2 & 22l & 4l^2 \end{bmatrix}.$$

(4.5.5)

If $\hat{\theta} = l\theta$ is introduced instead of $\theta$:

$$\mathbf{c}^e = \frac{cl}{420} \begin{bmatrix} 156 & -22 & 54 & 13 \\ -22 & 4 & -13 & -3 \\ 54 & -13 & 156 & 22 \\ 13 & -3 & 22 & 4 \end{bmatrix}.$$

(4.5.6)

Note that the above matrix is similar to that in Eq. (4.1.33) for the element mass matrix. For the entire structure:

$$\int_{x=0}^{x=L} \delta w (p_z)_{damping} dx = -\int_{x=0}^{x=L} \delta w \dot{w} c\, dx = -\delta \mathbf{q}^{\mathsf{T}} \mathbf{C} \dot{\mathbf{q}} = \delta \mathbf{q}^{\mathsf{T}} \mathbf{F}_{damping},$$

(4.5.7)

where $\mathbf{C}$ is the global damping matrix and

$$\mathbf{F}_{damping} = -\mathbf{C} \dot{\mathbf{q}}$$

(4.5.8)

is the global load vector due to damping. The global damping matrix is assembled in the same manner as the global stiffness matrix. The equation of motion including damping is

$$\mathbf{M} \ddot{\mathbf{q}} + \mathbf{K} \mathbf{q} = \mathbf{F} + \mathbf{F}_{damping},$$

(4.5.9)

from which it follows that

$$\mathbf{M} \ddot{\mathbf{q}} + \mathbf{C} \dot{\mathbf{q}} + \mathbf{K} \mathbf{q} = \mathbf{F}.$$

(4.5.10)

## PROBLEMS

**4.1** As shown in Figure 4.13, a slender body of length $L$, constant $EA$ and constant $m$, mass per unit length, is fixed at one end. To investigate longitudinal dynamics, the body is modeled using two-node uniaxial elements of equal length. Do the following using a three-element model:

(a) Construct the global mass matrix and the global stiffness matrix.

(b) Determine the eigenvalues and eigenvectors corresponding to free vibration. Scale such that the largest entry in each eigenvector is equal to unity.

(c) Determine the natural frequencies and natural modes (in terms of axial displacement).

(d) Compare the results with the exact natural frequencies and natural modes, given as follows:

$$\omega_n = \left(n - \frac{1}{2}\right)\pi\sqrt{\frac{EA}{mL^2}}, \quad \phi_n = \sin\left(n - \frac{1}{2}\right)\frac{\pi x}{L},$$

where $n = 1, 2, \ldots$ For comparison, plot the natural modes.

*Note:* Introduce $\bar{\lambda} = \omega^2 mL^2/EA$.

**Figure 4.13** For Problem 4.1

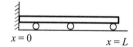   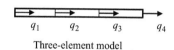

Three-element model

**4.2** Repeat Problem 4.1 using a model with six elements of equal length. Compare the results with those from the three-element model.

**4.3** A long slender body of length $L$ is constrained against torsional motion at $x = 0$ and $x = L$. The body of uniform cross-section is modeled with three two-node torsional elements of equal length.

(a) Construct the global mass matrix.

(b) Construct the global stiffness matrix.

(c) Determine the natural frequencies of the body.

(d) Sketch the natural modes.

**4.4** Repeat Problem 4.3 using a model with six elements of equal length.

**4.5** Carry out free bending vibration analysis of the slender body described in Example 4.1 in this chapter, using a model with four bending elements of equal length.

**4.6** Consider free bending vibration of a slender body clamped at $x = 0$ and free at $x = L$, as shown in Figure 4.14. For simplicity, assume that the body is of uniform cross-section. The body is modeled with two elements of equal length.

(a) Determine the natural frequencies and natural modes.

(b) Plot the natural modes. Scale each mode such that max displacement is equal to unity.

(c) Compare the results with the analytical solutions found in the open literature.

*Note:* Introduce variable $\hat{\theta} = l\theta$ and $\bar{\lambda} = \omega^2 \frac{mL^4}{EI_y}$.

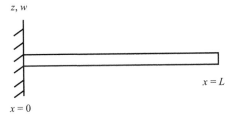

**Figure 4.14** For Problem 4.6

z, w

x = L

x = 0

**4.7** A slender body in level flight at a constant speed is modeled as an unconstrained beam of length $L$. For simplicity, assume a body of constant $m$ and $EI_y$.
   (a) Use a one-element model to determine natural frequencies and natural modes. Plot the natural modes (vs. $x/L$). Identify the two rigid-body modes. Scale the natural modes such that the largest value in each mode is equal to one.
   (b) Use a model with two elements of equal length to determine natural frequencies and natural modes. Plot the natural modes. Compare with the one-element model.

**4.8** A good scheme for numerical integration in time must conserve the total energy when it is applied to an undamped free vibration problem. Consider a single DOF system with the following equation of motion and the initial conditions:

$$m\ddot{q} + kq = F$$

where $m = 1.0$ kg, $k = 4\pi^2$ N/m, $F = 0$ and $q(0) = 0$, $\dot{q}(0) = 1.0$ m/s.
Use the trapezoidal rule to integrate up to $t = 8$ s with the following time increments:
Case 1: $\Delta t = 0.01$ s.
Case 2: $\Delta t = 0.05$ s.
Case 3: $\Delta t = 0.2$ s.
Case 4: $\Delta t = 0.5$ s.

   (a) Plot $q$ vs. time $t$ for the above four cases. Compare with exact $q$.
   (b) Plot non-dimensional error (NDE) in total energy vs. time for the four cases to check whether the total energy is conserved. Non-dimensionalize error as follows:
   NDE = (exact total energy − numerical total energy)/exact total energy.
   (c) How does the trapezoidal rule affect the frequency as $\Delta t$ increases? Does it increase or decrease?

**4.9** Repeat Problem 4.8 using the CDS.

**4.10** Use the trapezoidal rule to integrate the following equation of motion for a SDOF until the static equilibrium state is reached. The system is initially at rest:

$$m\ddot{q} + c\dot{q} + kq = F$$

where $m = 1.0$ kg, $c = 0.4\pi^2$N-s/m, $k = 4\pi^4$N/m, $F = 1.0$ N. Try the following time segment to plot $q$ vs. time $t$:
Case 1: $\Delta t = 0.01$ s.
Case 2: $\Delta t = 0.05$ s.
Case 3: $\Delta t = 0.2$ s.

**4.11** Repeat Problem 4.10 using the CDS.

# 5      Bending under Axial Force

In this chapter we consider the FE formulation for bending of slender bodies under a tensile or compressive axial force. In order to capture the effect of axial force, we look at the force and moment equilibrium in the deformed configuration, but still assuming small translational displacement and rotation of the cross-section. In the FE formulation, it is shown that the effect of axial force on bending manifests as an effective bending stiffness.

We first consider the case in which the direction of axial force does not change and remains always parallel to the body axis in the original undeformed configuration. It will be shown that the FE formulation of a slender body under compressive axial force results in a matrix equation for eigenvalue analysis from which we can determine the static buckling load and the buckling mode.

Subsequently, we consider the FE formulation for vibration analysis of slender bodies to investigate the effect of axial force on the natural frequencies and modes. The softening effect of a compressive force and the stiffening effect of a tensile force will be demonstrated via example problems.

Finally, we introduce the FE formulation of slender bodies subjected to a compressive follower force in which the direction of the applied force is always parallel to the body axis in the deformed configuration. In such a case, the force interacts with the structure as it undergoes bending deformation and the body experiences dynamic instability with sustained vibration if the applied force level is high enough. It will be shown that the FE formulation results in an equation for eigenvalue analysis which can be used to determine the magnitude of the follower force at which the system begins to exhibit dynamic instability.

## 5.1    Equilibrium Equation for Bending under Axial Force

In this section we will examine the effect of axial force on bending behavior of slender bodies. Previously, we have considered equilibrium neglecting the effect of deformation. However, in order to investigate the bending behavior of a slender body under axial force, it is necessary to examine equilibrium in the deformed configuration (even if the deformation involved is small). In order to derive the governing equation for bending under axial force, let's look at Figure 5.1, which shows a portion of the body in the undeformed configuration and in the deformed configuration in the $xz$-plane. A free body is formed by introducing cuts at $x$ and $x + dx$ in the undeformed configuration to consider equilibrium of the same body in the deformed configuration, as shown in Figure 5.2.

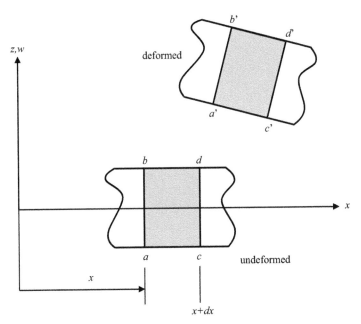

Figure 5.1 Portion of slender body in the undeformed configuration and the deformed configuration

Summing all forces acting in the $z$ direction on the free body:

$$(F \sin \theta)_x - (F \sin \theta)_{x+dx} - (V_z \cos \theta)_x + (V_z \cos \theta)_{x+dx} + p_z dx = 0, \qquad (5.1.1)$$

where

$$(F \sin \theta)_{x+dx} = (F \sin \theta)_x + \frac{\partial}{\partial x}(F \sin \theta)dx,$$

$$(V_z \cos \theta)_{x+dx} = (V_z \cos \theta)_x + \frac{\partial}{\partial x}(V_z \cos \theta)dx, \qquad (5.1.2)$$

and $p_z(x)$ is the applied force per unit length.

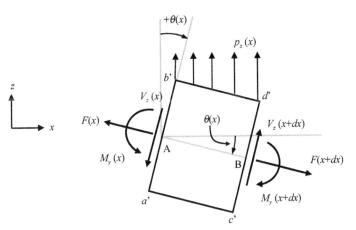

Figure 5.2 Free body diagram in the deformed configuration

Substituting Eq. (5.1.2) into Eq. (5.1.1):

$$\left[-\frac{\partial}{\partial x}(F\sin\theta)+\frac{\partial}{\partial x}(V_z\cos\theta)+p_z\right]dx=0. \tag{5.1.3}$$

For small $\theta$, $\sin\theta\approx\theta$, $\cos\theta\approx1$ and the above equation can be approximated as

$$\left[-\frac{\partial}{\partial}(F\theta)+\frac{\partial V_z}{\partial x}+p_z\right]dx=0\rightarrow-\frac{\partial}{\partial x}(F\theta)+\frac{\partial V_z}{\partial x}+p_z=0. \tag{5.1.4}$$

Also noting that $\theta=-\frac{\partial w}{\partial x}$, the above equation can be expressed as

$$\frac{\partial}{\partial x}\left(F\frac{\partial w}{\partial x}\right)+\frac{\partial V_z}{\partial x}+p_z=0. \tag{5.1.5}$$

For moment equilibrium, one may sum moments about point $B$ in Figure 5.2. Then

$$M_y(x+dx)-M_y(x)-V_z(x)\cdot|\overline{AB}|+p_z(x)dx\cdot\frac{1}{2}\left(|\overline{AB}|\cos\theta\right)=0. \tag{5.1.6}$$

Introducing

$$M_y(x+dx)=M_y(x)+\frac{\partial M_y}{\partial x}dx \tag{5.1.7}$$

and $|\overline{AB}|\cos\theta\approx dx$ for small rotation into Eq. (5.1.6):

$$\left(\frac{\partial M_y}{\partial x}-V_z\right)dx+p_z(x)\cdot\frac{1}{2}(dx)^2=0. \tag{5.1.8}$$

The second term in the above equation is of higher order and can be neglected. Accordingly:

$$\frac{\partial M_y}{\partial x}=V_z. \tag{5.1.9}$$

Placing Eq. (5.1.9) into Eq. (5.1.5):

$$\frac{\partial^2 M_y}{\partial x^2}+\frac{\partial}{\partial x}\left(F\frac{\partial w}{\partial x}\right)+p_z=0. \tag{5.1.10}$$

For equilibrium in the $x$ direction, we note that axial stress due to bending is linear through thickness and does not contribute to axial force. Accordingly, under the assumption of small deflection, it is adequate to look at the equilibrium in the unbent position.

As shown in Chapter 3, under the assumption of small rotation of the cross-section, the moment–curvature relation is

$$M_y(x)=-EI_y\frac{\partial^2 w}{\partial x^2}. \tag{5.1.11}$$

## 5.2    Finite Element Formulation

As a prelude to the FE formulation, we may follow the approach described in Chapter 1 for a slender body undergoing axial deformation and introduce virtual displacement $\delta w(x)$ to construct the following integral:

$$\int_{x=0}^{x=L} \left[ \frac{\partial^2 M_y}{\partial x^2} + \frac{\partial}{\partial x} \left( F \frac{\partial w}{\partial x} \right) + p_z \right] \delta w\, dx = 0, \tag{5.2.1}$$

which is equivalent to Eq. (5.1.10). The above integral can be transformed into a form involving the boundary terms by performing twice the integration by parts on the first term. The details are left as an exercise in Problem 6.3 at the end of Chapter 6. Instead, in this chapter we will take a different approach, considering the work done by the axial force as the body bends.

### 5.2.1    Slender Body under Compressive Tip Force

As an example of a slender body subjected to axial force, consider a column subjected to a compressive force $P$ applied always along the horizontal axis with no change in the direction, as shown in Figure 5.3. In the absence of any other load that can cause bending, the column initially remains straight and the length of the column shortens as the applied load increases. However, the column will bend or "buckle" laterally when the applied load reaches a certain critical value $P = P_{cr}$, called the "buckling load." This phenomenon is called bifurcation buckling, because the behavior of the column bifurcates from pure compression to bending. With bifurcation, the column is no longer straight.

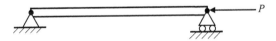

**Figure 5.3** Slender body subjected to a compressive tip force

We may now consider the work done by an axial force $P$ applied at the right end as the slender body bends. As shown in Figure 5.2, the cross-section of segment $dx$ rotates by angle $\theta$ as the slender body bends. Neglecting length change of the body axis, the net axial translation $d\Delta$ due to the rotation is then

$$d\Delta = dx - dx \cos\theta = dx(1 - \cos\theta). \tag{5.2.2}$$

For small rotation:

$$\cos\theta \simeq 1 - \frac{1}{2}\theta^2. \tag{5.2.3}$$

Accordingly:

$$d\Delta = dx(1 - \cos\theta) = \frac{1}{2}\theta^2 dx = \frac{1}{2}\left(-\frac{\partial w}{\partial x}\right)^2 dx. \tag{5.2.4}$$

The total translation of axial force $P$ is then

$$\Delta = \int d\Delta = \frac{1}{2} \int\limits_{x=0}^{x=L} \left(\frac{\partial w}{\partial x}\right)^2 dx. \tag{5.2.5}$$

The work done by axial force $P$ is then

$$W_P = P\Delta = \frac{1}{2} P \int\limits_{x=0}^{x=L} \left(\frac{\partial w}{\partial w}\right)^2 dx. \tag{5.2.6}$$

For the two-node bending element with 4 DOF described in Chapter 3:

$$w = \mathbf{Nd}, \tag{5.2.7}$$

where $\mathbf{N}$ is given in Eq. (3.2.10) and

$$\mathbf{d} = \begin{Bmatrix} w_1 \\ \theta_1 \\ w_2 \\ \theta_2 \end{Bmatrix}. \tag{5.2.8}$$

Then

$$\frac{\partial w}{\partial x} = \frac{\partial w}{\partial s}\frac{ds}{dx} = \frac{1}{l}\frac{\partial w}{\partial s} = \frac{1}{l}\frac{\partial}{\partial s}(\mathbf{Nd}) = \frac{1}{l}\mathbf{Ad} = \frac{1}{l}\mathbf{d}^T\mathbf{A}^T, \tag{5.2.9}$$

where

$$\mathbf{A} = \frac{\partial}{\partial s}(\mathbf{N}) = \begin{bmatrix} -6s + 6s^2 & (-1 + 4s - 3s^2)l & 6s - 6s^2 & (2s - 3s^2)l \end{bmatrix}. \tag{5.2.10}$$

Then, for the element:

$$\begin{aligned} W_P^e &= \frac{1}{2} P \int\limits_{x=x_1}^{x=x_2} \left(\frac{\partial w}{\partial x}\right)^2 dx = \frac{1}{2} P \int\limits_{s=0}^{s=1} \frac{1}{l^2} \mathbf{d}^T\mathbf{A}^T\mathbf{A}\mathbf{d}\, l\, ds \\ &= \frac{1}{2} P\mathbf{d}^T \left(\frac{1}{l} \int\limits_{s=0}^{s=1} \mathbf{A}^T\mathbf{A}\, ds\right) \mathbf{d} = \frac{1}{2} P\mathbf{d}^T\mathbf{k}_p^e\mathbf{d}, \end{aligned} \tag{5.2.11}$$

where

$$\mathbf{k}_P^e = \frac{1}{l} \int\limits_{s=0}^{s=1} \mathbf{A}^T\mathbf{A}\, ds \tag{5.2.12}$$

is a symmetric matrix. Introducing Eq. (5.2.10) into Eq. (5.2.12) and carrying out integration, one can show that

$$\mathbf{k}_P^e = \frac{1}{10l} \begin{bmatrix} 12 & -l & -12 & -l \\ -l & \dfrac{4l^2}{3} & l & -\dfrac{l^2}{3} \\ -12 & l & 12 & l \\ -l & -\dfrac{l^2}{3} & l & \dfrac{4l^2}{3} \end{bmatrix}. \tag{5.2.13}$$

For the entire body:

$$W_P = \frac{1}{2} P \mathbf{q}^T \mathbf{K}_P \mathbf{q}, \tag{5.2.14}$$

where $\mathbf{K}_P$ is assembled from $\mathbf{k}_P^e$. Then

$$\delta W_P = P \delta \mathbf{q}^T \mathbf{K}_P \mathbf{q}. \tag{5.2.15}$$

For bending of the slender body:

$$\delta U = \delta \mathbf{q}^T \mathbf{K}_B \mathbf{q}, \tag{5.2.16}$$

where $\mathbf{K}_B$ is the global bending stiffness matrix assembled from element bending stiffness matrices expressed in Eq. (3.2.26) in Chapter 3.

For equilibrium in the absence of any other applied loads:

$$\delta U - \delta W_P = 0 \rightarrow \delta \mathbf{q}^T (\mathbf{K}_B - P \mathbf{K}_P) \mathbf{q} = 0. \tag{5.2.17}$$

Thus

$$(\mathbf{K}_B - P \mathbf{K}_P) \mathbf{q} = \mathbf{0}, \tag{5.2.18}$$

which is a homogeneous equation. For static buckling, we look for nontrivial $\mathbf{q}$ (i.e. a bent configuration) and then

$$\det (\mathbf{K}_B - P \mathbf{K}_P) = 0, \tag{5.2.19}$$

where "det" stands for "determinant." The above equation represents an eigenvalue problem for $P$. The smallest eigenvalue that satisfies the above equation is the critical load or buckling load. The corresponding eigenvector leads to the buckling mode. For eigenvalue analysis, Eq. (5.2.18) can be expressed as

$$\mathbf{K}_B \mathbf{q} = P \mathbf{K}_P \mathbf{q}. \tag{5.2.20}$$

The above equation is in the standard form for eigenvalue analysis.

If the compressive load $P$ is smaller than the buckling load and the body is subjected to loads that cause bending:

$$(\mathbf{K}_B - P \mathbf{K}_P) \mathbf{q} = \mathbf{F}_{app} \rightarrow \mathbf{K}_{eff} \mathbf{q} = \mathbf{F}_{app}, \tag{5.2.21}$$

where $\mathbf{F}_{app}$ is the global load vector due to applied loads excluding $P$ and

$$\mathbf{K}_{eff} = \mathbf{K}_B - P \mathbf{K}_P : \text{effective bending stiffness matrix.} \tag{5.2.22}$$

As for the $\mathbf{k}_P^e$ matrix, we can show that introducing $\hat{\theta} = l\theta$:

$$\mathbf{k}_P^e = \frac{1}{10l}\begin{bmatrix} 12 & -1 & -12 & -1 \\ -1 & \dfrac{4}{3} & 1 & -\dfrac{1}{3} \\ -12 & 1 & 12 & 1 \\ -1 & -\dfrac{1}{3} & 1 & \dfrac{4}{3} \end{bmatrix}.$$  (5.2.23)

## Example 5.1

A slender column of constant $EI_y$ and length $L$ is simply supported with a fixed hinge at $x = 0$ and a roller hinge at $x = L$, as shown in Figure 5.3. The body is subjected to a compressive force $P$ applied at $x = L$. The direction of force $P$ does not change and is always in the horizontal direction. Use a model with two elements of equal length as shown in Figure 5.4 to do the following.

(a) Carry out the eigenvalue analysis to determine the buckling load and the buckling mode. Compare with the exact bucking load.
(b) Use the assumed $w$ of the element to plot the buckling mode with respect to $x/L$. Scale the magnitude of the mode such that max $w$ is equal to unity. Compare with the exact buckling mode.

*Note:* The exact solutions are $P_{cr} = \frac{\pi^2 EI_y}{L^2}$ for the buckling load and $w = \sin\frac{\pi x}{L}$ for the buckling mode.

**Figure 5.4** Two-element model for buckling analysis    $q_1 = 0$                    $q_5 = 0$

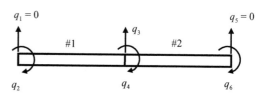

## Solution:

Introducing $\hat{\theta} = l\theta$ and $l = L/2$:

$$\mathbf{k}_B^e = \frac{2EI_y}{l^3}\begin{bmatrix} 6 & -3 & -6 & -3 \\ -3 & 2 & 3 & 1 \\ -6 & 3 & 6 & 3 \\ -3 & 1 & 3 & 2 \end{bmatrix} = \frac{16EI_y}{L^3}\begin{bmatrix} 6 & -3 & -6 & -3 \\ -3 & 2 & 3 & 1 \\ -6 & 3 & 6 & 3 \\ -3 & 1 & 3 & 2 \end{bmatrix},$$

$$\mathbf{k}_P^e = \frac{1}{10l}\begin{bmatrix} 12 & -1 & -12 & -1 \\ -1 & 4/3 & 1 & -1/3 \\ -12 & 1 & 12 & 1 \\ -1 & -1/3 & 1 & 4/3 \end{bmatrix} = \frac{1}{5L}\begin{bmatrix} 12 & -1 & -12 & -1 \\ -1 & 4/3 & 1 & -1/3 \\ -12 & 1 & 12 & 1 \\ -1 & -1/3 & 1 & 4/3 \end{bmatrix}.$$

Global stiffness matrix. After assembling the element matrices:

$$\mathbf{K}_B = \frac{16EI_y}{L^3} \begin{bmatrix} 6 & -3 & -6 & -3 & 0 & 0 \\ -3 & 2 & 3 & 1 & 0 & 0 \\ -6 & 3 & 12 & 0 & -6 & -3 \\ -3 & 1 & 0 & 4 & 3 & 1 \\ 0 & 0 & -6 & 3 & 6 & 3 \\ 0 & 0 & -3 & 1 & 3 & 2 \end{bmatrix},$$

$$\mathbf{K}_P = \frac{1}{5L} \begin{bmatrix} 12 & -1 & -12 & -1 & 0 & 0 \\ -1 & 4/3 & 1 & -1/3 & 0 & 0 \\ 12 & 1 & 24 & 0 & -12 & -1 \\ -1 & -1/3 & 0 & 8/3 & 1 & -1/3 \\ 0 & 0 & -12 & 1 & 12 & 1 \\ 0 & 0 & -1 & -1/3 & 1 & 4/3 \end{bmatrix}.$$

Deleting the first and fifth columns and rows to account for the geometric boundary conditions $(q_1 = q_5 = 0)$:

$$\mathbf{K}_B = \frac{16EI_y}{L^3} \begin{bmatrix} 2 & 3 & 1 & 0 \\ 3 & 12 & 0 & -3 \\ 1 & 0 & 4 & 1 \\ 0 & -3 & 1 & 2 \end{bmatrix}, \mathbf{K}_P = \frac{1}{5L} \begin{bmatrix} 4/3 & 1 & -1/3 & 0 \\ 1 & 24 & 0 & -1 \\ -1/3 & 0 & 8/3 & -1/3 \\ 0 & -1 & -1/3 & 4/3 \end{bmatrix}.$$

According to Eq. (5.2.20), the equation for eigenvalue analysis is

$$\begin{bmatrix} 2 & 3 & 1 & 0 \\ 3 & 12 & 0 & -3 \\ 1 & 0 & 4 & 1 \\ 0 & -3 & 1 & 2 \end{bmatrix} \begin{Bmatrix} q_2 \\ q_3 \\ q_4 \\ q_6 \end{Bmatrix} = \lambda \frac{1}{80} \begin{bmatrix} 4/3 & 1 & -1/3 & 0 \\ 1 & 24 & 0 & -1 \\ -1/3 & 0 & 8/3 & -1/3 \\ 0 & -1 & -1/3 & 4/3 \end{bmatrix} \begin{Bmatrix} q_2 \\ q_3 \\ q_4 \\ q_6 \end{Bmatrix},$$

where $\lambda = \frac{PL^2}{EI_y}$. For eigenvalue analysis, the above equation can be expressed as

$$\mathbf{A}\mathbf{x} = \lambda\mathbf{B}\mathbf{x},$$

where

$$\mathbf{A} = \begin{bmatrix} 2 & 3 & 1 & 0 \\ 3 & 12 & 0 & -3 \\ 1 & 0 & 4 & 1 \\ 0 & -3 & 1 & 2 \end{bmatrix}, \mathbf{x} = \begin{Bmatrix} \phi_2 \\ \phi_3 \\ \phi_4 \\ \phi_6 \end{Bmatrix}, \mathbf{B} = \frac{1}{80} \begin{bmatrix} 4/3 & 1 & -1/3 & 0 \\ 1 & 24 & 0 & -1 \\ -1/3 & 0 & 8/3 & -1/3 \\ 0 & -1 & -1/3 & 4/3 \end{bmatrix}.$$

One may then use the command [X, D] = eig(A, B) in Matlab to determine eigenvalues and eigenvectors. The output is as follows:

D: diagonal matrix of eigenvalues
X: full matrix in which the eigenvectors corresponding to the eigenvalues are stored columnwise.

Carrying out the eigenvalue analysis, we find that the lowest eigenvalue is

$$\lambda_1 = \frac{P_{cr}L^2}{EI_y} = 9.9438 \rightarrow P_{cr} = 9.9438\frac{EI_y}{L^2} = 1.0075\frac{\pi^2 EI_y}{L^2}.$$

This buckling load compares well with the exact solution. The corresponding eigenvector, after scaling, is

$$\begin{Bmatrix} q_2 \\ q_3 \\ q_4 \\ q_6 \end{Bmatrix} = \begin{Bmatrix} -1 \\ 0.6378 \\ 0 \\ 1 \end{Bmatrix}.$$

Expressing in an expanded form including all degrees of freedom:

$$\begin{Bmatrix} q_1 \\ q_2 \\ q_3 \\ q_4 \\ q_5 \\ q_6 \end{Bmatrix} = \begin{Bmatrix} 0 \\ -1 \\ 0.6378 \\ 0 \\ 0 \\ 1 \end{Bmatrix}.$$

The buckling mode can be obtained using the above eigenvector and assumed displacement for each element. We recall that for an element:

$$w = \hat{\mathbf{N}}\hat{\mathbf{d}}.$$

Accordingly, for element #1:

$$x = x_1 + ls = \frac{L}{2}s \rightarrow s = 2\frac{x}{L}, w = \lfloor N_1 \quad N_2 \quad N_3 \quad N_4 \rfloor \begin{Bmatrix} 0 \\ -1 \\ 0.6368 \\ 0 \end{Bmatrix}$$

and for element #2:

$$x = x_1 + ls = \frac{L}{2}(1+s) \rightarrow s = 2\frac{x}{L} - 1, w = \lfloor N_1 \quad N_2 \quad N_3 \quad N_4 \rfloor \begin{Bmatrix} 0.6368 \\ 0 \\ 0 \\ 1 \end{Bmatrix}.$$

Figure 5.5 shows the buckling mode which agrees well with the exact solution. The mode shape is scaled such that the maximum value is equal to unity. A model with more elements will further improve the solution.

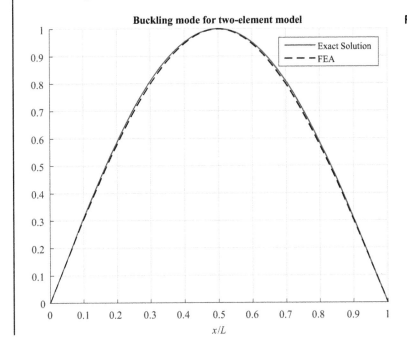

**Buckling mode for two-element model**

Exact Solution
- - - FEA

$x/L$

**Figure 5.5** Buckling mode

## 5.2.2  Slender Body under Non-uniform Axial Force

Consider now bending of a slender body subjected to a static axial force $F(x)$ which varies along the $x$-axis. As the slender body bends, the cross-section rotates by angle $\theta$. The net axial translation $d\Delta$ due to the rotation is given in Eq. (5.2.4). Accordingly, the work done by axial force $F$ is

$$W_F = - \int_{x=0}^{x=L} F d\Delta = -\frac{1}{2} \int_{x=0}^{x=L} F \left(\frac{\partial w}{\partial x}\right)^2 dx. \qquad (5.2.24)$$

Then, utilizing Eq. (5.2.9) for the 4-DOF element:

$$W_F^e = -\frac{1}{2} \int_{x=x_1}^{x=x_2} F \left(\frac{\partial w}{\partial x}\right)^2 dx = -\frac{1}{2} \int_{s=0}^{s=1} F \frac{1}{l^2} \mathbf{d}^\mathrm{T} \mathbf{A}^\mathrm{T} \mathbf{A} \mathbf{d} l ds$$

$$= -\frac{1}{2} \mathbf{d}^\mathrm{T} \left(\frac{1}{l} \int_{s=0}^{s=1} F \mathbf{A}^\mathrm{T} \mathbf{A} ds\right) \mathbf{d} = -\frac{1}{2} \mathbf{d}^\mathrm{T} \mathbf{k}_F^e \mathbf{d}, \qquad (5.2.25)$$

where

$$\mathbf{k}_F^e = \frac{1}{l} \int_{s=0}^{s=1} F \mathbf{A}^{\mathrm{T}} \mathbf{A} \, ds \qquad (5.2.26)$$

is a symmetric element stiffness matrix due to axial force. For the entire body:

$$W_F = -\frac{1}{2} \mathbf{q}^{\mathrm{T}} \mathbf{K}_F \mathbf{q}, \qquad (5.2.27)$$

where $\mathbf{K}_F$ is assembled from $\mathbf{k}_F^e$. The infinitesimal increment of $W_F$ can be expressed as

$$\delta W_F = -\delta \mathbf{q}^{\mathrm{T}} \mathbf{K}_F \mathbf{q} = \delta \mathbf{q}^{\mathrm{T}} \mathbf{F}_F, \qquad (5.2.28)$$

where

$$\mathbf{F}_F = -\mathbf{K}_F \mathbf{q}. \qquad (5.2.29)$$

Then the equation for the entire body is

$$\mathbf{K}_B \mathbf{q} = \mathbf{F} = \mathbf{F}_F + \mathbf{F}_{app} = -\mathbf{K}_F \mathbf{q} + \mathbf{F}_{app} \qquad (5.2.30)$$

or

$$\mathbf{K}_{eff} \mathbf{q} = \mathbf{F}_{app}, \qquad (5.2.31)$$

where

$$\mathbf{K}_{eff} = \mathbf{K}_B + \mathbf{K}_F. \qquad (5.2.32)$$

Note that $\mathbf{K}_B$ is the global bending stiffness matrix assembled from the element bending stiffness matrices and $\mathbf{F}_{app}$ is the global load vector due to applied loads excluding the axial force.

As an example, one may consider a column standing vertically under its own weight as shown in Figure 5.6. For constant $m$, mass per unit length, the axial force $F$ can be expressed as

$$F = -mg(L - x) = -mgL\left(1 - \frac{x}{L}\right). \qquad (5.2.33)$$

For an element, the mapping is

$$x = x_1 + ls$$

and

$$\mathbf{k}_F^e = \frac{1}{l} \int_{s=0}^{s=1} F \mathbf{A}^{\mathrm{T}} \mathbf{A} \, ds = -mgL \frac{1}{l} \int_{s=0}^{s=1} \left(1 - \frac{x_1 + ls}{L}\right) \mathbf{A}^{\mathrm{T}} \mathbf{A} \, ds = -mg \bar{\mathbf{k}}_F^e, \qquad (5.2.34)$$

where

$$\bar{\mathbf{k}}_F^e = \frac{L}{l} \int_{s=0}^{s=1} \left(1 - \frac{x_1 + ls}{L}\right) \mathbf{A}^T \mathbf{A} ds. \tag{5.2.35}$$

The global $\mathbf{K}_F$ matrix can then be assembled to construct the $\mathbf{K}_{eff}$ matrix shown in Eq. (5.2.32) such that

$$\mathbf{K}_{eff} = \mathbf{K}_B - mg\bar{\mathbf{K}}_F. \tag{5.2.36}$$

To determine the buckling load and modes under the weight, one can use Eq. (5.2.31) with $\mathbf{F}_{app} = 0$ such that

$$\mathbf{K}_{eff}\mathbf{q} = \mathbf{0} \rightarrow (\mathbf{K}_B - mg\bar{\mathbf{K}}_F)\mathbf{q} = \mathbf{0} \rightarrow \mathbf{K}_B\mathbf{q} = mg\bar{\mathbf{K}}_F\mathbf{q}, \tag{5.2.37}$$

which results in an equation for the eigenvalue problem.

 ## Example 5.2

Consider a slender column of length $L$ standing vertically under its own weight. The column is fixed at the bottom. The bending rigidity $EI_y$, cross-sectional area $A$, and mass $m$ per unit length are constant. For given $EI_y$ and $m$, determine the maximum column length that does not buckle. Use a model with one element.

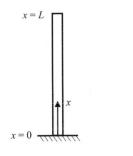

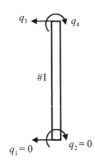

**Figure 5.6** A column under its own weight and the model with one element

## Solution:

For the one-element model, $l = L$. Accordingly, from Eq. (5.2.35):

$$\bar{\mathbf{k}}_F^1 = \int_{s=0}^{s=1} \left(1 - \frac{x_1 + ls}{L}\right) \mathbf{A}^T \mathbf{A} ds = \int_{s=0}^{s=1} (1 - s) \mathbf{A}^T \mathbf{A} ds.$$

Carrying out the integration:

$$\bar{\mathbf{k}}_F^1 = \frac{1}{60} \begin{bmatrix} 36 & 0 & -36 & -6 \\ 0 & 6 & 0 & -1 \\ -36 & 0 & 36 & 6 \\ -6 & -1 & 6 & 2 \end{bmatrix}.$$

Also, from Eq. (3.3.10) with $l = L$:

$$\hat{k}_B^1 = \frac{2EI_y}{L^3} \begin{bmatrix} 6 & -3 & -6 & -3 \\ -3 & 2 & 3 & 1 \\ -6 & 3 & 6 & 3 \\ -3 & 1 & 3 & 2 \end{bmatrix}.$$

According to Eq. (5.2.37), after deleting the first and second columns and the first and second rows of the above two matrices, the equation for eigenvalue analysis is

$$\frac{2EI_y}{L^3} \begin{bmatrix} 6 & 3 \\ 3 & 2 \end{bmatrix} \begin{Bmatrix} q_3 \\ q_4 \end{Bmatrix} = mg \frac{1}{60} \begin{bmatrix} 36 & 6 \\ 6 & 2 \end{bmatrix} \begin{Bmatrix} q_3 \\ q_4 \end{Bmatrix}$$

or

$$\begin{bmatrix} 6 & 3 \\ 3 & 2 \end{bmatrix} \begin{Bmatrix} q_3 \\ q_4 \end{Bmatrix} = \frac{mgL^3}{EI_y} \frac{1}{120} \begin{bmatrix} 36 & 6 \\ 6 & 2 \end{bmatrix} \begin{Bmatrix} q_3 \\ q_4 \end{Bmatrix}.$$

The above equation can be written as

$$\begin{bmatrix} 6 & 3 \\ 3 & 2 \end{bmatrix} \begin{Bmatrix} q_3 \\ q_4 \end{Bmatrix} = \lambda \frac{1}{120} \begin{bmatrix} 36 & 6 \\ 6 & 2 \end{bmatrix} \begin{Bmatrix} q_3 \\ q_4 \end{Bmatrix}$$

where

$$\lambda = \frac{mgL^3}{EI_y}.$$

Carrying out the eigenvalue analysis, we obtain two eigenvalues. Selecting the smaller of the two eigenvalues:

$$\lambda_1 = \frac{mgL^3}{EI_y} = 7.889 \rightarrow L_{max} = \sqrt[3]{7.889 \frac{EI_y}{mg}}.$$

It can be shown that, using a model with two elements of equal length:

$$L_{max} = \sqrt[3]{7.857 \frac{EI_y}{mg}}.$$

Finding solutions using multiple elements is left as a problem at the end of this chapter.

## 5.3   Bending Vibration under Axial Force

The presence of axial force manifests as effective bending stiffness. For dynamic problems, the equation of motion can then be symbolically expressed as

$$\mathbf{M}\ddot{\mathbf{q}} + \mathbf{K}_{eff}\mathbf{q} = \mathbf{F}. \tag{5.3.1}$$

For free vibration problems:

$$\mathbf{M}\ddot{\mathbf{q}} + \mathbf{K}_{eff}\mathbf{q} = \mathbf{0}. \tag{5.3.2}$$

As shown in Chapter 4, the solution to the above homogeneous equation is of the following form:

$$\mathbf{q} = \boldsymbol{\varphi}e^{\pm\omega t}. \tag{5.3.3}$$

Substituting the above solution into Eq. (5.3.2):

$$\left(\mathbf{K}_{eff} - \omega^2\mathbf{M}\right)\boldsymbol{\varphi}e^{\pm i\omega t} = \mathbf{0} \rightarrow \left(\mathbf{K}_{eff} - \omega^2\mathbf{M}\right)\boldsymbol{\varphi} = \mathbf{0} \tag{5.3.4}$$

or

$$\mathbf{K}_{eff}\boldsymbol{\varphi} = \omega^2\mathbf{M}\boldsymbol{\varphi} \rightarrow \mathbf{K}_{eff}\boldsymbol{\varphi} = \lambda\mathbf{M}\boldsymbol{\varphi}, \tag{5.3.5}$$

where $\lambda = \omega^2$. Equation (5.3.5) is an equation for eigenvalue analysis for determination of the natural frequencies and natural modes.

## Example 5.3

As shown in Figure 5.3, a slender body simply supported with a fixed hinge at $x = 0$ and a roller hinge at $x = L$ is subjected to a compressive axial force $P$ at $x = L$. For simplicity, assume the body is of uniform cross-section with constant $EI_y$. The body is modeled with two elements of equal length (Figure 5.7).

Introduce

$$\hat{\theta} = l\theta, \ \lambda = \omega^2\frac{mL^4}{EI_y}, \ \bar{\omega} = \frac{\omega}{\sqrt{\dfrac{EI_y}{mL^4}}},$$

$$\bar{P} = \frac{P}{\left(\dfrac{\pi^2 EI_y}{L^2}\right)} : \text{axial force relative to the static buckling load.}$$

$q_1 = 0$    $q_5 = 0$    **Figure 5.7** Two-element model for free vibration of simply supported body under compression

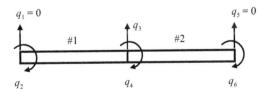

(a) Determine the natural frequencies for $\bar{P} = 0, 0.25, 0.5, 0.75$.

(b) Compare the results of free bending vibration with the analytical solution given as follows:

$$\omega_r = (r\pi)^2 \sqrt{\frac{EI_y}{mL^4}} \sqrt{1 - \frac{\bar{P}}{r^2}} : \text{natural frequency of the } r\text{th mode}$$

$$\phi_r(x) = \sin\frac{r\pi x}{L} : \text{mode shape of the } r\text{th mode}.$$

## Solution:

According to Example 4.1 of Chapter 4, the $6 \times 6$ global mass matrix reduces to the $4 \times 4$ matrix after deleting columns and rows in accordance with the geometric boundary conditions $(q_1 = q_5 = 0)$ such that

$$\mathbf{M} = \frac{mL}{840}\begin{bmatrix} 4 & -13 & -3 & 0 \\ -13 & 312 & 0 & 13 \\ -3 & 0 & 8 & -3 \\ 0 & 13 & -3 & 4 \end{bmatrix} = mL\bar{\mathbf{M}},$$

where

$$\bar{\mathbf{M}} = \frac{1}{840}\begin{bmatrix} 4 & -13 & -3 & 0 \\ -13 & 312 & 0 & 13 \\ -3 & 0 & 8 & 3 \\ 0 & 13 & -3 & 4 \end{bmatrix}.$$

From Example 5.1 of this chapter, the effective bending stiffness matrix after applying the geometric boundary condition is

$$\mathbf{K}_{eff} = \mathbf{K}_B - P\mathbf{K}_P = \frac{16EI_y}{L^3}\begin{bmatrix} 2 & 3 & 1 & 0 \\ 3 & 12 & 0 & -3 \\ 1 & 0 & 4 & 1 \\ 0 & -3 & 1 & 2 \end{bmatrix} - \frac{P}{5L}\begin{bmatrix} 4/3 & 1 & -1/3 & 0 \\ 1 & 24 & 0 & -1 \\ -1/3 & 0 & 8/3 & -1/3 \\ 0 & -1 & -1/3 & 4/3 \end{bmatrix}$$

or

$$\mathbf{K}_{eff} = \frac{EI_y}{L^3}\left(16\begin{bmatrix} 2 & 3 & 1 & 0 \\ 3 & 12 & 0 & -3 \\ 1 & 0 & 4 & 1 \\ 0 & -3 & 1 & 2 \end{bmatrix} - \bar{P}\frac{\pi^2}{5}\begin{bmatrix} 4/3 & 1 & -1/3 & 0 \\ 1 & 24 & 0 & -1 \\ -1/3 & 0 & 8/3 & -1/3 \\ 0 & -1 & -1/3 & 4/3 \end{bmatrix}\right) = \frac{EI_y}{L^3}\bar{\mathbf{K}}_{eff},$$

where

$$\bar{\mathbf{K}}_{eff} = \left( 16 \begin{bmatrix} 2 & 3 & 1 & 0 \\ 3 & 12 & 0 & -3 \\ 1 & 0 & 4 & 1 \\ 0 & -3 & 1 & 2 \end{bmatrix} - \bar{P}\frac{\pi^2}{5} \begin{bmatrix} 4/3 & 1 & -1/3 & 0 \\ 1 & 24 & 0 & -1 \\ -1/3 & 0 & 8/3 & -1/3 \\ 0 & -1 & -1/3 & 4/3 \end{bmatrix} \right).$$

According to Eqs (5.3.4) and (5.3.5):

$$\left( \frac{EI_y}{L^3}\bar{\mathbf{K}}_{eff} - \omega^2 mL\bar{\mathbf{M}} \right)\varphi = 0 \rightarrow \frac{EI_y}{L^3}\left( \bar{\mathbf{K}}_{eff} - \omega^2 \frac{mL^4}{EI_y}\bar{\mathbf{M}} \right)\varphi = 0$$

$$\rightarrow \left( \bar{\mathbf{K}}_{eff} - \omega^2 \frac{mL^4}{EI_y}\bar{\mathbf{M}} \right)\varphi = 0 \rightarrow \left( \bar{\mathbf{K}}_{eff} - \bar{\lambda}\bar{\mathbf{M}} \right)\varphi = 0 \rightarrow \bar{\mathbf{K}}_{eff}\varphi = \bar{\lambda}\bar{\mathbf{M}}\varphi,$$

where

$$\varphi = \begin{Bmatrix} \phi_2 \\ \phi_3 \\ \phi_4 \\ \phi_6 \end{Bmatrix}, \bar{\lambda} = \omega^2 \frac{mL^4}{EI_y}.$$

Carrying out the eigenvalue analysis, we obtain four eigenvalues. Note that

$$\bar{\lambda} = \omega^2 \frac{mL^4}{EI_y} \rightarrow \omega^2 = \bar{\lambda}\frac{EI_y}{mL^4} \rightarrow \omega = \sqrt{\bar{\lambda}\frac{EI_y}{mL^4}} \rightarrow \bar{\omega} = \sqrt{\bar{\lambda}}.$$

We can find that, corresponding to $\bar{P} = 0, 0.25, 0.5, 0.75$, the first two non-dimensional natural frequencies are

$$\bar{\omega}_1 = 9.9086, 8.5917, 7.0325, 5.0095$$
$$\bar{\omega}_2 = 43.8178, 42.6767, 41.5043, 40.2978.$$

For comparison, the exact solutions are given as follows:

$$\bar{\omega}_1 = 9.8696, 8.5473, 6.9789, 4.9348$$
$$\bar{\omega}_2 = 39.4784, 38.2248, 36.9287, 35.5854.$$

The first natural frequencies obtained by the two-element model are fairly accurate. However, the second natural frequencies are not as accurate, indicating that more elements are needed to improve the accuracy.

## Example 5.4

As shown in Figure 5.8, a cantilevered beam, clamped at $x = 0$ and free at $x = L$, is rotating around the $z$-axis at a constant rotating speed of $\Omega$ (rad/s).

**Figure 5.8** A slender body rotating at a constant speed

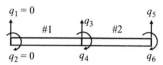

**Figure 5.9** Two-element model for a rotating cantilevered body

For simplicity, assume that the beam is of uniform cross-section with constant $m$ and $EI_y$. Carry out free bending vibration analysis using a model with two elements of equal length, as shown in Figure 5.9.

(a) Construct the element stiffness matrix due to axial force for each element.
(b) Determine the natural frequencies vs. the rotating speed.

## Solution:

(a) As shown in Chapter 1, for a uniform cross-section, the axial force due to rotation is

$$F = \frac{1}{2} m\Omega^2 \left(L^2 - x^2\right) = \frac{1}{2} m\Omega^2 L^2 \left[1 - \left(\frac{x}{L}\right)^2\right].$$

According to the mapping $x = x_1 + ls$ and then from Eq. (5.2.26):

$$\mathbf{k}_F^e = \frac{1}{l} \int_{s=0}^{s=1} F \mathbf{A}^T \mathbf{A} \, ds = \frac{1}{2l} m\Omega^2 L^2 \int_{s=0}^{s=1} \left[1 - \left(\frac{x}{L}\right)^2\right] \mathbf{A}^T \mathbf{A} \, ds$$

$$= \frac{1}{2l} m\Omega^2 L^2 \int_{s=0}^{s=1} \left[1 - \left(\frac{x_1 + ls}{L}\right)^2\right] \mathbf{A}^T \mathbf{A} \, ds.$$

For element #1 with $x_1 = 0$ and $l = L/2$:

$$\mathbf{k}_F^1 = m\Omega^2 L \int_{s=0}^{s=1} \left[1 - \frac{1}{4} s^2\right] \mathbf{A}^T \mathbf{A} \, ds.$$

For element #2 with $x_1 = L/2$:

$$\mathbf{k}_F^2 = m\Omega^2 L \int_{s=0}^{s=1} \left[1 - \frac{1}{4}(1+s)^2\right] \mathbf{A}^T \mathbf{A} \, ds.$$

Carrying out the integrations:

$$\mathbf{k}_F^1 = m\Omega^2 L \begin{bmatrix} 39/35 & -23/280 & -39/35 & -3/28 \\ -23/280 & 9/70 & 23/280 & -5/168 \\ -39/35 & 23/280 & 39/35 & 3/28 \\ -3/28 & -5/168 & 3/28 & 47/420 \end{bmatrix},$$

$$\mathbf{k}_F^2 = m\Omega^2 L \begin{bmatrix} 18/35 & -1/140 & -18/35 & -23/280 \\ -1/140 & 11/140 & 1/140 & -11/840 \\ -18/35 & 1/140 & 18/35 & 23/280 \\ -23/280 & -11/840 & 23/280 & 1/35 \end{bmatrix}.$$

The above two matrices are assembled into the 6 × 6 global matrix.

(b) In Example 4.1 in Chapter 4, the global mass matrix and the global stiffness matrix for the two-element model are given as

$$\mathbf{M}_{6\times 6} = \frac{mL}{840} \begin{bmatrix} 156 & -22 & 54 & 13 & 0 & 0 \\ -22 & 4 & -13 & -3 & 0 & 0 \\ 54 & -13 & 312 & 0 & 54 & 13 \\ 13 & -3 & 0 & 8 & -13 & -3 \\ 0 & 0 & 54 & -13 & 156 & 22 \\ 0 & 0 & 13 & -3 & 22 & 4 \end{bmatrix},$$

$$\mathbf{K}_{6\times 6} = \frac{16EI_y}{L^3} \begin{bmatrix} 6 & -3 & -6 & -3 & 0 & 0 \\ -3 & 2 & 3 & 1 & 0 & 0 \\ -6 & 3 & 12 & 0 & -6 & -3 \\ -3 & 1 & 0 & 4 & 3 & 1 \\ 0 & 0 & -6 & 3 & 6 & 3 \\ 0 & 0 & -3 & 1 & 3 & 2 \end{bmatrix}.$$

Deleting the first two columns and two rows corresponding to the geometric boundary conditions ($q_1 = q_2 = 0$), the global mass and stiffness matrices reduce to

$$\mathbf{M} = \frac{mL}{840} \begin{bmatrix} 312 & 0 & 54 & 13 \\ 0 & 8 & -13 & -3 \\ 54 & -13 & 156 & 22 \\ 13 & -3 & 22 & 4 \end{bmatrix} = mL\bar{\mathbf{M}}$$

with

$$\bar{\mathbf{M}} = \frac{1}{840} \begin{bmatrix} 312 & 0 & 54 & 13 \\ 0 & 8 & -13 & -3 \\ 54 & -13 & 156 & 22 \\ 13 & -3 & 22 & 4 \end{bmatrix},$$

$$\mathbf{K}_B = \frac{16EI_y}{L^3} \begin{bmatrix} 12 & 0 & -6 & -3 \\ 0 & 4 & 3 & 1 \\ -6 & 3 & 6 & 3 \\ -3 & 1 & 3 & 1 \end{bmatrix} = \frac{EI_y}{L^3} \bar{\mathbf{K}}_B$$

with

$$\bar{\mathbf{K}}_B = 16 \begin{bmatrix} 12 & 0 & -6 & -3 \\ 0 & 4 & 3 & 1 \\ -6 & 3 & 6 & 3 \\ -3 & 1 & 3 & 1 \end{bmatrix},$$

and

$$\mathbf{K}_F = m\Omega^2 L \begin{bmatrix} 1.6286 & 0.1000 & -0.5143 & -0.0821 \\ 0.1000 & 0.1905 & 0.0071 & -0.0131 \\ -0.5143 & 0.0071 & 0.5143 & 0.0821 \\ -0.0821 & -0.0131 & 0.0821 & 0.0286 \end{bmatrix} = m\Omega^2 L \bar{\mathbf{K}}_F$$

with

$$\bar{\mathbf{K}}_F = \begin{bmatrix} 1.6286 & 0.1000 & -0.5143 & -0.0821 \\ 0.1000 & 0.1905 & 0.0071 & -0.0131 \\ -0.5143 & 0.0071 & 0.5143 & 0.0821 \\ -0.0821 & -0.0131 & 0.0821 & 0.0286 \end{bmatrix}.$$

Accordingly:

$$\mathbf{K}_{eff} = \mathbf{K}_B + \mathbf{K}_F = \frac{EI_y}{L^3} \bar{\mathbf{K}}_B + m\Omega^2 L \bar{\mathbf{K}}_F = \frac{EI_y}{L^3} \left( \bar{\mathbf{K}}_B + \frac{m\Omega^2 L^4}{EI_y} \bar{\mathbf{K}}_F \right)$$

$$\to \mathbf{K}_{eff} = \frac{EI_y}{L^3} \left( \bar{\mathbf{K}}_B + \bar{\Omega}^2 \bar{\mathbf{K}}_F \right),$$

where

$$\bar{\Omega}^2 = \frac{m\Omega^2 L^4}{EI_y} \to \bar{\Omega} = \frac{\Omega}{\sqrt{\dfrac{EI_y}{mL^4}}}.$$

According to Eq. (5.3.4) for free vibration:

$$\left(\mathbf{K}_{eff} - \omega^2 \mathbf{M}\right)\boldsymbol{\varphi} = \mathbf{0} \to \left[ \frac{EI_y}{L^3} \left( \bar{\mathbf{K}}_B + \bar{\Omega}^2 \bar{\mathbf{K}}_F \right) - \omega^2 mL\bar{\mathbf{M}} \right]\boldsymbol{\varphi} = \mathbf{0}$$

$$\to \frac{EI_y}{L^3} \left[ \left( \bar{\mathbf{K}}_B + \bar{\Omega}^2 \bar{\mathbf{K}}_F \right) - \omega^2 \frac{mL^4}{EI_y} \bar{\mathbf{M}} \right]\boldsymbol{\varphi} = \mathbf{0} \to \left[ \left( \bar{\mathbf{K}}_B + \bar{\Omega}^2 \bar{\mathbf{K}}_F \right) - \omega^2 \frac{mL^4}{EI_y} \bar{\mathbf{M}} \right]\boldsymbol{\varphi} = \mathbf{0}.$$

The last of the above equations can be expressed as

$$\left[\bar{\mathbf{K}}_{eff} - \bar{\lambda}\mathbf{M}\right]\boldsymbol{\varphi} = 0 \rightarrow \bar{\mathbf{K}}_{eff}\boldsymbol{\varphi} = \bar{\lambda}\mathbf{M}\boldsymbol{\varphi},$$

which is a standard equation for eigenvalue analysis where

$$\bar{\mathbf{K}}_{eff} = \bar{\mathbf{K}}_B + \bar{\Omega}^2\bar{\mathbf{K}}_F, \quad \bar{\lambda} = \omega^2\frac{mL^4}{EI_y}.$$

Figure 5.10 shows the first natural frequency vs. the rotating speed. They are normalized with respect to the first natural frequency of the non-rotating body such that

$$\hat{\Omega} = \frac{\Omega}{(\omega_1)_{NR}} : \text{normalized rotating speed}$$

$$\hat{\omega} = \frac{\omega_1}{(\omega_1)_{NR}} : \text{normalized first natural frequency}$$

$$(\omega_1)_{NR} = 3.5177\sqrt{\frac{EI_y}{mL^4}} : \text{first natural frequency of non-rotating body.}$$

As the rotating speed increases, the first natural frequency increases, reflecting the stiffening effect of the centrifugal force.

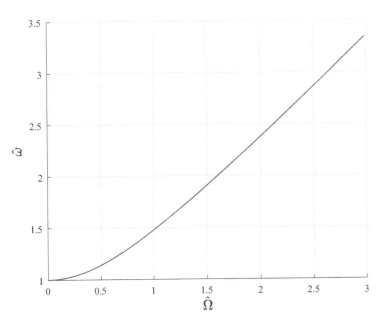

Figure 5.10 Normalized first natural frequency vs. normalized rotating speed

## 5.4    Slender Body under Compressive Follower Force

Consider a clamped-free column subjected to a compressive force as shown in Figure 5.11. Force $P$ is always parallel to the body axis in the deformed configuration and is called a follower force. This may represent a rocket on a test stand with end thrust. This is also known as the garden hose problem or fireman's hose problem. It is known that the column does not buckle statically but becomes dynamically unstable if the force is large enough. Assuming small rotation of the cross-section, the transverse component and the longitudinal component of the follower force can be expressed as

$$P\sin\theta \approx P\theta, P\cos\theta \approx P. \tag{5.4.1}$$

The incremental work done by the transverse component is

$$\delta W_f = (P\theta\delta w)_{x=L}. \tag{5.4.2}$$

**Figure 5.11** Clamped-free column subjected to a follower force

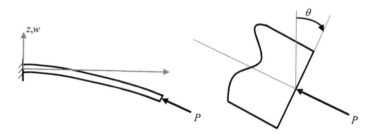

For the two-element model shown in Figure 5.12:

$$\delta W_f = (P\theta\delta w)_{x=L} = \left(P\frac{1}{l}\hat\theta\delta w\right)_{x=L} = P\frac{1}{l}q_6\delta q_5 = \delta q_5(F_5)_P, \tag{5.4.3}$$

where

$$\hat\theta = l\theta, \quad (F_5)_P = P\frac{1}{l}q_6. \tag{5.4.4}$$

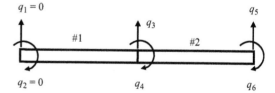

**Figure 5.12** Two-element model for the follower force problem

The incremental work done by the follower force can be symbolically expressed as

$$\delta W_f = \delta\mathbf{q}^T\mathbf{F}_f. \tag{5.4.5}$$

For the two-element model:

$$\mathbf{F}_f = \frac{P}{l}\begin{Bmatrix} 0 \\ 0 \\ 0 \\ 0 \\ q_6 \\ 0 \end{Bmatrix} = P(\mathbf{K}_P)_{nonsymmetric}\mathbf{q}, \tag{5.4.6}$$

where

$$(\mathbf{K}_P)_{nonsymmetric} = \frac{1}{l}\begin{bmatrix} 0 & 0 & 0 & 0 & 0 & 0 \\ 0 & 0 & 0 & 0 & 0 & 0 \\ 0 & 0 & 0 & 0 & 0 & 0 \\ 0 & 0 & 0 & 0 & 0 & 0 \\ 0 & 0 & 0 & 0 & 0 & 1 \\ 0 & 0 & 0 & 0 & 0 & 0 \end{bmatrix}. \tag{5.4.7}$$

The equation of motion is then

$$\mathbf{M}\ddot{\mathbf{q}} + \left(\mathbf{K}_B - P(\mathbf{K}_P)_{symmetric}\right)\mathbf{q} = \mathbf{F}_f. \tag{5.4.8}$$

Note that $(\mathbf{K}_P)_{symmetric} = \mathbf{K}_P$ was introduced in Section 5.2.1. Substituting Eq. (5.4.6) into Eq. (5.4.8) and rearranging:

$$\mathbf{M}\ddot{\mathbf{q}} + \left(\mathbf{K}_B - P\mathbf{K}_{P\text{-}follower}\right)\mathbf{q} = \mathbf{0}, \tag{5.4.9}$$

where

$$\mathbf{K}_{P\text{-}follower} = (\mathbf{K}_P)_{symmetric} + (\mathbf{K}_P)_{nonsymmetric} \tag{5.4.10}$$

or

$$\mathbf{M}\ddot{\mathbf{q}} + \mathbf{K}_{eff}\mathbf{q} = \mathbf{0}, \tag{5.4.11}$$

where

$$\mathbf{K}_{eff} = \mathbf{K}_B - P\mathbf{K}_{P\text{-}follower}. \tag{5.4.12}$$

Equation (5.4.11) is a homogeneous equation representing free vibration following initial disturbances. The solution to Eq. (5.4.11) is of the form

$$\mathbf{q} = \boldsymbol{\varphi}e^{\lambda t}. \tag{5.4.13}$$

Placing Eq. (5.4.13) into Eq. (5.4.11) yields

$$\left(\mathbf{K}_{eff} + \lambda^2\mathbf{M}\right)\boldsymbol{\varphi} = \mathbf{0}. \tag{5.4.14}$$

For nontrivial $\varphi$:

$$\det\left(\mathbf{K}_{eff} + \lambda^2 \mathbf{M}\right) = 0, \tag{5.4.15}$$

which can be used to solve for $\lambda = \lambda_1, \lambda_2, \ldots, \lambda_N$. Alternatively, Eq. (5.4.14) can be expressed as

$$\mathbf{K}_{eff}\,\varphi = \Lambda \mathbf{M}\varphi, \tag{5.4.16}$$

where

$$\Lambda = -\lambda^2. \tag{5.4.17}$$

Equation (5.4.16) is in standard form for eigenvalue analysis. For a given compressive force $P$ one can find eigenvalues $\Lambda_k$ from Eq. (5.4.16) and then determine $\lambda_k$ from Eq. (5.4.17). Then $\lambda_k$ can be expressed as

$$\lambda_k = \mathrm{Re}(\lambda_k) + i\mathrm{Im}(\lambda_k) = \sigma_k \pm i\omega_k \tag{5.4.18}$$

and

$$e^{\lambda_k t} = e^{(\sigma_k \pm i\omega_k)t} = e^{\sigma_k t}e^{\pm i\omega_k t} = e^{\sigma_k t}(\cos\omega_k t \pm i\sin\omega_k t), \tag{5.4.19}$$

where $e^{\sigma_k t}$ indicates change of amplitude with time and $(\cos\omega_k t \pm i\sin\omega_k t)$ represents oscillation with frequency $\omega_k$.

*Stability criteria:* If $\sigma_k = \mathrm{Re}(\lambda_k) > 0$ for any $k$, the amplitude of vibration excited by initial disturbance will grow in time. The system is then said to be dynamically unstable.

One can carry out the eigenvalue analyses for varying values of $P$. For small $P$ $\sigma_k = \mathrm{Re}(\lambda_k) = 0$ for all $k$, and the system is stable. As $P$ increases, $\sigma_k = \mathrm{Re}(\lambda_k)$ becomes positive at a critical value $(P_{cr})$, indicating the onset of amplitude growth with time. Two lowest frequencies $\omega_1$ and $\omega_2$ approach each other as $P$ increases, and they meet or coalesce at $P_{cr}$. This is called frequency coalescence.

## ◥ Example 5.5

For simplicity, assume that the body is of uniform cross-section. Use a model with two elements of equal length. For convenience, introduce

$$\hat{\theta} = l\theta,\ \bar{P} = \frac{PL^2}{EI_y},\ \bar{\lambda} = \lambda\sqrt{\frac{mL^4}{EI_y}} = \bar{\sigma} \pm i\bar{\omega}.$$

Plot $\bar{\sigma} = \mathrm{Re}(\bar{\lambda})$ vs. $\bar{P}$ and $\bar{\omega}_1, \bar{\omega}_2$ vs. $\bar{P}$ and determine the critical load $\bar{P}_{cr}$ at which the body becomes dynamically unstable.

### Solution:

For the two-element model, $\bar{\sigma} = \mathrm{Re}(\bar{\lambda})$ vs. $\bar{P}$ and $\bar{\omega}_1, \bar{\omega}_2$ vs. $\bar{P}$ are plotted as shown in Figure 5.13, from which $\bar{P}_{cr}$ is found to be slightly over 20. Figure 5.14 shows frequency coalescence as the follower force increases and the system becomes dynamically unstable.

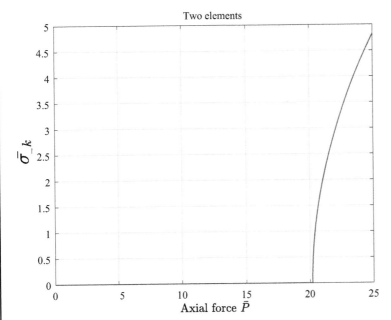

**Figure 5.13** Real part of the non-dimensional eigenvalues of the follower force problem

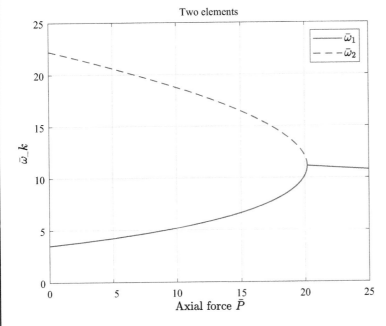

**Figure 5.14** Frequency coalescence of the follower force problem

## PROBLEMS

**5.1** Figure 5.15 shows a buckled position of a slender body, clamped at $x = 0$ and free at $x = L$, under a compressive force $P$ applied at the free end. The direction of load $P$ does not change and is always parallel to the body axis in the original undeformed configuration. The body is of uniform cross-section with constant $EI_y$.

*Note:* Introduce $\bar{P} = \frac{PL^2}{EI_y}$.

**Figure 5.15** For Problem 5.1

(a) Use a model with one element to determine the static buckling load and buckling mode. Scale the mode such that the maximum value of $w$ is equal to unity. Compare with the exact solution given as

$$P_{cr} = \frac{\pi^2 EI_y}{4L^2}.$$

(b) Repeat using a model with two elements of equal length.
(c) Repeat using a model with three elements of equal length.

**5.2** Carry out the buckling analysis described in Example 5.1 using a model with three elements of equal length.

**5.3** A slender body of constant $EI_y$ and length $L$ is clamped at $x = 0$ and supported by a roller hinge at $x = L$, as shown in Figure 5.16. The body is subjected to a compressive axial force $P$ at $x = L$. The direction of $P$ does not change and remains parallel to the $x$-axis. Use a model with two elements of equal length to do as follows:

(a) The static buckling load can be expressed as $P_{cr} = CEI_y/L^2$. Determine $C$.
(b) Sketch the buckling mode. Scale the mode such that the maximum value of $w$ is equal to unity.

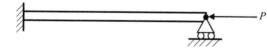

Figure 5.16  For Problem 5.3

**5.4** Repeat Problem 5.3 using a model with four elements of equal length.
**5.5** Repeat the problem described in Example 5.2 using a model with two elements of equal length as shown in Figure 5.17.
**5.6** Repeat the problem described in Example 5.2 using a model with three elements of equal length.

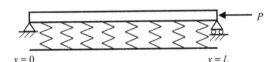

**Figure 5.17** For Problem 5.5

$q_5$ $q_6$

#2

$q_3$ $q_4$

#1

$q_1 = 0$     $q_2 = 0$

**5.7** As shown in Figure 5.18, a simply supported column is placed on an elastic foundation modeled as a distributed spring with spring constant $k$ per unit length.

(a) Use a model with six elements of equal length to determine the buckling load and the mode for the non-dimensional ratio defined as

$$\bar{k} = \frac{kL}{\pi^4 \left(\dfrac{EI_y}{L^3}\right)} = \frac{kL^4}{\pi^4 EI_y} = 0, 1, 2, \ldots, 8.$$

Note that these ratios represent the relative stiffness between the elastic foundation and the column bending rigidity. Plot $\bar{P}_{cr}$ vs. $\bar{k}$, where

$$\bar{P}_{cr} = \frac{P}{\left(\dfrac{\pi^2 EI_y}{L^2}\right)}.$$

(b) The buckling mode may change as the relative stiffness ratio defined in (a) increases. At what value of the relative stiffness ratio does the buckling mode change? Plot the buckling modes corresponding to the range of relative stiffness ratios considered. Scale such that the maximum value of $w$ in each mode is equal to unity.

$P$     **Figure 5.18** For Problem 5.7

$x = 0$                    $x = L$

**5.8** Repeat the problem described in Example 5.3 using a model with four elements of equal length.

**5.9** Repeat the problem described in Example 5.4 using a model with three elements of equal length.

**5.10** Repeat the problem described in Example 5.5 using a model with four elements of equal length. In addition, carry out eigenvalue analysis to check whether the slender body buckles statically under the follower force. Note that, for static buckling to occur, there must be a real eigenvalue.

**5.11** A slender body in level flight is modeled as a column free at both $x = 0$ and $x = L$. The body is subjected to thrust $P$ applied at $x = L$. The direction of load $P$ remains always in parallel with the body axis in the deformed configuration. For simplicity, assume that the body is of uniform cross-section with constant $EI_y$.

*Note:*

(1) Assume zero-gravity condition.

(2) Axial force is expressed as $F(x) = -Px/L$.

(3) Introduce $\hat{\theta} = l\theta$, $\bar{P} = PL^2/EI_y$, and $\bar{\lambda} = \lambda\sqrt{mL^4/EI_y} = \bar{\sigma} + i\bar{\omega}$.

(a) Using a model with three elements of equal length, carry out free bending vibration analysis for $P = 0$ to determine natural frequencies and modes. Plot the modes. Scale the modes such that max displacement is equal to unity.

(b) Using a model with three elements of equal length, check whether there exists a critical load $\bar{P}_{cr}$ at which the body becomes dynamically unstable.

**5.12** Repeat Problem 5.11 using a model with five elements of equal length.

**5.13** A slender body of constant $EI_y$ and length $L$ is fixed at $x = 0$ and supported by a roller hinge at $x = L$. The body is subjected to a compressive axial force $P$. The direction of $P$ is always parallel to the body axis in the deformed configuration.

(a) Can the body experience dynamic instability?

(b) Can the body buckle statically? What is the buckling load?

**5.14** Consider a slender body of non-uniform cross-section clamped at $x = 0$ and free at $x = L$, as shown in Figure 5.19. The body is subjected to a compressive force $P$ applied at the free end. The direction of load $P$ does not change and is always parallel to the body axis in the original undeformed configuration. The area moment of inertia is given as

$$I_y = I_0\left(1 - \frac{x}{2L}\right),$$

where $I_0$ is a given constant value.

(a) Use a model with two elements of equal length to determine the static buckling load and the buckling mode.

(b) Use a model with four elements of equal length to determine the static buckling load and the buckling mode.

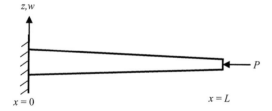

Figure 5.19 For Problem 5.14

# 6 Virtual Displacement and Virtual Work

The mathematical statement for analysis of solids and structures includes the equilibrium equation, strain–displacement relation, and constitutive equation or stress–strain relation. In addition, the force boundary condition and geometric boundary condition are prescribed over a given geometry.

In this chapter, we introduce the concept of an arbitrary virtual displacement which may also be considered as an arbitrary weight function. This will be used to express the equilibrium equation for 3D solids and structures in a scalar integral form. Subsequently, the divergence theorem is applied to transform the scalar integral into another form to which the force boundary condition can be introduced. This results in the statement for the principle of virtual work involving internal virtual work and external virtual work. The internal virtual work and the external virtual work will then be expressed in matrix form so that they can be used for the FE formulation in later chapters.

We then consider plane stress and plane strain problems in which the principle of virtual work can be expressed in 2D domains in accordance with simplifying conditions. In the last section, the Lagrange equation is derived within the context of deformable solid bodies, starting from the principle of virtual work.

In the problems at the end of this chapter, we introduce proper simplifying assumptions to the virtual work statement for 3D solids in order to derive the virtual work statements for slender bodies undergoing uniaxial deformation and pure bending deformation. We can also construct the virtual work statement for slender bodies in bending under axial force starting from the equilibrium equation derived in Chapter 5, as indicated in the problems.

## 6.1 Mathematical Description of Deformable Solids

Considering the fundamental equations provided in Appendix 1, the mathematical description for analysis of solids and structures subjected to applied loads can be summarized as follows.

**Equilibrium equation:**
$$\frac{\partial \boldsymbol{\sigma}_x}{\partial x} + \frac{\partial \boldsymbol{\sigma}_y}{\partial y} + \frac{\partial \boldsymbol{\sigma}_z}{\partial z} + \mathbf{F}_B = 0. \tag{6.1.1}$$

**Stress–strain relation:**
$$\boldsymbol{\sigma} = \mathbf{C}(\boldsymbol{\varepsilon} - \boldsymbol{\varepsilon}^o). \tag{6.1.2}$$

**Strain–displacement relation:**
$$\varepsilon_{xx} = \frac{\partial u}{\partial x}, \text{etc.} \tag{6.1.3}$$

**165**

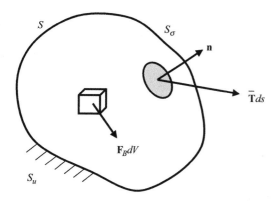

Figure 6.1 Solid body under body force, surface force and prescribed displacement

**Boundary conditions:**

(1) Force boundary condition: $\mathbf{T} = \bar{\mathbf{T}}$ on $S_\sigma$.
(2) Geometric boundary condition: $\mathbf{u} = \bar{\mathbf{u}}$ on $S_u$.

$\bar{\mathbf{T}}$ and $\bar{\mathbf{u}}$ are prescribed values, and $S_\sigma$ is the part of surface $S$ on which the traction vector is prescribed, while $S_u$ is the part of the surface on which displacement is prescribed as shown in Figure 6.1.

## 6.2    Virtual Displacement and Work for 3D Solids

We will now develop a scalar equation which is equivalent to equilibrium in volume $V$ and incorporates the force boundary conditions on $S_\sigma$. Starting from the equilibrium in Eq. (6.1.1), introduce an arbitrary function $\delta\mathbf{u}$, called a virtual displacement, and construct the following scalar integral:

$$\int_V \left( \frac{\partial \boldsymbol{\sigma}_x}{\partial x} + \frac{\partial \boldsymbol{\sigma}_y}{\partial y} + \frac{\partial \boldsymbol{\sigma}_z}{\partial z} + \mathbf{F}_B \right) \cdot \delta\mathbf{u} \, dV = 0. \tag{6.2.1}$$

For arbitrary $\delta\mathbf{u}$, Eq. (6.2.1) is equivalent to Eq. (6.1.1). To show this, set

$$\frac{\partial \boldsymbol{\sigma}_x}{\partial x} + \frac{\partial \boldsymbol{\sigma}_y}{\partial y} + \frac{\partial \boldsymbol{\sigma}_z}{\partial z} + \mathbf{F}_B = \mathbf{g}. \tag{6.2.2}$$

Equation (6.2.1) is then expressed as

$$\int_V \mathbf{g} \cdot \delta\mathbf{u} \, dV = 0. \tag{6.2.3}$$

Since Eq. (6.2.3) holds for arbitrary $\delta\mathbf{u}$, it holds for the following $\delta\mathbf{u}$ chosen as

$$\delta\mathbf{u} = \varepsilon\mathbf{g}, \tag{6.2.4}$$

where $\varepsilon$ is a constant. Placing Eq. (6.2.4) into Eq. (6.2.3):

$$\varepsilon \int_V \mathbf{g} \cdot \mathbf{g} \, dV = 0 \rightarrow \mathbf{g} = \mathbf{0}. \tag{6.2.5}$$

### 6.2.1    Principle of Virtual Work

Equation (6.2.1) can now be transformed to a form to which the force boundary condition can be applied. For this, we need to use the divergence theorem stated as follows.

**Divergence Theorem**

This theorem relates a volume integral to a surface integral. In Figure 6.2:

$$\mathbf{h}(x, y, z) = L\mathbf{i} + M\mathbf{j} + N\mathbf{k} : \text{vector field function of position,} \qquad (6.2.6)$$

$$\mathbf{n} = l\mathbf{i} + m\mathbf{j} + n\mathbf{k} : \text{unit vector outward normal to the surface,} \qquad (6.2.7)$$

$$\nabla = \mathbf{i}\frac{\partial}{\partial x} + \mathbf{j}\frac{\partial}{\partial y} + \mathbf{k}\frac{\partial}{\partial z} : \text{gradient operator.} \qquad (6.2.8)$$

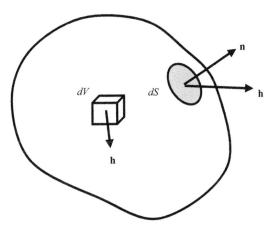

**Figure 6.2** Vector field function $\mathbf{h}$ in the volume and on the surface of a solid body

The divergence theorem states that

$$\int_V \nabla \cdot \mathbf{h}\, dV = \int_S \mathbf{h} \cdot \mathbf{n}\, dS \qquad (6.2.9)$$

or

$$\int_V \left( \frac{\partial L}{\partial x} + \frac{\partial M}{\partial y} + \frac{\partial N}{\partial z} \right) dV = \int_S (Ll + Mm + Nn)\, dS. \qquad (6.2.10)$$

To apply the divergence theorem, Eq. (6.2.1) can be expressed as

$$\int_V \left[ \frac{\partial}{\partial x}(\boldsymbol{\sigma}_x \cdot \delta\mathbf{u}) + \frac{\partial}{\partial y}(\boldsymbol{\sigma}_y \cdot \delta\mathbf{u}) + \frac{\partial}{\partial z}(\boldsymbol{\sigma}_z \cdot \delta\mathbf{u}) \right] dV$$

$$- \int_V \left( \boldsymbol{\sigma}_x \cdot \frac{\partial \delta\mathbf{u}}{x} + \boldsymbol{\sigma}_y \cdot \frac{\partial \delta\mathbf{u}}{\partial y} + \boldsymbol{\sigma}_z \cdot \frac{\partial \delta\mathbf{u}}{\partial z} \right) dV + \int_V \mathbf{F}_B \cdot \delta\mathbf{u}\, dV = 0. \qquad (6.2.11)$$

Applying the divergence theorem to the first integral in the above equation:

$$I_1 = \int_V \left[ \underbrace{\frac{\partial}{\partial x}(\boldsymbol{\sigma}_x \cdot \delta\mathbf{u})}_{L} + \underbrace{\frac{\partial}{\partial y}(\boldsymbol{\sigma}_y \cdot \delta\mathbf{u})}_{M} + \underbrace{\frac{\partial}{\partial z}(\boldsymbol{\sigma}_z \cdot \delta\mathbf{u})}_{N} \right] dV$$

$$= \int_V \left[ (\boldsymbol{\sigma}_x \cdot \delta\mathbf{u})l + (\boldsymbol{\sigma}_y \cdot \delta\mathbf{u})m + (\boldsymbol{\sigma}_z \cdot \delta\mathbf{u})n \right] dS$$

$$= \int_S (\boldsymbol{\sigma}_x l + \boldsymbol{\sigma}_y m + \boldsymbol{\sigma}_z n) \cdot \delta\mathbf{u} dS. \tag{6.2.12}$$

In Appendix 1 we show that

$$\mathbf{T} = \boldsymbol{\sigma}_x l + \boldsymbol{\sigma}_y m + \boldsymbol{\sigma}_z n. \tag{6.2.13}$$

Applying the above equation to Eq. (6.2.12):

$$I_1 = \int_S \delta\mathbf{u} \cdot \mathbf{T} dS = \int_{S_\sigma} \delta\mathbf{u} \cdot \mathbf{T} dS + \int_{S_u} \delta\mathbf{u} \cdot \mathbf{T} dS. \tag{6.2.14}$$

Applying the force boundary condition $\mathbf{T} = \bar{\mathbf{T}}$ on $S_\sigma$:

$$I_1 = \int_{S_\sigma} \delta\mathbf{u} \cdot \bar{\mathbf{T}} dS + \int_{S_u} \delta\mathbf{u} \cdot \mathbf{T} dS. \tag{6.2.15}$$

Placing Eq. (6.2.15) into Eq. (6.2.11) and changing the sign:

$$\int_V \left( \boldsymbol{\sigma}_x \cdot \frac{\partial \delta\mathbf{u}}{\partial x} + \boldsymbol{\sigma}_y \cdot \frac{\partial \delta\mathbf{u}}{\partial y} + \boldsymbol{\sigma}_z \cdot \frac{\partial \delta\mathbf{u}}{\partial z} \right) dV - \int_V \delta\mathbf{u} \cdot \mathbf{F}_B dV - \int_{S_\sigma} \delta\mathbf{u} \cdot \bar{\mathbf{T}} dS - \int_{S_u} \delta\mathbf{u} \cdot \mathbf{T} dS = 0.$$

$$\tag{6.2.16}$$

The last term in the above equation accounts for the reaction force on the surface on which displacement is prescribed. One may now define internal virtual work $\delta U$ and external virtual work $\delta W$ as follows:

$$\delta U = \int_V \left( \boldsymbol{\sigma}_x \cdot \frac{\partial \delta\mathbf{u}}{\partial x} + \boldsymbol{\sigma}_y \cdot \frac{\partial \delta\mathbf{u}}{\partial y} + \boldsymbol{\sigma}_z \cdot \frac{\partial \delta\mathbf{u}}{\partial z} \right) dV, \tag{6.2.17}$$

$$\delta W = \int_V \delta\mathbf{u} \cdot \mathbf{F}_B dV + \int_{S_\sigma} \delta\mathbf{u} \cdot \bar{\mathbf{T}} dS + \int_{S_u} \delta\mathbf{u} \cdot \mathbf{T} dS. \tag{6.2.18}$$

Equation (6.2.16) is then expressed as

$$\delta U - \delta W = 0. \tag{6.2.19}$$

If we set $\delta\mathbf{u} = 0$ on $S_u$:

$$\delta W = \int_V \delta\mathbf{u} \cdot \mathbf{F}_B dV + \int_{S_\sigma} \delta\mathbf{u} \cdot \bar{\mathbf{T}} dS \tag{6.2.20}$$

and the reaction force term disappears. The equivalence between the two blocks shown below is known as the principle of virtual work (PVW).

Equilibrium:

$$\frac{\partial \boldsymbol{\sigma}_x}{\partial x} + \frac{\partial \boldsymbol{\sigma}_y}{\partial y} + \frac{\partial \boldsymbol{\sigma}_z}{\partial z} + \mathbf{F}_B = \mathbf{0}$$

Force B.C.:

$$\mathbf{T} = \bar{\mathbf{T}} \text{ on } S_\sigma$$

$\rightleftarrows$  $\boxed{\delta U - \delta W = 0.}$

## 6.2.2 Virtual Work Expression in Matrix Form

For the FE formulation, it is convenient to express $\delta U$ and $\delta W$ in matrix form as follows.

### External Virtual Work in Matrix Form

For the body force term:

$$\int_V \delta \mathbf{u} \cdot \mathbf{F}_B dV = \int_V (\delta u \mathbf{i} + \delta v \mathbf{j} + \delta w \mathbf{k}) \cdot (X_B \mathbf{i} + Y_B \mathbf{j} + Z_B \mathbf{k}) dV$$

$$= \int_V (\delta u X_B + \delta v Y_B + \delta w Z_B) dV \qquad (6.2.21)$$

$$= \int_V \lfloor \delta u \quad \delta v \quad \delta w \rfloor \begin{Bmatrix} X_B \\ Y_B \\ Z_B \end{Bmatrix} dV = \int_V \delta \mathbf{u}^\mathrm{T} \mathbf{F}_B dV,$$

where

$$\delta \mathbf{u} = \begin{Bmatrix} \delta u \\ \delta v \\ \delta w \end{Bmatrix}, \quad \mathbf{F}_B = \begin{Bmatrix} X_B \\ Y_B \\ Z_B \end{Bmatrix}. \qquad (6.2.22)$$

For the surface traction term:

$$\int_{S_\sigma} \delta \mathbf{u} \cdot \bar{\mathbf{T}} dS = \int_{S_\sigma} (\delta u \mathbf{i} + \delta v \mathbf{j} + \delta w \mathbf{k}) \cdot (\bar{T}_x \mathbf{i} + \bar{T}_y \mathbf{j} + \bar{T}_z \mathbf{k}) dS$$

$$= \int_{S_\sigma} (\delta u \bar{T}_x + \delta v \bar{T}_y + \delta w \bar{T}_z) dS \qquad (6.2.23)$$

$$= \int_{S_\sigma} \lfloor \delta u \quad \delta v \quad \delta w \rfloor \begin{Bmatrix} \bar{T}_x \\ \bar{T}_y \\ \bar{T}_z \end{Bmatrix} dS = \int_{S_\sigma} \delta \mathbf{u}^\mathrm{T} \bar{\mathbf{T}} dS,$$

where

$$\bar{\mathbf{T}} = \begin{Bmatrix} \bar{T}_x \\ \bar{T}_y \\ \bar{T}_z \end{Bmatrix} : \text{applied traction vector on } S_\sigma. \qquad (6.2.24)$$

Similarly:

$$\int_{S_u} \delta \mathbf{u} \cdot \mathbf{T} dS = \int_{S_u} \lfloor \delta u \quad \delta v \quad \delta w \rfloor \begin{Bmatrix} T_x \\ T_y \\ T_z \end{Bmatrix} dS = \int_{S_u} \delta \mathbf{u}^{\mathsf{T}} \mathbf{T} dS, \tag{6.2.25}$$

where $\mathbf{T}$ is the reaction traction vector corresponding to the applied displacement on $S_u$.

**Internal Virtual Work in Matrix Form**

Recall that

$$\delta U = \int_V \left( \boldsymbol{\sigma}_x \cdot \frac{\partial \delta \mathbf{u}}{\partial x} + \boldsymbol{\sigma}_y \cdot \frac{\partial \delta \mathbf{u}}{\partial y} + \boldsymbol{\sigma}_z \cdot \frac{\partial \delta \mathbf{u}}{\partial z} \right) dV, \tag{6.2.26}$$

where

$$\delta \mathbf{u} = \delta u \mathbf{i} + \delta v \mathbf{j} + \delta w \mathbf{k}. \tag{6.2.27}$$

Then

$$\begin{aligned} \boldsymbol{\sigma}_x \cdot \frac{\partial \delta \mathbf{u}}{\partial x} &= (\sigma_{xx}\mathbf{i} + \sigma_{xy}\mathbf{j} + \sigma_{xz}\mathbf{k}) \cdot \left( \frac{\partial \delta u}{\partial x}\mathbf{i} + \frac{\partial \delta v}{\partial x}\mathbf{j} + \frac{\partial \delta w}{\partial x}\mathbf{k} \right) \\ &= \sigma_{xx}\frac{\partial \delta u}{\partial x} + \sigma_{xy}\frac{\partial \delta v}{\partial x} + \sigma_{xz}\frac{\partial \delta w}{\partial x}. \end{aligned} \tag{6.2.28}$$

Similarly

$$\boldsymbol{\sigma}_y \cdot \frac{\partial \delta \mathbf{u}}{\partial y} = \sigma_{yx}\frac{\partial \delta u}{\partial y} + \sigma_{yy}\frac{\partial \delta v}{\partial y} + \sigma_{yz}\frac{\partial \delta w}{\partial y}, \tag{6.2.29}$$

$$\boldsymbol{\sigma}_z \cdot \frac{\partial \delta \mathbf{u}}{\partial z} = \sigma_{zx}\frac{\partial \delta u}{\partial z} + \sigma_{zy}\frac{\partial \delta v}{\partial z} + \sigma_{zz}\frac{\partial \delta w}{\partial z}. \tag{6.2.30}$$

Placing Eqs (6.2.28), (6.2.29), and (6.2.30) into Eq. (6.2.26) and using the symmetry of shear stress components:

$$\begin{aligned} \delta U = \int_V \Big[ &\sigma_{xx}\frac{\partial \delta u}{\partial x} + \sigma_{yy}\frac{\partial \delta v}{\partial y} + \sigma_{zz}\frac{\partial \delta w}{\partial z} + \sigma_{xy}\left( \frac{\partial \delta u}{\partial y} + \frac{\partial \delta v}{\partial x} \right) \\ &+ \sigma_{yz}\left( \frac{\partial \delta v}{\partial z} + \frac{\partial \delta w}{\partial y} \right) + \sigma_{zx}\left( \frac{\partial \delta w}{\partial x} + \frac{\partial \delta u}{\partial z} \right) \Big] dV. \end{aligned} \tag{6.2.31}$$

For convenience, we introduce "virtual strain" components defined as

$$\begin{aligned} \delta \varepsilon_{xx} &= \frac{\partial \delta u}{\partial x}, \delta \varepsilon_{yy} = \frac{\partial \delta v}{\partial y}, \delta \varepsilon_{zz} = \frac{\partial \delta w}{\partial z}, \\ \delta \varepsilon_{xy} &= \frac{\partial \delta u}{\partial y} + \frac{\partial \delta v}{\partial x}, \delta \varepsilon_{yz} = \frac{\partial \delta v}{\partial z} + \frac{\partial \delta w}{\partial y}, \delta \varepsilon_{zx} = \frac{\partial \delta w}{\partial x} + \frac{\partial \delta u}{\partial z}. \end{aligned} \tag{6.2.32}$$

Equation (6.2.31) can then be expressed as

$$\delta U = \int_V \left( \sigma_{xx}\delta\varepsilon_{xx} + \sigma_{yy}\delta\varepsilon_{yy} + \sigma_{zz}\delta\varepsilon_{zz} + \sigma_{yz}\delta\varepsilon_{yz} + \sigma_{zx}\delta\varepsilon_{zx} + \sigma_{xy}\delta\varepsilon_{xy} \right) dV, \tag{6.2.33}$$

which can be written in matrix form as

$$\delta U = \int_V \delta\boldsymbol{\varepsilon}^{\mathrm{T}}\boldsymbol{\sigma}\, dV, \tag{6.2.34}$$

where $\delta\boldsymbol{\varepsilon}$ and $\boldsymbol{\sigma}$ are

$$\delta\boldsymbol{\varepsilon} = \begin{Bmatrix} \delta\varepsilon_{xx} \\ \delta\varepsilon_{yy} \\ \delta\varepsilon_{zz} \\ \delta\varepsilon_{xy} \\ \delta\varepsilon_{yz} \\ \delta\varepsilon_{zx} \end{Bmatrix}, \boldsymbol{\sigma} = \begin{Bmatrix} \sigma_{xx} \\ \sigma_{yy} \\ \sigma_{zz} \\ \sigma_{xy} \\ \sigma_{yz} \\ \sigma_{zx} \end{Bmatrix}. \tag{6.2.35}$$

From Appendix 1, the stress–strain relationship is

$$\boldsymbol{\sigma} = \mathbf{C}(\boldsymbol{\varepsilon} - \boldsymbol{\varepsilon}^o). \tag{6.2.36}$$

Substituting the above equation into Eq. (6.2.34):

$$\delta U = \int_V \delta\boldsymbol{\varepsilon}^{\mathrm{T}}\mathbf{C}(\boldsymbol{\varepsilon} - \boldsymbol{\varepsilon}^o) dV = \int_V \delta\boldsymbol{\varepsilon}^{\mathrm{T}}\mathbf{C}\boldsymbol{\varepsilon}\, dV - \int_V \delta\boldsymbol{\varepsilon}^{\mathrm{T}}\mathbf{C}\boldsymbol{\varepsilon}^o dV. \tag{6.2.37}$$

### Infinitesimally Incremental Displacement

If an infinitesimally incremental displacement is chosen as a virtual displacement, $\delta W$ is the incremental work done by applied loads and $\delta U$ is the incremental strain energy. Noting that, for a given temperature change, $\delta\boldsymbol{\varepsilon}^o = 0$:

$$\delta U = \int_V \delta\boldsymbol{\varepsilon}^{\mathrm{T}}\mathbf{C}(\boldsymbol{\varepsilon} - \boldsymbol{\varepsilon}^o) dV = \int_V \delta(\boldsymbol{\varepsilon} - \boldsymbol{\varepsilon}^o)^{\mathrm{T}}\mathbf{C}(\boldsymbol{\varepsilon} - \boldsymbol{\varepsilon}^o) dV. \tag{6.2.38}$$

Integrating from $\boldsymbol{\varepsilon} = \mathbf{0}$ to $\boldsymbol{\varepsilon} = \hat{\boldsymbol{\varepsilon}}$ over a change in state:

$$U(\hat{\boldsymbol{\varepsilon}}) = \frac{1}{2}\int_V (\hat{\boldsymbol{\varepsilon}} - \boldsymbol{\varepsilon}^o)^{\mathrm{T}}\mathbf{C}(\hat{\boldsymbol{\varepsilon}} - \boldsymbol{\varepsilon}^o) dV. \tag{6.2.39}$$

Dropping the "hat" for generality, strain energy can then be expressed as

$$U(\boldsymbol{\varepsilon}) = \frac{1}{2}\int_V (\boldsymbol{\varepsilon} - \boldsymbol{\varepsilon}^o)^{\mathrm{T}}\mathbf{C}(\boldsymbol{\varepsilon} - \boldsymbol{\varepsilon}^o) dV. \tag{6.2.40}$$

## 6.3    2D Problems

When the geometry and loading conditions meet certain conditions, the mathematical equations governing structural behavior can be simplified to allow the formulation in a 2D domain. Examples are the plane stress state and the plane strain state problems described in the following.

### 6.3.1    Plane Stress State Problems

Consider a thin flat plate as shown in Figure 6.3 in which the plane formed by the $x$ and $y$-axes coincides with the midplane of the plate and the $z$-axis is normal to the midplane. Suppose that there are no variations in geometry and temperature through the thickness, the top and bottom surfaces are stress free, and applied loads are in-plane with no traction and body force in the $z$ direction is such that

$$\overline{T}_z = \mathbf{0}, Z_B = 0 \tag{6.3.1}$$

and

$$\frac{\partial}{\partial z}\left(\overline{T}_x, \overline{T}_y, X_B, Y_B\right) = 0. \tag{6.3.2}$$

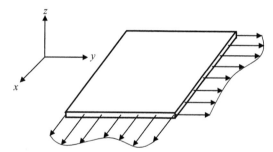

**Figure 6.3** Thin plate under in-plane loadings

### Plane Stress State

Since $\sigma_{zx} = \sigma_{zy} = \sigma_{zz} = 0$ on the top and bottom surfaces and considering the conditions regarding the geometry and applied loads, one may assume $\sigma_{zx} = \sigma_{zy} = \sigma_{zz} = 0$ through the thickness and

$$\sigma_{xx} = \sigma_{xx}(x, y), \quad \sigma_{yy} = \sigma_{yy}(x, y), \quad \sigma_{xy} = \sigma_{xy}(x, y). \tag{6.3.3}$$

The above stress state, called a "plane stress state," is independent of the $z$-coordinate.

### Stress–Strain Relation

As shown in Appendix 1, strain is related to stress and temperature change such that

$$\boldsymbol{\varepsilon} = \hat{\mathbf{C}}\boldsymbol{\sigma} + \boldsymbol{\varepsilon}^o. \tag{6.3.4}$$

For an isotropic material:

$$\varepsilon_{xx} = \frac{1}{E}\left(\sigma_{xx} - \nu\sigma_{yy} - \nu\sigma_{zz}\right) + \alpha\Delta T,$$

$$\varepsilon_{yy} = \frac{1}{E}\left(\sigma_{yy} - \nu\sigma_{xx} - \nu\sigma_{zz}\right) + \alpha\Delta T,$$

$$\varepsilon_{zz} = \frac{1}{E}\left(\sigma_{zz} - \nu\sigma_{xx} - \nu\sigma_{yy}\right) + \alpha\Delta T,$$

$$\varepsilon_{xy} = \frac{1}{G}\sigma_{xy}, \quad \varepsilon_{yz} = \frac{1}{G}\sigma_{yz} = 0, \quad \varepsilon_{zx} = \frac{1}{G}\sigma_{zx} = 0.$$

(6.3.5)

Written in matrix form:

$$\left\{\begin{array}{c} \varepsilon_{xx} \\ \varepsilon_{yy} \\ \varepsilon_{xy} \end{array}\right\} = \begin{bmatrix} \dfrac{1}{E} & -\dfrac{\nu}{E} & 0 \\ -\dfrac{\nu}{E} & \dfrac{1}{E} & 0 \\ 0 & 0 & \dfrac{1}{G} \end{bmatrix} \left\{\begin{array}{c} \sigma_{xx} \\ \sigma_{yy} \\ \sigma_{xy} \end{array}\right\} + \left\{\begin{array}{c} \alpha\Delta T \\ \alpha\Delta T \\ 0 \end{array}\right\}, G = \frac{E}{2(1+\nu)}. \quad (6.3.6)$$

Since stress and temperature change are independent of the $z$-coordinate, strain is also independent of $z$. We also note from the third row in Eq. (6.3.5) that

$$\varepsilon_{zz} = -\frac{\nu}{E}\left(\sigma_{xx} + \sigma_{yy}\right) + \alpha\Delta T \qquad (6.3.7)$$

is redundant and does not appear in Eq. (6.3.6). In matrix form, Eq. (6.3.6) can be expressed as

$$\boldsymbol{\varepsilon} = \hat{\mathbf{C}}\boldsymbol{\sigma} + \boldsymbol{\varepsilon}^o. \qquad (6.3.8)$$

The above equation is symbolically identical to Eq. (6.3.4). However, in Eq. (6.3.8) total strain, stress, and thermal strain are $3 \times 1$ column vectors, and $\hat{\mathbf{C}}$ is a $3 \times 3$ matrix as shown in Eq. (6.3.6). From Eq. (6.3.6):

$$\left\{\begin{array}{c} \sigma_{xx} \\ \sigma_{yy} \\ \sigma_{zz} \end{array}\right\} = \frac{E}{(1-\nu^2)} \begin{bmatrix} 1 & \nu & 0 \\ \nu & 1 & 0 \\ 0 & 0 & \dfrac{1-\nu}{2} \end{bmatrix} \left\{\begin{array}{c} \varepsilon_{xx} - \alpha\Delta T \\ \varepsilon_{yy} - \alpha\Delta T \\ \varepsilon_{xy} \end{array}\right\} \qquad (6.3.9)$$

or

$$\boldsymbol{\sigma} = \mathbf{C}(\boldsymbol{\varepsilon} - \boldsymbol{\varepsilon}^o), \qquad (6.3.10)$$

where

$$\mathbf{C} = \hat{\mathbf{C}}^{-1}. \qquad (6.3.11)$$

## Displacement Field

One may observe that in-plane displacements $u, v$ are independent of the $z$-coordinate and can be expressed as

$$u = u(x, y),$$
$$v = v(x, y). \tag{6.3.12}$$

Then

$$\varepsilon_{xx} = \frac{\partial u}{\partial x} = \varepsilon_{xx}(x, y), \quad \varepsilon_{yy} = \frac{\partial v}{\partial y} = \varepsilon_{yy}(x, y),$$

$$\varepsilon_{xy} = \frac{\partial u}{\partial y} + \frac{\partial v}{\partial x} = \varepsilon_{xy}(x, y). \tag{6.3.13}$$

The three in-plane strains are independent of $z$, consistent with the previous conclusion drawn from the strain–stress relation.

### Virtual Work Expression for the Plane Stress Problem

For plane stress states, $\sigma_{zx} = \sigma_{zy} = \sigma_{zz} = 0$ and Eq. (6.2.33) is reduced to

$$\delta U = \int_V \left( \sigma_{xx}\delta\varepsilon_{xx} + \sigma_{yy}\delta\varepsilon_{yy} + \sigma_{xy}\delta\varepsilon_{xy} \right) dV. \tag{6.3.14}$$

In matrix form:

$$\delta U = \int_V \lfloor \delta\varepsilon_{xx} \quad \delta\varepsilon_{yy} \quad \delta\varepsilon_{xy} \rfloor \left\{ \begin{array}{c} \sigma_{xx} \\ \sigma_{yy} \\ \sigma_{xy} \end{array} \right\} dV = \int_V \delta\boldsymbol{\varepsilon}^\mathrm{T}\boldsymbol{\sigma} dV = \int_A \delta\boldsymbol{\varepsilon}^\mathrm{T}\boldsymbol{\sigma} t dA, \tag{6.3.15}$$

where $t$ is the thickness and $A$ stands for the planform area. The integration is now over an area instead of a volume. Introducing Eq. (6.3.10) into Eq. (6.3.15):

$$\delta U = \int_A \delta\boldsymbol{\varepsilon}^\mathrm{T}\mathbf{C}(\boldsymbol{\varepsilon} - \boldsymbol{\varepsilon}^o) t dA = \int_A \delta\boldsymbol{\varepsilon}^\mathrm{T}\mathbf{C}\boldsymbol{\varepsilon} t dA - \int_A \delta\boldsymbol{\varepsilon}^\mathrm{T}\mathbf{C}\boldsymbol{\varepsilon}^o t dA. \tag{6.3.16}$$

For the body force term in the external virtual work expression:

$$\int_V \delta\mathbf{u}^\mathrm{T}\mathbf{F}_B dV = \int_V [\delta u \quad \delta v \quad \delta w] \left\{ \begin{array}{c} X_B \\ Y_B \\ Z_B = 0 \end{array} \right\} dV = \int_A [\delta u \quad \delta v] \left\{ \begin{array}{c} X_B \\ Y_B \end{array} \right\} t dA. \tag{6.3.17}$$

Once again, integration is over the area in the $xy$-plane. For the surface traction term:

$$\int_{S_\sigma} \delta\mathbf{u}^\mathrm{T}\bar{\mathbf{T}} dS = \int_{S_\sigma} [\delta u \quad \delta v \quad \delta w] \left\{ \begin{array}{c} \bar{T}_x \\ \bar{T}_y \\ \bar{T}_z = 0 \end{array} \right\} dS = \int_{l_\sigma} [\delta u \quad \delta v] \left\{ \begin{array}{c} \bar{T}_x \\ \bar{T}_y \end{array} \right\} t dl. \tag{6.3.18}$$

In the above equation, integration is now over "$l$" which is a coordinate along the boundary of the 2D domain.

## 6.3.2 Plane Strain State Problems

Consider a long prismatic body with a uniform cross-section as shown in Figure 6.4. The body is very long in the $z$ direction such that the $z$ dimension of the body is much greater than the dimension in the $xy$-plane.

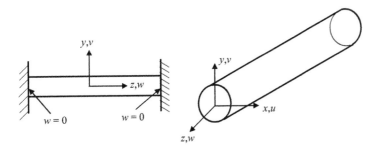

Figure 6.4 Long body of uniform cross-section under plane strain states

The geometry, applied loads, and temperature changes are uniform along the $z$-axis and there are no applied tractions or body forces in the $z$ direction. In addition, the body is constrained such that $w = 0$ at both ends.

### Strain–Displacement Relations

One may assume that the displacement field is uniform along the $z$-axis such that

$$u = u(x, y), \quad v = v(x, y), \quad w = 0. \tag{6.3.19}$$

Then, according to the strain–displacement relationship:

$$\varepsilon_{zz} = \frac{\partial w}{\partial z} = 0, \quad \varepsilon_{zx} = \frac{\partial w}{\partial z} + \frac{\partial u}{\partial z} = 0, \quad \varepsilon_{yz} = \frac{\partial v}{\partial z} + \frac{\partial w}{\partial y} = 0,$$

$$\varepsilon_{xx} = \frac{\partial u}{\partial x} = \varepsilon_{xx}(x, y), \quad \varepsilon_{yy} = \frac{\partial v}{\partial y} = \varepsilon_{yy}(x, y), \quad \varepsilon_{xy} = \frac{\partial u}{\partial y} + \frac{\partial v}{\partial x} = \varepsilon_{xy}(x, y). \tag{6.3.20}$$

The above strain state is called a "plane strain state" and is independent of the $z$-axis.

### Stress–Strain Relation

For an isotropic material:

$$\varepsilon_{zz} = \frac{1}{E}\left(\sigma_{zz} - \nu\sigma_{xx} - \nu\sigma_{yy}\right) + \alpha\Delta T. \tag{6.3.21}$$

Setting $\varepsilon_{zz} = 0$,

$$\sigma_{zz} = \nu\left(\sigma_{xx} + \sigma_{yy}\right) - E\alpha\Delta T. \tag{6.3.22}$$

Recall that

$$\varepsilon_{xx} = \frac{1}{E}\left(\sigma_{xx} - v\sigma_{yy} - v\sigma_{zz}\right) + \alpha\Delta T. \tag{6.3.23}$$

Placing Eq. (6.3.22) into Eq. (6.3.23):

$$\begin{aligned}\varepsilon_{xx} &= \frac{1}{E}\left[\sigma_{xx} - v\sigma_{yy} - v^2\left(\sigma_{xx} + \sigma_{yy}\right) + vE\alpha\Delta T\right] + \alpha\Delta T \\ &= \frac{1-v^2}{E}\sigma_{xx} - \frac{v(1+v)}{E}\sigma_{yy} + (1+v)\alpha\Delta T.\end{aligned} \tag{6.3.24}$$

Similarly, one can show that

$$\varepsilon_{yy} = \frac{1-v^2}{E}\sigma_{yy} - \frac{v(1+v)}{E}\sigma_{xx} + (1+v)\alpha\Delta T. \tag{6.3.25}$$

For shear strain and stress:

$$\varepsilon_{xy} = \frac{1}{G}\sigma_{xy}. \tag{6.3.26}$$

From Eqs (6.3.24)–6.3.26), one can obtain

$$\left\{\begin{array}{c}\sigma_{xx} \\ \sigma_{yy} \\ \sigma_{xy}\end{array}\right\} = \frac{E}{(1+v)(1-2v)}\begin{bmatrix}1-v & v & 0 \\ v & 1-v & 0 \\ 0 & 0 & \frac{1-2v}{2}\end{bmatrix}\left\{\begin{array}{c}\varepsilon_{xx} - (1+v)\alpha\Delta T \\ \varepsilon_{yy} - (1+v)\alpha\Delta T \\ \varepsilon_{xy}\end{array}\right\}. \tag{6.3.27}$$

The above equation can be expressed symbolically as

$$\boldsymbol{\sigma} = \mathbf{C}(\boldsymbol{\varepsilon} - \boldsymbol{\varepsilon}^o), \tag{6.3.28}$$

where

$$\boldsymbol{\varepsilon}^o = \left\{\begin{array}{c}(1+v)\alpha\Delta T \\ (1+v)\alpha\Delta T \\ 0\end{array}\right\}. \tag{6.3.29}$$

### Virtual Work Expression for the Plane Strain Problem

For the plane strain state, we may choose

$$\delta u = \delta u(x,y), \quad \delta v = \delta v(x,y) \quad \delta w = 0. \tag{6.3.30}$$

Then

$$\delta\varepsilon_{zz} = \delta\varepsilon_{zy} = \delta\varepsilon_{zx} = 0 \tag{6.3.31}$$

and Eq. (6.2.33) reduces to

$$\delta U = \int_V \left( \sigma_{xx}\delta\varepsilon_{xx} + \sigma_{yy}\delta\varepsilon_{yy} + \sigma_{xy}\delta\varepsilon_{xy} \right)dV. \tag{6.3.32}$$

The above expression is identical to that for the plane stress state case. In matrix form

$$\delta U = \int_V \lfloor \delta\varepsilon_{xx} \quad \delta\varepsilon_{yy} \quad \delta\varepsilon_{xy} \rfloor \left\{ \begin{array}{c} \sigma_{xx} \\ \sigma_{yy} \\ \sigma_{xy} \end{array} \right\} dV = \int_V \delta\boldsymbol{\varepsilon}^T \boldsymbol{\sigma} dV = \int_A \delta\boldsymbol{\varepsilon}^T \boldsymbol{\sigma} t dA, \tag{6.3.33}$$

where $t$ is the body length in the $z$ direction and $A$ stands for the uniform cross-sectional area. The integration is now over an area instead of a volume. Introducing Eq. (6.3.28):

$$\delta U = \int_A \delta\boldsymbol{\varepsilon}^T \mathbf{C}(\boldsymbol{\varepsilon} - \boldsymbol{\varepsilon}^o) t dA = \int_A \delta\boldsymbol{\varepsilon}^T \mathbf{C}\boldsymbol{\varepsilon} t dA - \int_A \delta\boldsymbol{\varepsilon}^T \mathbf{C}\boldsymbol{\varepsilon}^o t dA. \tag{6.3.34}$$

For the body force term in the external virtual work:

$$\int_V \delta\mathbf{u}^T \mathbf{F}_B dV = \int_V \lfloor \delta u \quad \delta v \quad \delta w \rfloor \left\{ \begin{array}{c} X_B \\ Y_B \\ Z_B = 0 \end{array} \right\} dV = \int_A \lfloor \delta u \quad \delta v \rfloor \left\{ \begin{array}{c} X_B \\ Y_B \end{array} \right\} t dA. \tag{6.3.35}$$

For the surface traction term:

$$\int_{S_\sigma} \delta\mathbf{u}^T \bar{\mathbf{T}} dS = \int_{S_\sigma} \lfloor \delta u \quad \delta v \quad \delta w \rfloor \left\{ \begin{array}{c} \bar{T}_x \\ \bar{T}_y \\ \bar{T}_z = 0 \end{array} \right\} dS = \int_{l_\sigma} \lfloor \delta u \quad \delta v \rfloor \left\{ \begin{array}{c} \bar{T}_x \\ \bar{T}_y \end{array} \right\} t dl. \tag{6.3.36}$$

We note that in the above equation the integration is over coordinate "$l$" along the boundary line of the 2D domain of a uniform cross-section.

## 6.4    Lagrange Equation

In Chapter 4, the Lagrange equation was introduced without derivation to construct the equation of motion for slender bodies. In this section, we will derive the Lagrange equation starting from the statement of the principle of virtual work. Treating inertia force as a body force:

$$\mathbf{F}_B dV = -(\rho dV)\ddot{\mathbf{u}}, \tag{6.4.1}$$

where $\rho$ is the mass per volume. Introducing the above equation into the first term on the right-hand side of Eq. (6.2.18):

$$-\int_V \delta\mathbf{u} \cdot \mathbf{F}_B dV = \int_V \delta\mathbf{u} \cdot \ddot{\mathbf{u}}\rho dV = \int_V \delta\mathbf{u} \cdot \ddot{\mathbf{u}} dm, \tag{6.4.2}$$

where $dm = \rho dV$ is the mass of infinitesimal volume $dV$. Suppose $\mathbf{u}$ is a function of $q_1, q_2, \ldots, q_N$, called generalized coordinates, such that

$$\mathbf{u} = \mathbf{u}(q_1, q_2, \ldots, q_N). \tag{6.4.3}$$

In the FE formulation, $q_1, q_2, \ldots, q_N$ are the nodal DOF. Then

$$\dot{\mathbf{u}} = \frac{\partial \mathbf{u}}{\partial q_1} \dot{q}_1 + \frac{\partial \mathbf{u}}{\partial q_2} \dot{q}_2 + \cdots + \frac{\partial \mathbf{u}}{\partial q_N} \dot{q}_N. \tag{6.4.4}$$

From the above equation, we find that

$$\frac{\partial \dot{\mathbf{u}}}{\partial \dot{q}_k} = \frac{\partial \mathbf{u}}{\partial q_k} \quad (k = 1, 2, \ldots, N). \tag{6.4.5}$$

We may also set

$$\delta \mathbf{u} = \frac{\partial \mathbf{u}}{\partial q_1} \delta q_1 + \frac{\partial \mathbf{u}}{\partial q_2} \delta q_2 + \cdots + \frac{\partial \mathbf{u}}{\partial q_N} \delta q_N. \tag{6.4.6}$$

Substituting Eq. (6.4.6) into the last of Eq. (6.4.2):

$$dm \ddot{\mathbf{u}} \cdot \delta \mathbf{u} = dm \ddot{\mathbf{u}} \cdot \left( \frac{\partial \mathbf{u}}{\partial q_1} \delta q_1 + \frac{\partial \mathbf{u}}{\partial q_2} \delta q_2 + \cdots + \frac{\partial \mathbf{u}}{\partial q_N} \delta q_N \right) = dm \sum_{k=1}^{N} \left( \ddot{\mathbf{u}} \cdot \frac{\partial \mathbf{u}}{\partial q_k} \delta q_k \right). \tag{6.4.7}$$

We also note that

$$\ddot{\mathbf{u}} \cdot \frac{\partial \mathbf{u}}{\partial q_k} = \frac{d}{dt} \left( \dot{\mathbf{u}} \cdot \frac{\partial \mathbf{u}}{\partial q_k} \right) - \dot{\mathbf{u}} \cdot \frac{d}{dt} \left( \frac{\partial \mathbf{u}}{\partial q_k} \right). \tag{6.4.8}$$

Introducing Eq. (6.4.5) into the first term on the right-hand side of the above equation:

$$\ddot{\mathbf{u}} \cdot \frac{\partial \mathbf{u}}{\partial q_k} = \frac{d}{dt} \left( \dot{\mathbf{u}} \cdot \frac{\partial \dot{\mathbf{u}}}{\partial \dot{q}_k} \right) - \dot{\mathbf{u}} \cdot \frac{\partial \dot{\mathbf{u}}}{\partial q_k} = \frac{d}{dt} \left( \frac{\partial}{\partial \dot{q}_k} \left( \frac{1}{2} \dot{\mathbf{u}} \cdot \dot{\mathbf{u}} \right) \right) - \frac{\partial}{\partial q_k} \left( \frac{1}{2} \dot{\mathbf{u}} \cdot \dot{\mathbf{u}} \right). \tag{6.4.9}$$

Introducing the above equation into Eq. (6.4.7):

$$dm \ddot{\mathbf{u}} \cdot \delta \mathbf{u} = dm \sum_{k=1}^{N} \ddot{\mathbf{u}} \cdot \frac{\partial \mathbf{u}}{\partial q_k} \delta q_k = \sum_{k=1}^{N} \left[ \frac{d}{dt} \left( \frac{\partial}{\partial \dot{q}_k} \left( \frac{1}{2} dm \dot{\mathbf{u}} \cdot \dot{\mathbf{u}} \right) \right) - \frac{\partial}{\partial q_k} \left( \frac{1}{2} dm \dot{\mathbf{u}} \cdot \dot{\mathbf{u}} \right) \right] \delta q_k. \tag{6.4.10}$$

Accordingly, the last term in Eq. (6.4.2) can be expressed as

$$\int_V \delta \mathbf{u} \cdot \ddot{\mathbf{u}} \, dm = \int_V \sum_{k=1}^{N} \left[ \frac{d}{dt} \left( \frac{\partial}{\partial \dot{q}_k} \left( \frac{1}{2} dm \dot{\mathbf{u}} \cdot \dot{\mathbf{u}} \right) \right) - \frac{\partial}{\partial q_k} \left( \frac{1}{2} dm \dot{\mathbf{u}} \cdot \dot{\mathbf{u}} \right) \right] \delta q_k$$

$$= \sum_{k=1}^{N} \left[ \frac{d}{dt} \left( \frac{\partial T}{\partial \dot{q}_k} \right) - \frac{\partial T}{\partial q_k} \right] \delta q_k, \tag{6.4.11}$$

where

$$T = \int_V \left( \frac{1}{2} dm \dot{\mathbf{u}} \cdot \dot{\mathbf{u}} \right) = \frac{1}{2} \int_V (\dot{\mathbf{u}} \cdot \dot{\mathbf{u}} \rho dV) \tag{6.4.12}$$

is the kinetic energy of the entire body. For the strain energy

$$U = U(q_1, q_2, \ldots, q_N). \tag{6.4.13}$$

We may then set

$$\delta U = \frac{\partial U}{\partial q_1} \delta q_1 + \frac{\partial U}{\partial q_2} \delta q_2 + \cdots + \frac{\partial U}{\partial q_N} \delta q_N = \sum_{k=1}^N \frac{\partial U}{\partial q_k} \delta q_k. \tag{6.4.14}$$

Substituting Eq. (6.4.6) into Eq. (6.2.20):

$$\delta W_{applied} = \sum_{k=1}^N F_k \delta q_k, \tag{6.4.15}$$

where

$$F_k = \int_V \mathbf{F}_B \cdot \frac{\partial \mathbf{u}}{\partial q_k} dV + \int_{S_\sigma} \bar{\mathbf{T}} \cdot \frac{\partial \mathbf{u}}{\partial q_k} dS. \tag{6.4.16}$$

The body force in the above equation excludes the inertia force. Collecting Eqs (6.4.11), (6.4.14), and (6.4.16):

$$\delta U - \delta W = \sum_{k=1}^N \left[ \frac{d}{dt} \left( \frac{\partial T}{\partial \dot{q}_k} \right) - \frac{\partial T}{\partial q_k} + \frac{\partial U}{\partial q_k} - F_k \right] \delta q_k = 0. \tag{6.4.17}$$

It follows that for arbitrary $\delta q_k$:

$$\frac{d}{dt} \left( \frac{\partial T}{\partial \dot{q}_k} \right) - \frac{\partial T}{\partial q_k} + \frac{\partial U}{\partial q_k} - F_k = 0, \quad (k = 1, 2, \ldots, N), \tag{6.4.18}$$

which is the Lagrange equation.

## PROBLEMS

**6.1** The slender body under uniaxial forces as shown in Figure 6.5 was described in Chapter 1.

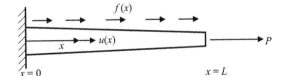

Figure 6.5 For Problem 6.1

(a) Show that Eq. (6.2.33) is simplified to

$$\delta U = \int_V \sigma_{xx}\delta\varepsilon_{xx}dV = \int_{x=0}^{x=L} \sigma_{xx}\delta\varepsilon_{xx}Adx.$$

(b) For the external virtual work, show that

$$\delta W = \int_V \delta u X_B dV + \int_S \delta u T_x dS = \int_V \delta u f dx + (\delta u P)_{x=L},$$

where

$$f = \int_A X_B dydz, \quad P = \int_{x=L} T_x dydz = \int_{x=L} \sigma_{xx}dydz.$$

We observe that the expressions for $\delta U$ and $\delta W$ are identical to those given in Chapter 1.

**6.2** Consider a cantilevered slender body as shown in Figure 6.6.

In the figure, $w(x)$ is the transverse displacement in the $z$ direction, $p_z(x)$ is the applied force per unit length, $P_L$ is the force applied at $x = L$, and $M_L$ is the moment applied at $x = L$. There is no body force applied. Starting from the virtual work for 3D solids, introduce proper simplifying assumptions to show that

$$\delta U = \int_{x=0}^{x=L} EI_y \frac{\partial^2 \delta w}{\partial x^2}\frac{\partial^2 w}{\partial x^2}dx, \quad \delta W = (P_L\delta w)_{x=L} + (M_L\delta\theta)_{x=L} + \int_{x=0}^{x=L} \delta w p_z dx,$$

where

$$p_z(x) = \int \sigma_{zz}dy, \quad P_L = \int_{x=L} \sigma_{xz}dydz,$$

$$(F)_{x=L} = \int_{x=L} \sigma_{xx}dydz = 0, \quad (M_L)_{x=L} = \int_{x=L} z\sigma_{xx}dydz.$$

The above expressions can be used for the FE formulation of beam bending with the two-node element described in Chapter 3.

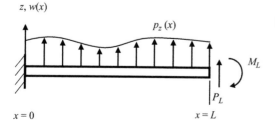

$z, w(x)$

$p_z(x)$

$M_L$

$P_L$

$x = 0$    $x = L$

**Figure 6.6** For Problem 6.2

**6.3** Consider bending of slender bodies subjected to axial force as described in Chapter 5. As shown in Eq. (5.1.10), the equilibrium equation is

$$\frac{\partial^2 M_y}{\partial x^2} + \frac{\partial}{\partial x}\left(F\frac{\partial w}{\partial x}\right) + p_z = 0.$$

(a) Introducing virtual displacement $\delta w(x)$, we may construct a scalar equation as follows:

$$\int_{x=0}^{x=L} \left[\frac{\partial^2 M_y}{\partial x^2} + \frac{\partial}{\partial x}\left(F\frac{\partial w}{\partial x}\right) + p_z\right]\delta w\, dx = 0.$$

Apply the integration by parts twice to show that the first term in the above equation can be expressed as

$$I_1 = \int_{x=0}^{x=L} \frac{\partial^2 M_y}{\partial x^2}\delta w\, dx = (V_z\delta w)_{x=0}^{x=L} + (M_y\delta\theta)_{x=0}^{x=L} + \int_{x=0}^{x=L} M_y\frac{\partial^2(\delta w)}{\partial x^2}\, dx,$$

where

$$V_z = \frac{\partial M_y}{\partial x}, \quad \delta\theta = -\frac{\partial(\delta w)}{\partial x}.$$

(b) Apply the integration by parts formula once for the second term in the first equation in part (a) to show that

$$I_2 = \int_{x=0}^{x=L} \frac{\partial}{\partial x}\left(F\frac{\partial w}{\partial x}\right)\delta w\, dx = \left(\delta w F\frac{\partial w}{\partial x}\right)_{x=0}^{x=L} - \int_{x=0}^{x=L} F\frac{\partial(\delta w)}{\partial x}\frac{\partial w}{\partial x}\, dx.$$

(c) Show that, using the result in parts (a) and (b), the first equation in part (a) can be expressed as

$$-\int_{x=0}^{x=L} M_y\frac{\partial^2\delta w}{\partial x^2}\, dx + \int_{x=0}^{x=L} F\frac{\partial\delta w}{\partial x}\frac{\partial w}{\partial x}\, dx - \int_{x=0}^{x=L} \delta w p_z\, dx - \delta W_B = 0,$$

where $\delta W_B$ is the boundary term, defined as follows:

$$\delta W_B = (M_y\delta\theta)_{x=L} - (M_y\delta\theta)_{x=0} + \left[\left(V_z + F\frac{\partial w}{\partial x}\right)\delta w\right]_{x=L} - \left[\left(V_z + F\frac{\partial w}{\partial x}\right)\delta w\right]_{x=0}.$$

Introducing Eq. (5.1.11), the virtual work can now be expressed as

$$\int_{x=0}^{x=L} EI_y\frac{\partial^2\delta w}{\partial x^2}\frac{\partial^2 w}{\partial x^2}\, dx + \int_{x=0}^{x=L} F\frac{\partial\delta w}{\partial x}\frac{\partial w}{\partial x}\, dx - \int_{x=0}^{x=L} \delta w p_z\, dx - \delta W_B = 0.$$

We note that when $F = 0$ the above expression is identical to that derived in Problem 6.2 for pure bending.

**6.4** For a simply supported column under compressive axial force as shown in Figure 6.7, $F = -P$, $p_z = 0$, and the boundary conditions are as follows: at $x = 0$, $w = 0$ and $M_y = 0$ and at $x = L$, $w = 0$ and $M_y = 0$.

(a) Apply boundary conditions of zero moment at $x = 0$ and $x = L$ and, corresponding to $w = 0$ at $x = 0$ and $x = L$, set $\delta w = 0$ at $x = 0$ and $x = L$ to show that $\delta W_B = 0$. The last equation in Problem 6.3 is then simplified as

$$\int_{x=0}^{x=L} EI_y \frac{\partial^2 \delta w}{\partial x^2} \frac{\partial^2 w}{\partial x^2} dx - P \int_{x=0}^{x=L} \frac{\partial \delta w}{\partial x} \frac{\partial w}{\partial x} dx = 0.$$

(b) Using a model with two bending elements of equal length, describe how the FE formulation can be applied to derive the matrix equation for eigenvalue analysis which can be used to determine the static buckling load and mode as discussed in Chapter 5.

**Figure 6.7** For Problem 6.4    $z, w(x)$

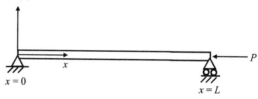

**6.5** For a column clamped against bending at both ends and subjected to compressive axial force as shown in Figure 6.8, the boundary conditions are as follows: at $x = 0$, $w = 0$ and $\theta = 0$ and at $x = L$, $w = 0$ and $\theta = 0$.

(a) Show that, for virtual displacement that satisfies geometric constraints, $\delta W_B = 0$ and the last equation in Problem 6.3 is simplified as

$$\int_{x=0}^{x=L} EI_y \frac{\partial^2 \delta w}{\partial x^2} \frac{\partial^2 w}{\partial x^2} dx - P \int_{x=0}^{x=L} \frac{\partial \delta w}{\partial x} \frac{\partial w}{\partial x} dx = 0.$$

(b) Using a model with two elements of equal length, describe how the FE formulation can be applied to derive the equation for eigenvalue analysis which can be used to determine the static buckling load and mode as discussed in Chapter 5.

**Figure 6.8** For Problem 6.5    $z, w(x)$

**6.6** Figure 6.9 shows a clamped-free column in a buckled position with the boundary conditions given as follows.

$$\text{At } x = 0, w = 0, \theta = -\frac{\partial w}{\partial x} = 0.$$

$$\text{At } x = L, M_y = 0, V_z = -P\sin\theta \cong -P\theta = P\frac{\partial w}{\partial x} \rightarrow V_z - P\frac{\partial w}{\partial x} = 0 \rightarrow V_z + F\frac{\partial w}{\partial x} = 0.$$

(a) Apply boundary conditions on force and moment and choose virtual displacement that satisfies the geometric constraint to show that $\delta W_B = 0$. The last equation in Problem 6.3 is then simplified as

$$\int_{x=0}^{x=L} EI_y \frac{\partial^2 \delta w}{\partial x^2}\frac{\partial^2 w}{\partial x^2} dx - P\int_{x=0}^{x=L}\frac{\partial \delta w}{\partial x}\frac{\partial w}{\partial x} dx = 0.$$

(b) Using a model with two elements of equal length, describe how the FE formulation can be applied to derive the equation for eigenvalue analysis which can be used to determine the static buckling load and mode as discussed in Chapter 5.

**Figure 6.9** For Problem 6.6

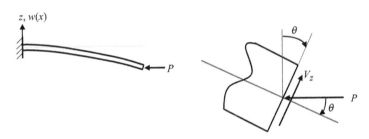

$z, w(x)$

$P$

$\theta$

$V_z$

$P$

$\theta$

**6.7** Show that, for the clamped free column subjected to a follower force as described in Chapter 5, we obtain the following expression from Problem 6.3:

$$\delta W_B = \left[\left(F\frac{\partial w}{\partial x}\right)\delta w\right]_{x=L} = (P\theta\delta w)_{x=L}.$$

# 7 Mapping, Shape Functions, and Numerical Integration

In solid and structural mechanics, we deal with a body in two configurations: original undeformed configuration and deformed configuration. In the FE formulation, the body is divided into elements of various types. This chapter describes mapping functions for the description of element geometry in the undeformed configuration and shape functions for the description of displacement and thus deformed geometry in the 2D and 3D domains. This is followed by discussions on integration in the mapped domains and numerical integration.

We introduce the "isoparametric" formulation in which mapping functions and shape functions are identical, beginning with the review of the mapping and shape functions of the two-node element in the 1D domains. We then consider mapping and shape functions of finite elements in the 2D domains to show that quadrilateral elements in the physical plane are mapped into a square in the mapped plane while triangular elements are mapped into a right triangle. This is followed by construction of mapping and shape functions in the 3D domains for elements with shapes of hexahedra and tetrahedra and various node numbers. The isoparametric formulation guarantees that the geometry and assumed displacement are continuous along the element boundary when two adjacent elements of arbitrary shape are joined at the nodal points of the common boundary.

In the isoparametric FE formulation, integrands are dependent on the coordinates in the mapped domain. Accordingly, we discuss how integrations over the physical domains can be transformed into integrations over the mapped domains. This includes integration along a line, over an area and a volume, and an arbitrarily oriented surface in 3D space.

The last section covers the methods of numerical integration. We first introduce the Gaussian quadrature for a line integral, which can then be extended for integration over areas and volumes with the use of mapping functions. This is followed by discussions on numerical integrations over triangular areas and tetrahedral volumes. These numerical integration rules will be used to construct the element stiffness matrices and load vectors in later chapters.

## 7.1 Mapping and Shape Functions of 1D Elements

In Chapter 1, we introduced a mapping between the $x$ coordinate and a non-dimensional $s$ coordinate. We also expressed the assumed displacement as a function of the $s$ coordinate. Recall that for the two-node element in Chapter 1 (see Figure 7.1):

$$\text{Mapping}: x = (1 - s)x_1 + sx_2. \tag{7.1.1}$$

$$\text{Assumed displacement}: u = (1 - s)u_1 + su_2. \tag{7.1.2}$$

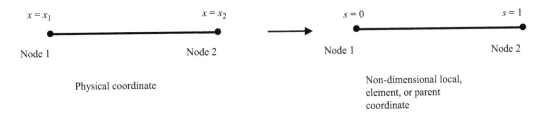

**Figure 7.1** One-dimensional two-node element mapped into the $s$-coordinate system

Using Eqs (7.1.1) and (7.1.2), the integrations to generate the element stiffness matrix and load vector can be carried out in the mapped domain. So, the limits of integrations are always from $s = 0$ to $s = 1$, regardless of the location of the element in the 1D domain.

Instead of the $s$-coordinate, let us introduce the $\xi$-coordinate that varies from $-1$ to $+1$, as shown in Figure 7.2. We will use $\xi$ henceforth for convenience because of the quadrature rules for numerical integration that will be described later in this chapter.

$\xi = -1$          $\xi = 1$

Node 1        Node 2

**Figure 7.2** One-dimensional two-node element mapped into the $\xi$-coordinate system

$$\text{Mapping}: x = a_1 + a_2\xi. \tag{7.1.3}$$

$$\text{At node 1}: (\xi = -1), x_1 = a_1 - a_2. \tag{7.1.4}$$

$$\text{At node 2}: (\xi = +1), x_2 = a_1 + a_2. \tag{7.1.5}$$

From Eqs (7.1.4) and (7.1.5):

$$a_1 = \frac{x_1 + x_2}{2}, \quad a_2 = \frac{x_2 - x_1}{2}. \tag{7.1.6}$$

Substituting Eq. (7.1.6) into Eq. (7.1.3):

$$x = \frac{x_1 + x_2}{2} + \frac{x_2 - x_1}{2}\xi = \frac{1}{2}(1 - \xi)x_1 + \frac{1}{2}(1 + \xi)x_2$$

$$= N_1(\xi)x_1 + N_2(\xi)x_2 = \sum_{i=1}^{2} N_i(\xi)x_i, \tag{7.1.7}$$

where

$$N_1(\xi) = \frac{1}{2}(1 - \xi), \quad N_2(\xi) = \frac{1}{2}(1 + \xi) \tag{7.1.8}$$

are called "mapping" functions. Note that $N_1 = 1$ at node 1 and 0 at node 2, while $N_2 = 1$ at node 2 and 0 at node 1.

*Assumed displacement:* For the displacement assumed to be linear in $\xi$ in the element, we can show that

$$u = \sum_{i=1}^{2} N_i(\xi)u_i, \tag{7.1.9}$$

where $N_1(\xi)$ and $N_2(\xi)$ are now called "shape" functions. Note that $N_1(\xi) + N_2(\xi) = 1$. In the isoparametric formulation, shape functions for the assumed displacement are identical to mapping functions for the geometry. They are also called "interpolation" functions for geometry and displacement.

## 7.2 Mapping and Shape Functions of 2D Elements

For FE modeling of 2D problems, the domain or area may be divided into a mesh of many elements of quadrilateral and triangular shapes, as shown in Figure 7.3.

### 7.2.1 Four-Node Quadrilateral Element

Figure 7.4 shows the four-node element with the nodes placed at the vertices. For the node numbering, one may place node number 1 at any corner and then increase node numbers in the anticlockwise direction. In the physical $xy$-plane, the element sides may not be parallel with the coordinate axes. The element is mapped into a square in the $\xi\eta$-plane, where $-1 \le \xi \le 1$ and $-1 \le \eta \le 1$. The non-dimensional $\xi, \eta$-coordinates are called local coordinates, element coordinates, parent coordinates, or natural coordinates. Figure 7.4 also shows natural coordinates inscribed in the element in the physical plane.

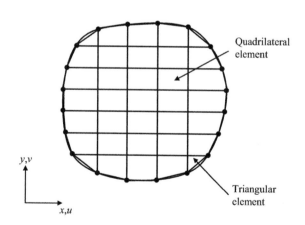

Quadrilateral element

Triangular element

**Figure 7.3** 2D domain divided into a mesh of quadrilateral and triangular elements

### Geometry

For the four-node element, we may express the mapping as

$$x(\xi, \eta) = a_1 + a_2\xi + a_3\eta + a_4\xi\eta. \tag{7.2.1}$$

We note that, for $\eta = \pm 1$, $x$ is linear in $\xi$ with two nodes to match while for $\xi = \pm 1$, $x$ is linear in $\eta$ with two nodes to match. The polynomial terms in the above equation are bilinear

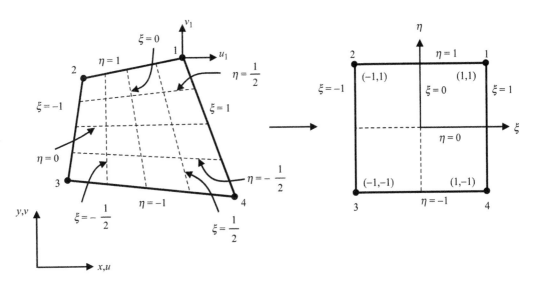

**Figure 7.4** Four-node element in the physical plane (left) and the mapped plane (right)

and balanced in $\xi$ and $\eta$, since the number of nodes in the $\xi$, $\eta$ directions are equal. The four coefficients $(a_1, a_2, a_3, a_4)$ can be related to the nodal coordinates by matching the above equation at the four nodes. At node $i$:

$$x_i = a_1 + a_2\xi_i + a_3\eta_i + a_4\xi_i\eta_i \quad (i = 1, 2, 3, 4) \tag{7.2.2}$$

or, written in matrix form:

$$\begin{Bmatrix} x_1 \\ x_2 \\ x_3 \\ x_4 \end{Bmatrix} = \begin{bmatrix} 1 & 1 & 1 & 1 \\ 1 & -1 & 1 & -1 \\ 1 & -1 & -1 & 1 \\ 1 & 1 & -1 & -1 \end{bmatrix} \begin{Bmatrix} a_1 \\ a_2 \\ a_3 \\ a_4 \end{Bmatrix} = \mathbf{A}\begin{Bmatrix} a_1 \\ a_2 \\ a_3 \\ a_4 \end{Bmatrix}, \quad \mathbf{A} = \begin{bmatrix} 1 & 1 & 1 & 1 \\ 1 & -1 & 1 & -1 \\ 1 & -1 & -1 & 1 \\ 1 & 1 & -1 & -1 \end{bmatrix}. \tag{7.2.3}$$

Inverting:

$$\begin{Bmatrix} a_1 \\ a_2 \\ a_3 \\ a_4 \end{Bmatrix} = \mathbf{A}^{-1}\begin{Bmatrix} x_1 \\ x_2 \\ x_3 \\ x_4 \end{Bmatrix}. \tag{7.2.4}$$

Substituting Eq. (7.2.4) into Eq. (7.2.1):

$$x = \lfloor 1 \ \ \xi \ \ \eta \ \ \xi\eta\rfloor\begin{Bmatrix} a_1 \\ a_2 \\ a_3 \\ a_4 \end{Bmatrix} = \lfloor 1 \ \ \xi \ \ \eta \ \ \xi\eta\rfloor\mathbf{A}^{-1}\begin{Bmatrix} x_1 \\ x_2 \\ x_3 \\ x_4 \end{Bmatrix} = \lfloor N_1 \ \ N_2 \ \ N_3 \ \ N_4\rfloor\begin{Bmatrix} x_1 \\ x_2 \\ x_3 \\ x_4 \end{Bmatrix},$$

$$\tag{7.2.5}$$

where

$$N_1 = \frac{1}{4}(1 + \xi)(1 + \eta), \quad N_2 = \frac{1}{4}(1 - \xi)(1 + \eta),$$
$$N_3 = \frac{1}{4}(1 - \xi)(1 - \eta), \quad N_4 = \frac{1}{4}(1 + \xi)(1 - \eta) \tag{7.2.6}$$

are the mapping functions. Note that $N_i$ is equal to unity at node $i$ and zero at all other nodes. Equation (7.2.5) can also be expressed as

$$x = \sum_{i=1}^{4} N_i x_i. \tag{7.2.7}$$

Similarly, the mapping for $y$ is expressed as

$$y = \sum_{i=1}^{4} N_i y_i. \tag{7.2.8}$$

## Assumed Displacement

The displacement $u$ in the $x$ direction can be assumed as

$$u = b_1 + b_2 \xi + b_3 \eta + b_4 \xi \eta, \tag{7.2.9}$$

which is linear in $\xi$ along the $\eta = \pm 1$ boundaries and linear in $\eta$ along the $\xi = \pm 1$ boundaries. Once again, polynomial terms in the above equation are bilinear and balanced in $\xi$ and $\eta$, since the number of nodes in the $\xi$, $\eta$ directions are equal. At node $i$:

$$u_i = b_1 + b_2 \xi_i + b_3 \eta_i + b_4 \xi_i \eta_i \quad (i = 1, 2, 3, 4). \tag{7.2.10}$$

Then, following the similar procedure that led to the last of Eq. (7.2.5):

$$u = \lfloor N_1 \quad N_2 \quad N_3 \quad N_4 \rfloor \begin{Bmatrix} u_1 \\ u_2 \\ u_3 \\ u_4 \end{Bmatrix}, \tag{7.2.11}$$

where the shape functions $N_1, N_2, N_3, N_4$ are identical to the mapping functions in Eq. (7.2.6) in accordance with the isoparametric formulation. A similar expression holds for displacement component $v$ and thus for the assumed displacement:

$$u = \sum_{i=1}^{4} N_i u_i, \quad v = \sum_{i=1}^{4} N_i v_i. \tag{7.2.12}$$

Note that

$$\sum_{i=1}^{4} N_i = 1, \tag{7.2.13}$$

which can be confirmed by considering an element undergoing rigid-body translation.

**Construction of Shape Functions by Inspection**

Noting that $N_i = 1$ at node $i$ and 0 at all other nodes, one may construct the shape functions by inspection as shown in Figure 7.5.

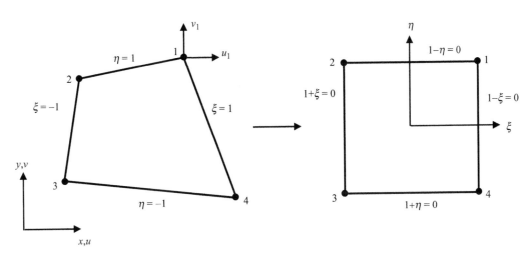

**Figure 7.5** Four-node element with alternative equations for the boundary lines in the mapped plane

Note that in Figure 7.5 the equations for the element boundaries in the mapped plane are expressed in a different, but equivalent, form. For example, the equation of the line connecting nodes 2 and 3 is expressed as $1 + \xi = 0$ while the equation for the line connecting nodes 3 and 4 is written as $1 + \eta = 0$.

To construct $N_1$, we note that $N_1 = 0$ at nodes 2, 3, and 4, and we then choose the minimum number of element boundary lines that do not contain node 1. They are the line connecting nodes 2 and 3, and the line connecting nodes 3 and 4. We may then take the left-hand side of the equations corresponding to these lines and try for $N_1$ as follows:

$$N_1 = \alpha_1(1 + \xi)(1 + \eta). \tag{7.2.14}$$

Note that $N_1$ is zero at all nodes except node 1 and that $N_1$ is bilinear in $\xi$ and $\eta$. Since $N_1 = 1$ at node 1 ($\xi = 1, \eta = 1$), we find that

$$\alpha_1 = \frac{1}{4}. \tag{7.2.15}$$

Following a similar procedure, we can find that

$$N_2 = \frac{1}{4}(1 - \xi)(1 + \eta), N_3 = \frac{1}{4}(1 - \xi)(1 - \eta), N_4 = \frac{1}{4}(1 + \xi)(1 - \eta), \tag{7.2.16}$$

which are indeed identical to those in Eq. (7.2.6). Written in expanded form, the assumed displacement $u$ is

$$u = \frac{1}{4}(1+\xi)(1+\eta)u_1 + \frac{1}{4}(1-\xi)(1+\eta)u_2 + \frac{1}{4}(1-\xi)(1-\eta)u_3 + \frac{1}{4}(1+\xi)(1-\eta)u_4.$$

$$(7.2.17)$$

Now consider the assumed displacement along the element boundaries. For example, along the side connecting nodes 1 and 2, $\eta = 1$ and then Eq. (7.2.17) reduces to

$$u = \frac{1}{2}(1+\xi)u_1 + \frac{1}{2}(1-\xi)u_2,$$

$$(7.2.18)$$

which is linear in $\xi$, while along the side connecting nodes 3 and 4, $\eta = -1$ and

$$u = \frac{1}{2}(1-\xi)u_3 + \frac{1}{2}(1+\xi)u_4,$$

$$(7.2.19)$$

which is linear in $\xi$. Similarly, along the side connecting nodes 2 and 3 and the side connecting nodes 1 and 4, $u$ is linear in $\eta$. This ensures that the assumed displacement is continuous when two adjacent elements of arbitrary shape share a common boundary between two nodes. A similar observation holds for the displacement component $v$ in Eq. (7.2.12).

## 7.2.2 Quadrilateral Elements of Higher Order

For construction of the mapping and shape functions, it is helpful to look at polynomial functions in increasing order as shown below.

Constant          $1$

Linear            $\xi \quad \eta$

Quadratic         $\xi^2 \quad \xi\eta \quad \eta^2$

Cubic             $\xi^3 \quad \xi^2\eta \quad \xi\eta^2 \quad \eta^3$

Quartic           $\xi^4 \quad \xi^3\eta \quad \xi^2\eta^2 \quad \xi\eta^3 \quad \eta^4$

Quintic           $\xi^5 \quad \xi^4\eta \quad \xi^3\eta^2 \quad \xi^2\eta^3 \quad \xi\eta^4 \quad \eta^5$

Sextic            $\xi^6 \quad \xi^5\eta \quad \xi^4\eta^2 \quad \xi^3\eta^3 \quad \xi^2\eta^4 \quad \xi\eta^5 \quad \eta^6$

We will now consider the construction of the mapping and shape functions of the nine-node element and eight-node element, following the procedure described in the previous section. Extensions to even higher-order elements are left as an exercise in the problems at the end of this chapter.

### Nine-Node Element

As shown in Figure 7.6, there are three nodes along each of the four sides. Along the $\eta = \pm 1$ boundaries, assumed displacement $u, v$ are quadratic in $\xi$ to match the three nodes. The element has 18 DOF.

There are three types of nodes: corner nodes, midside nodes, and a center node. For example, to construct $N_1$, we choose the minimum number of lines that pass through all

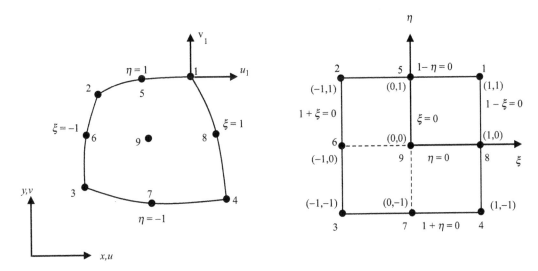

**Figure 7.6** Nine-node element in the 2D physical and mapped planes

other nodes except node 1. They are the $1 + \xi = 0$ line, $1 + \eta = 0$ line, $\xi = 0$ line, and $\eta = 0$ line. Taking and multiplying the left-hand side of these equations and noting that $N_1 = 1$ at node 1, we can find that

$$N_1 = \frac{1}{4}(1 + \xi)(1 + \eta)\xi\eta. \tag{7.2.20}$$

Following a similar procedure, we can determine all other mapping and shape functions. For example, for the midside node:

$$N_5 = \frac{1}{2}(1 + \xi)(1 - \xi)\eta(1 + \eta) \tag{7.2.21}$$

and for the center node:

$$N_9 = (1 - \xi)(1 + \xi)(1 + \eta)(1 - \eta). \tag{7.2.22}$$

Checking the polynomials in the shape functions, we can confirm that

$$u = b_1 + b_2\xi + b_3\eta + b_4\xi^2 + b_5\xi\eta + b_6\eta^2 + b_7\xi^2\eta + b_8\xi\eta^2 + b_9\xi^2\eta^2, \tag{7.2.23}$$

which is symmetric and balanced in $\xi$ and $\eta$. Accordingly, for the nine-node element:

$$x = \sum_{i=1}^{9} N_i x_i, \quad y = \sum_{i=1}^{9} N_i y_i,$$

$$\tag{7.2.24}$$

$$u = \sum_{i=1}^{9} N_i u_i, \quad v = \sum_{i=1}^{9} N_i v_i.$$

## Eight-Node Element

As shown in Figure 7.7, there are three nodes along each of the four sides. Along the $\eta = \pm 1$ boundaries, assumed displacements $u$, $v$ are quadratic in $\xi$ to match the three nodes. This element has 16 DOF. There are two types of nodes in this element: four corner nodes and four midside nodes.

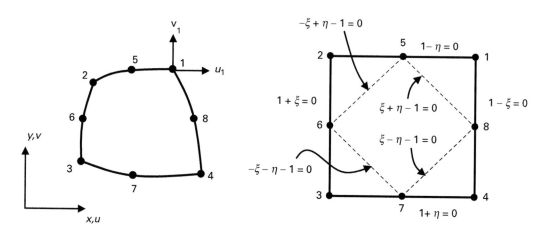

**Figure 7.7** Eight-node element in 2D plane

To construct $N_1$, we choose the minimum number of lines that pass through all other nodes except node 1. They are the $1 + \xi = 0$ line, $1 + \eta = 0$ line, and $\xi + \eta - 1 = 0$ line. Taking and multiplying the left-hand side of these equations, and noting that $N_1 = 1$ at node 1, we can find that

$$N_1 = \frac{1}{4}(1 + \xi)(1 + \eta)(\xi + \eta - 1).$$

We note that $N_1 = 0$ at all other nodes except node 1. Following similar procedures, it can be shown that for corner nodes:

$$N_1 = \frac{1}{4}(1 + \xi)(1 + \eta)(\xi + \eta - 1),\ N_2 = \frac{1}{4}(1 - \xi)(1 + \eta)(-\xi + \eta - 1),$$
$$N_3 = \frac{1}{4}(1 - \xi)(1 - \eta)(-\xi - \eta - 1),\ N_4 = \frac{1}{4}(1 + \xi)(1 - \eta)(\xi - \eta - 1)$$

(7.2.25)

and for midside nodes:

$$N_5 = \frac{1}{2}(1 - \xi)(1 + \xi)(1 + \eta),\ N_6 = \frac{1}{2}(1 - \xi)(1 - \eta)(1 + \eta),$$
$$N_7 = \frac{1}{2}(1 - \xi)(1 + \xi)(1 - \eta),\ N_8 = \frac{1}{2}(1 + \xi)(1 - \eta)(1 + \eta).$$

(7.2.26)

Checking the polynomials in the shape functions, we can confirm that

$$u = b_1 + b_2\xi + b_3\eta + b_4\xi^2 + b_5\xi\eta + b_6\eta^2 + b_7\xi^2\eta + b_8\xi\eta^2, \qquad (7.2.27)$$

which is symmetric in $\xi, \eta$ and balanced. A similar observation holds for assumed displacement $v$. Accordingly, for the eight-node element:

$$x = \sum_{i=1}^{8} N_i x_i, \quad y = \sum_{i=1}^{8} N_i y_i,$$

$$\qquad (7.2.28)$$

$$u = \sum_{i=1}^{8} N_i u_i, \quad v = \sum_{i=1}^{8} N_i v_i.$$

## 7.2.3 Triangular Elements

Triangular elements in the physical plane map into a right triangle in the $\xi\eta$-plane as shown in Figures 7.8 and 7.9.

### Three-Node Element

This element has 6 DOF. Following the procedure described in the previous section, it can be shown that

$$N_1 = \alpha_1(1 - \xi - \eta), \quad N_2 = \alpha_2\xi, \quad N_3 = \alpha_3\eta. \qquad (7.2.29)$$

They are linear in $\xi$ and $\eta$, and $\alpha_1 = \alpha_2 = \alpha_3 = 1$. We also note that $\sum N_i = 1$.

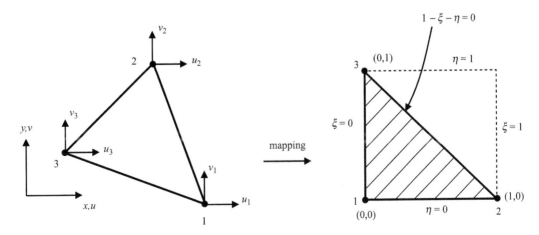

**Figure 7.8** Three-node element in the 2D physical plane and mapped plane

**Six-Node Element**

This element has 12 DOF and element boundaries in the physical domain can be curved, although Figure 7.9 shows straight boundaries. By inspection:

$$N_1 = 2(1 - \xi - \eta)\left(\frac{1}{2} - \xi - \eta\right), \quad N_2 = 2\xi\left(\xi - \frac{1}{2}\right), \quad N_3 = 2\eta\left(\eta - \frac{1}{2}\right),$$

$$N_4 = 4\xi(1 - \xi - \eta), \quad N_5 = 4\xi\eta, \quad N_6 = 4\eta(1 - \xi - \eta). \tag{7.2.30}$$

We confirm that $\sum_{i=1}^{6} N_i = 1$ and

$$u = \sum_{i=1}^{6} N_i u_i = a_1 + a_2\xi + a_3\eta + a_4\xi^2 + a_5\xi\eta + a_6\eta^2. \tag{7.2.31}$$

Accordingly, the assumed displacement is quadratic.

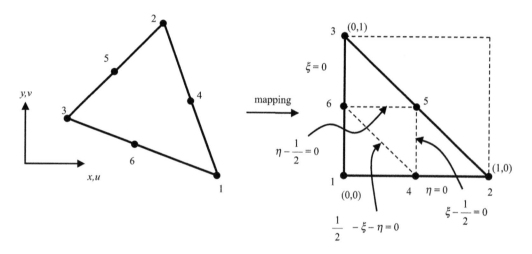

**Figure 7.9** Six-node element in the 2D physical and mapped planes

## 7.2.4 Property of Shape Functions

The assumed displacement must be able to represent a constant strain state that is the simplest nonzero strain state. The displacement field corresponding to a constant strain state in the 2D domain can be expressed as

$$u = a_1 + a_2 x + a_3 y, \quad v = b_1 + b_2 x + b_3 y. \tag{7.2.32}$$

To examine whether an element can represent this state, let's assign the displacement corresponding to the constant strain state to each node. For example, the assumed displacement $u$ in the element is

$$u = \sum N_i u_i = \sum N_i(a_1 + a_2 x_i + a_3 y_i)$$
$$= a_1 \sum N_i + a_2 \sum N_i x_i + a_3 \sum N_i y_i. \qquad (7.2.33)$$

For an isoparametric element:

$$\sum N_i = 1, \quad \sum N_i x_i = x, \quad \sum N_i y_i = y. \qquad (7.2.34)$$

Then, Eq. (7.2.33) becomes

$$u = a_1 + a_2 x + a_3 y \qquad (7.2.35)$$

and the displacement within the element corresponds to the constant strain state. Accordingly, the element can represent a constant strain state. Extension of the above discussion to isoparametric elements in the 3D space is straightforward.

## 7.3    Mapping and Shape Functions of 3D Elements

In this section, we will consider the mapping and shape functions for the eight-node hexahedral or brick element, the 20-node hexahedral element, and the four-node tetrahedral element.

### Eight-Node Hexahedron or Brick Element

Figure 7.10 shows the eight-node element in the 3D physical and mapped spaces. For mapping:

$$x = \sum_{i=1}^{8} N_i(\xi, \eta, \zeta) x_i, \quad y = \sum_{i=1}^{8} N_i(\xi, \eta, \zeta) y_i, \quad z = \sum_{i=1}^{8} N_i(\xi, \eta, \zeta) z_i. \qquad (7.3.1)$$

For assumed displacement:

$$u = \sum_{i=1}^{8} N_i(\xi, \eta, \zeta) u_i, \quad v = \sum_{i=1}^{8} N_i(\xi, \eta, \zeta) v_i, \quad w = \sum_{i=1}^{8} N_i(\xi, \eta, \zeta) w_i. \qquad (7.3.2)$$

There is only one type of node, located at the vertices or corners. Noting that $N_1 = 1$ at node 1 and 0 at all other nodes, we seek the smallest set of the element boundary surfaces in the mapped space that contain all other nodes except node 1. They are $1 + \xi = 0, 1 + \eta = 0$, and $1 + \zeta = 0$. We then take the product of the left-hand side of these equations and scale the result to construct $N_1$ as

$$N_1 = \frac{1}{8}(1 + \xi)(1 + \eta)(1 + \zeta). \qquad (7.3.3)$$

Similar expressions hold for other shape functions. The assumed displacement is trilinear as follows:

$$u = a_1 + a_2 \xi + a_3 \eta + a_4 \zeta + a_5 \xi \eta + a_6 \eta \zeta + a_7 \zeta \xi + a_8 \xi \eta \zeta. \qquad (7.3.4)$$

**Figure 7.10** Eight-node element in the 3D physical and mapped spaces

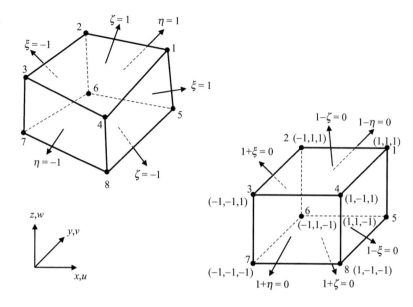

On the $\zeta = 1$ surface:

$$u = \sum_{i=1}^{8} N_i(\xi, \eta, \, +1)u_i. \tag{7.3.5}$$

Then $N_5 = N_6 = N_7 = N_8 = 0$ and

$$
\begin{aligned}
u = {} & \frac{1}{4}(1 + \xi)(1 + \eta)u_1 + \frac{1}{4}(1 - \xi)(1 + \eta)u_2 \\
& + \frac{1}{4}(1 - \xi)(1 - \eta)u_3 + \frac{1}{4}(1 + \xi)(1 - \eta)u_4.
\end{aligned} \tag{7.3.6}
$$

This guarantees continuity of the assumed displacement over the element boundary surfaces where two elements are joined. Note also that along the line connecting nodes 1 and 2, $\eta = 1$ and

$$u = \frac{1}{2}(1 + \xi)u_1 + \frac{1}{2}(1 - \xi)u_2. \tag{7.3.7}$$

Symbolically, the assumed displacement can be expressed as

$$\mathbf{u} = \mathbf{Nd}, \tag{7.3.8}$$

where

$$\mathbf{u} = \left\{ \begin{array}{c} u \\ v \\ w \end{array} \right\} \tag{7.3.9}$$

is the displacement vector, $\mathbf{N}$ is the $3 \times 24$ matrix of shape functions, and $\mathbf{d}$ is the $24 \times 1$ element DOF vector such that

$$\mathbf{d}^{\mathrm{T}} = \lfloor u_1 \quad v_1 \quad w_1 \quad u_2 \quad v_2 \quad w_2 \quad \cdots \quad \cdots \quad u_8 \quad v_8 \quad w_8 \rfloor. \tag{7.3.10}$$

## 20-Node Hexahedral Element

Figure 7.11 shows the 20-node element in the 3D physical and mapped spaces. For mapping:

$$x = \sum_{i=1}^{20} N_i(\xi, \eta, \zeta) x_i, \quad y = \sum_{i=1}^{20} N_i(\xi, \eta, \zeta) y_i, \quad z = \sum_{i=1}^{20} N_i(\xi, \eta, \zeta) z_i. \tag{7.3.11}$$

For assumed displacement:

$$u = \sum_{i=1}^{20} N_i(\xi, \eta, \zeta) u_i, \quad v = \sum_{i=1}^{20} N_i(\xi, \eta, \zeta) v_i, \quad w = \sum_{i=1}^{20} N_i(\xi, \eta, \zeta) w_i. \tag{7.3.12}$$

There are two types of nodes in this element: corner nodes and midside nodes.

To construct $N_i$ by inspection, we look for the equations of surfaces that constitute the smallest set of the element boundary and internal surfaces in the mapped space that contain all other nodes except node $i$. We then take the product of the left-hand side of these equations and scale the result such that $N_i = 1$ at node $i$ and 0 at all other nodes. As an example for a corner node, it can be shown that

$$N_4 = \frac{1}{8}(1+\xi)(1-\eta)(1+\zeta)(\xi - \eta + \zeta - 2), \tag{7.3.13}$$

where $\xi - \eta + \zeta - 2 = 0$ is the equation for the shaded plane in Figure 7.11. Similarly, as an example for a midside node, it can be shown that

$$N_7 = \frac{1}{4}(1+\xi)(1-\xi)(1-\eta)(1+\zeta). \tag{7.3.14}$$

**Figure 7.11** 20-node element in the 3D physical and mapped spaces

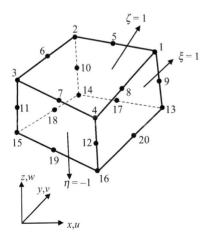
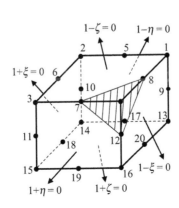

### Four-Node Tetrahedral Element

Consider the four-node tetrahedral element in the 3D physical and mapped spaces as shown in Figure 7.12. The nodes are located at the vertices of the tetrahedron. It can be shown by inspection that

$$N_1 = 1 - \xi - \eta - \zeta,$$
$$N_2 = \xi, N_3 = \eta, N_4 = \zeta. \tag{7.3.15}$$

Note that $1 - \xi - \eta - \zeta = 0$ is the equation formed by connecting nodes 2, 3, and 4 in Figure 7.12.

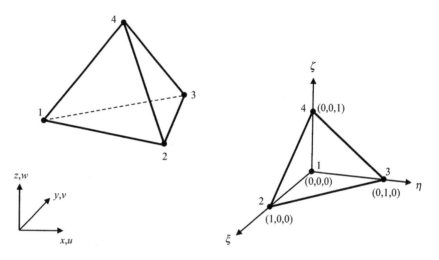

**Figure 7.12** Four-node tetrahedron element in the 3D physical and mapped spaces

## 7.4    Integration in Mapped Domains

In the FE formulation, integrals must be evaluated for the construction of the element stiffness matrices and load vectors. The integrals, originally expressed in the physical domains, can be transformed into those in the mapped domain using the mapping as shown in the following sections. Integration can then be carried out analytically or via numerical integration methods, to be discussed later.

### 7.4.1    Integration along a Line in the 2D Domain

A line integral can be expressed as

$$I = \int f_1 dl, \tag{7.4.1}$$

where $f_1$ is an integrand and $l$ is a coordinate along the line. Consider an element with two nodes along the boundary, as shown in the left of Figure 7.13. We note that coordinates along the line connecting nodes 1 and 4 can be expressed as

$$x = \frac{1}{2}(1 - \eta)x_4 + \frac{1}{2}(1 + \eta)x_1, \ y = \frac{1}{2}(1 - \eta)y_4 + \frac{1}{2}(1 + \eta)y_1 \tag{7.4.2}$$

and then

$$dl = \sqrt{(dx)^2 + (dy)^2} = \sqrt{\left(\frac{dx}{d\eta}\right)^2 + \left(\frac{dy}{d\eta}\right)^2} \, d\eta = J_1 d\eta, \tag{7.4.3}$$

where

$$J_1 = \sqrt{\left(\frac{dx}{d\eta}\right)^2 + \left(\frac{dy}{d\eta}\right)^2} : \text{length scale between } dl \text{ and } d\eta. \tag{7.4.4}$$

From Eq. (7.4.2):

$$\frac{dx}{d\eta} = \frac{1}{2}(x_1 - x_4), \frac{dy}{d\eta} = \frac{1}{2}(y_1 - y_4). \tag{7.4.5}$$

Then

$$I = \int f_1 dl = \int_{-1}^{1} f_1 J_1 d\eta, \tag{7.4.6}$$

where $f_1$ is expressed as a function of $\eta$ according to Eq. (7.4.2).

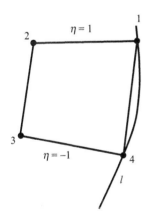 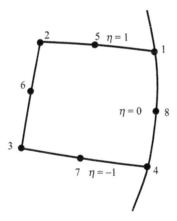

**Figure 7.13** Element boundaries along which line integration is carried out

For the element with three nodes along the boundary as shown on the right-hand side of Figure 7.13:

$$x = -\frac{1}{2}\eta(1 - \eta)x_4 + (1 - \eta)(1 + \eta)x_8 + \frac{1}{2}\eta(1 + \eta)x_1,$$

$$y = -\frac{1}{2}\eta(1 - \eta)y_4 + (1 - \eta)(1 + \eta)y_8 + \frac{1}{2}\eta(1 + \eta)y_1. \tag{7.4.7}$$

The length scale $J_1$ is determined using Eq. (7.4.4).

## 7.4.2  Integration over a 2D Element Area

Integration over the area of a 2D element in the physical plane can be expressed as

$$I = \int_{A_e} f_2 dA, \tag{7.4.8}$$

where $A_e$ is the element area. The above integration can be transformed into the integration in the mapped plane. For the element in Figure 7.14:

$$x = \sum_{i=1} N_i(\xi, \eta) x_i, \quad y = \sum_{i=1} N_i(\xi, \eta) y_i. \tag{7.4.9}$$

**Figure 7.14** Infinitesimal area in the physical plane

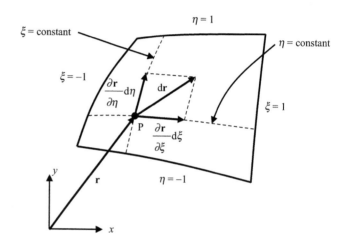

The position vector of a generic point P is

$$\mathbf{r} = x\mathbf{i} + y\mathbf{j} \tag{7.4.10}$$

and

$$d\mathbf{r} = \frac{\partial \mathbf{r}}{\partial \xi} d\xi + \frac{\partial \mathbf{r}}{\partial \eta} d\eta. \tag{7.4.11}$$

An infinitesimal area $dA$ in the physical plane can then be expressed as

$$dA = \left| \frac{\partial \mathbf{r}}{\partial \xi} d\xi \times \frac{\partial \mathbf{r}}{\partial \eta} d\eta \right| = \left| \frac{\partial \mathbf{r}}{\partial \xi} \times \frac{\partial \mathbf{r}}{\partial \eta} \right| d\xi d\eta = J_2 d\xi d\eta. \tag{7.4.12}$$

The area scale $J_2$ is evaluated from the mapping as

$$J_2 = \left| \frac{\partial \mathbf{r}}{\partial \xi} \times \frac{\partial \mathbf{r}}{\partial \eta} \right| = \left| \left( \frac{\partial x}{\partial \xi} \mathbf{i} + \frac{\partial y}{\partial \xi} \mathbf{j} \right) \times \left( \frac{\partial x}{\partial \eta} \mathbf{i} + \frac{\partial y}{\partial \eta} \mathbf{j} \right) \right|$$

$$= \left| \mathbf{k} \left( \frac{\partial x}{\partial \xi} \frac{\partial y}{\partial \eta} - \frac{\partial x}{\partial \eta} \frac{\partial y}{\partial \xi} \right) \right| = \begin{vmatrix} \dfrac{\partial x}{\partial \xi} & \dfrac{\partial y}{\partial \xi} \\[2mm] \dfrac{\partial x}{\partial \eta} & \dfrac{\partial y}{\partial \eta} \end{vmatrix} = \det \mathbf{J}_2, \tag{7.4.13}$$

where "det" stands for "determinant" and

$$\mathbf{J}_2 = \begin{bmatrix} \dfrac{\partial x}{\partial \xi} & \dfrac{\partial y}{\partial \xi} \\[2mm] \dfrac{\partial x}{\partial \eta} & \dfrac{\partial y}{\partial \eta} \end{bmatrix} : \text{Jacobian matrix in the 2D plane.} \tag{7.4.14}$$

Accordingly, for integration over a quadrilateral element:

$$I = \int f_2 dA = \int_{-1}^{1} \int_{-1}^{1} f_2 J_2 d\xi d\eta. \tag{7.4.15}$$

In the last of the above equation, $f_2$ is now expressed as a function of $\xi$ and $\eta$. For triangular elements, the integration is transformed into the right triangle in the mapped plane.

### 7.4.3  Integration over a Volume

Integration over an element in 3D space can be expressed as

$$I = \int_{V_e} f_3 dV, \tag{7.4.16}$$

where $V_e$ is the element volume. The above integration can be transformed into the integration in the mapped space using the mapping for a 3D element expressed as

$$x = \sum N_i(\xi, \eta, \zeta) x_i, \quad y = \sum N_i(\xi, \eta, \zeta) y_i, \quad z = \sum N_i(\xi, \eta, \zeta) z_i. \tag{7.4.17}$$

As shown in Figure 7.15, the position vector of a point in 3D space is

$$\mathbf{r} = x\mathbf{i} + y\mathbf{j} + z\mathbf{k} = \mathbf{r}(\xi, \eta, \zeta) \tag{7.4.18}$$

and

$$d\mathbf{r} = \frac{\partial \mathbf{r}}{\partial \xi} d\xi + \frac{\partial \mathbf{r}}{\partial \eta} d\eta + \frac{\partial \mathbf{r}}{\partial \zeta} d\zeta. \tag{7.4.19}$$

According to the triple scalar product formula:

$$dV = \frac{\partial \mathbf{r}}{\partial \xi} d\xi \cdot \left( \frac{\partial \mathbf{r}}{\partial \eta} d\eta \times \frac{\partial \mathbf{r}}{\partial \zeta} d\zeta \right) = \frac{\partial \mathbf{r}}{\partial \xi} \cdot \left( \frac{\partial \mathbf{r}}{\partial \eta} \times \frac{\partial \mathbf{r}}{\partial \zeta} \right) d\xi d\eta d\zeta = J_3 d\xi d\eta d\zeta, \tag{7.4.20}$$

**Figure 7.15** Infinitesimal volume in the physical 3D space

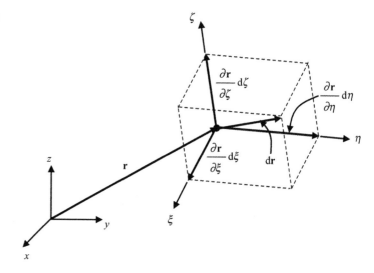

where

$$J_3 = \frac{\partial \mathbf{r}}{\partial \xi} \cdot \left( \frac{\partial \mathbf{r}}{\partial \eta} \times \frac{\partial \mathbf{r}}{\partial \zeta} \right) : \text{volume scale} \tag{7.4.21}$$

and

$$\frac{\partial \mathbf{r}}{\partial \xi} = \frac{\partial x}{\partial \xi}\mathbf{i} + \frac{\partial y}{\partial \xi}\mathbf{j} + \frac{\partial z}{\partial \xi}\mathbf{k},$$

$$\frac{\partial \mathbf{r}}{\partial \eta} = \frac{\partial x}{\partial \eta}\mathbf{i} + \frac{\partial y}{\partial \eta}\mathbf{j} + \frac{\partial z}{\partial \eta}\mathbf{k}, \tag{7.4.22}$$

$$\frac{\partial \mathbf{r}}{\partial \zeta} = \frac{\partial x}{\partial \zeta}\mathbf{i} + \frac{\partial y}{\partial \zeta}\mathbf{j} + \frac{\partial z}{\partial \zeta}\mathbf{k}.$$

Then, it can be shown that

$$J_3 = \begin{vmatrix} \dfrac{\partial x}{\partial \xi} & \dfrac{\partial y}{\partial \xi} & \dfrac{\partial z}{\partial \xi} \\[2mm] \dfrac{\partial x}{\partial \eta} & \dfrac{\partial y}{\partial \eta} & \dfrac{\partial z}{\partial \eta} \\[2mm] \dfrac{\partial x}{\partial \zeta} & \dfrac{\partial y}{\partial \zeta} & \dfrac{\partial z}{\partial \zeta} \end{vmatrix} \tag{7.4.23}$$

is the determinant of the Jacobian matrix in 3D space. For a hexahedral element, the integration is over the cube in the mapped space, such that

$$\int_{V_e} f_3 dV = \int_{-1}^{1} \int_{-1}^{1} \int_{-1}^{1} f_3 J_3 d\xi d\eta d\zeta. \tag{7.4.24}$$

For the tetrahedral element, the integration is over the right tetrahedron in the mapped space.

### 7.4.4 Integration over a Surface in 3D Space

Integration over a surface in 3D space can be expressed as

$$I = \int_{S_e} f_S dS,$$ (7.4.25)

where $S_e$ is the surface area of a 3D element. For example, consider integration over the $\zeta = 1$ surface of an element, as shown in Figure 7.16.

The position vector of a point on the $\zeta = 1$ surface is $\mathbf{r} = \mathbf{r}(\xi, \eta, \zeta = 1)$ and thus $d\zeta = 0$. Accordingly:

$$d\mathbf{r} = \frac{\partial \mathbf{r}}{\partial \xi} d\xi + \frac{\partial \mathbf{r}}{\partial \eta} d\eta.$$ (7.4.26)

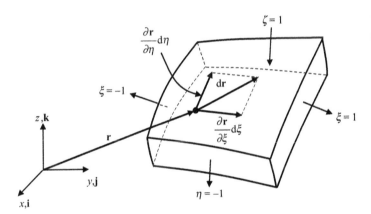

**Figure 7.16** Infinitesimal area over a surface in the physical 3D space

For an infinitesimal area $dS$ on the $\zeta = 1$ surface as shown in the figure:

$$dS = \left| \frac{\partial \mathbf{r}}{\partial \xi} d\xi \times \frac{\partial \mathbf{r}}{\partial \eta} d\eta \right|_{\zeta=1} = \left| \frac{\partial \mathbf{r}}{\partial \xi} \times \frac{\partial \mathbf{r}}{\partial \eta} \right|_{\zeta=1} d\xi d\eta = g d\xi d\eta,$$ (7.4.27)

where

$$g = \left| \frac{\partial \mathbf{r}}{\partial \xi} \times \frac{\partial \mathbf{r}}{\partial \eta} \right|_{\zeta=1} \quad : \text{area scale.}$$ (7.4.28)

Recall that

$$\mathbf{r} = x\mathbf{i} + y\mathbf{j} + z\mathbf{k} \rightarrow \frac{\partial \mathbf{r}}{\partial \xi} = \frac{\partial x}{\partial \xi}\mathbf{i} + \frac{\partial y}{\partial \xi}\mathbf{j} + \frac{\partial z}{\partial \xi}\mathbf{k}, \quad \frac{\partial \mathbf{r}}{\partial \eta} = \frac{\partial x}{\partial \eta}\mathbf{i} + \frac{\partial y}{\partial \eta}\mathbf{j} + \frac{\partial z}{\partial \eta}\mathbf{k}.$$ (7.4.29)

Then

$$I = \int_{S_{\sigma e}} f_S dS = \int f_S g d\xi d\eta. \tag{7.4.30}$$

## 7.5    Numerical Integration

Various numerical methods are available for integration in the mapped domain. In this section, we will consider Gaussian quadrature for integration over lines, squares, cuboids of equal edge lengths, and integration rules over triangles and tetrahedra.

### 7.5.1    Gaussian Quadrature

Gaussian quadrature is a method for evaluating integrals of functions numerically. For the special case of polynomial integrands, the integration can be exact through careful choice of the number of "quadrature points." We also note that an arbitrary function can be approximated by a combination of polynomial functions.

**Numerical Integration in One Dimension**

Numerical integration of the function $f(\xi)$ over the 1D domain bounded by $-1$ and $1$ can be expressed as

$$\int_{-1}^{1} f(\xi)d\xi = \sum_{i=1}^{n} f(a_i) W_i, \tag{7.5.1}$$

where
$a_i$: sampling point (or integration point) at which the integrand is evaluated
$W_i$: weight of sampling point $i$
$n$: number of sampling points.

Placing $f(\xi) = 1$ into Eq. (7.5.1):

$$\sum W_i = 2, \tag{7.5.2}$$

which states that the sum of the weights is equal to 2. Consider $f(\xi)$, which is expressed or approximated as a polynomial function such that

$$f(\xi) = c_0 + c_1\xi + c_2\xi^2 + c_3\xi^3 + c_4\xi^4 + c_5\xi^5 + c_6\xi^6 + c_7\xi^7 \cdots. \tag{7.5.3}$$

The exact integration of the above polynomial function is

$$\int_{-1}^{1} f(\xi)d\xi = 2c_0 + \frac{2}{3}c_2 + \frac{2}{5}c_4 + \frac{2}{7}c_6 + \cdots. \tag{7.5.4}$$

Note that the integration produces zero for all odd power terms. One may then determine the sampling points and the weights as follows, considering integration of individual terms in Eq. (7.5.3).

(1) One-point rule ($n = 1$):

$$\int_{-1}^{1} f(\xi)d\xi = f(a_1)W_1 = f(0)W_1,$$ (7.5.5)

in which we set $a_1 = 0$ as the sampling point so that integration of any odd function is equal to zero. Applying the one-point integration rule to the polynomial function in Eq. (7.5.3), we obtain

$$f(0)W_1 = c_0 W_1.$$ (7.5.6)

Comparing the right-hand side of the above equation with the exact integration in Eq. (7.5.4):

$$c_0 W_1 = 2c_0 \rightarrow W = 2,$$ (7.5.7)

which is consistent with Eq. (7.5.2). For the $\xi$ term in Eq. (7.5.3), the one-point rule produces zero which matches with the exact integration. However, for the $\xi^2$ term, the one-point rule produces zero which does not agree with 2/3 from the exact integration. Accordingly, the one-point rule can integrate exactly up to the first-order polynomial function.

(2) Two-point rule ($n = 2$):

$$\int_{-1}^{1} f(\xi)d\xi = f(a_1)W_1 + f(a_2)W_2.$$ (7.5.8)

As shown in Figure 7.17, we may set $a_1 = -a, a_2 = a$ and $W_1 = W_2 = W$ such that

$$\int_{-1}^{1} f(\xi)d\xi = [f(-a) + f(a)]W.$$ (7.5.9)

The above choices ensure that integration of any odd function is equal to zero. Applying the two-point integration rule up to the fifth-order polynomial function in Eq. (7.5.3), we obtain

$$\int_{-1}^{1} f(\xi)d\xi = [f(-a) + f(a)]W = 2Wc_0 + 2a^2 Wc_2 + 2a^4 Wc_4.$$ (7.5.10)

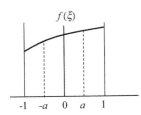

Figure 7.17 Sampling or integration points to be determined for the two-point rule

Comparing the right-hand side of the above equation with the exact integration in Eq. (7.5.4):

$$2Wc_0 = 2c_0 \rightarrow W = 1,$$

$$2a^2 Wc_2 = \frac{2}{3}c_2 \rightarrow a^2 = \frac{1}{3} \rightarrow a = \sqrt{\frac{1}{3}}. \tag{7.5.11}$$

The sum of the two weights is equal to 2, in agreement with Eq. (7.5.2). For the $\xi^4$ term, the two-point rule produces 2/9, which does not agree with 2/5 from the exact integration. Accordingly, the two-point rule can integrate exactly up to the third-order polynomial function.

(3) Three-point rule ($n = 3$):

$$\int_{-1}^{1} f(\xi)d\xi = f(a_1)W_1 + f(a_2)W_2 + f(a_3)W_3. \tag{7.5.12}$$

We may set $a_1 = -a, a_2 = 0, a_3 = a, W_3 = W_1$, which guarantees that integration of any odd function results in zero. Extending the procedure described for the two-point rule to the three-point rule, we can show that

$$a = \sqrt{\frac{3}{5}}, W_1 = W_3 = \frac{5}{9}, W_2 = \frac{8}{9}.$$

Accordingly, for the three-point rule in one dimension:

$$(a_1, a_2, a_3) = \left(-\sqrt{3/5},\ 0,\ \sqrt{3/5}\right), \quad (W_1, W_2, W_3) = (5/9, 8/9, 5/9). \tag{7.5.13}$$

It can be shown that the three-point rule can integrate exactly up to the fifth-order polynomial function.

(4) Four-point rule ($n = 4$):

$$\int_{-1}^{1} f(\xi)d\xi = f(a_1)W_1 + f(a_2)W_2 + f(a_3)W_3 + f(a_4)W_4. \tag{7.5.14}$$

We may set $a_1 = -b, a_2 = -a, a_3 = a, a_4 = b, W_4 = W_1, W_3 = W_2$. Then

$$\int_{-1}^{1} f(\xi)d\xi = [f(-b) + f(b)]W_1 + [f(-a) + f(a)]W_2. \tag{7.5.15}$$

Applying the four-point rule to the polynomial function in Eq. (7.5.3) results in nonlinear algebraic equations which can be solved via an iterative method such as the Newton–Raphson method for preset accuracy. Alternatively, for the four-point rule or higher-order integration rules, we may work with the Legendre polynomials to determine the sampling points and the weights, without using an iterative method. Table 7.1 summarizes the integration rules. The rules can be specified to arbitrarily large numbers of points, but the table shows only the first five. In general, we can show that the $n$-point rule integrates exactly a polynomial function of order $(2n - 1)$.

**Table 7.1 Gauss quadrature points**

| $n$ Number of points | $\xi_i$ Integration point | $w_i$ Quadrature weight |
| --- | --- | --- |
| 1 | 0 | 2 |
| 2 | $\pm\sqrt{\dfrac{1}{3}}$ | 1 |
| 3 | 0 | 8/9 |
| | $\pm\sqrt{\dfrac{3}{5}}$ | 5/9 |
| 4 | $\pm\sqrt{\dfrac{1}{7}\left(3-2\sqrt{\dfrac{6}{5}}\right)}$ | $\dfrac{18+\sqrt{30}}{36}$ |
| | $\pm\sqrt{\dfrac{1}{7}\left(3+2\sqrt{\dfrac{6}{5}}\right)}$ | $\dfrac{18-\sqrt{30}}{36}$ |
| 5 | 0 | 128/225 |
| | $\pm\dfrac{1}{3}\sqrt{5-2\sqrt{\dfrac{10}{7}}}$ | $\dfrac{322+13\sqrt{70}}{900}$ |
| | $\pm\dfrac{1}{3}\sqrt{5+2\sqrt{\dfrac{10}{7}}}$ | $\dfrac{322-13\sqrt{70}}{900}$ |

## Example 7.1

Carry out the following integration using numerical integration:

$$I = \int_{x=0}^{x=R} \sqrt{R^2 - x^2}\,dx.$$

(a) If we introduce linear mapping for the entire interval:

$$x = \frac{1}{2}(1-\xi)x_1 + \frac{1}{2}(1+\xi)x_2 = \frac{1}{2}(1+\xi)R.$$

Introducing the above mapping into the integral:

$$I = \frac{1}{2}R^2 \int_{\xi=-1}^{\xi=1} \sqrt{1-\frac{1}{4}(1+\xi)^2}\,d\xi = \frac{1}{4}R^2 \int_{\xi=-1}^{\xi=1} \sqrt{4-(1+\xi)^2}\,d\xi.$$

(b) If the three-point rule is applied:

$$I = \frac{1}{4}R^2\left[\sqrt{4-\left(1-\sqrt{\frac{3}{5}}\right)^2}\cdot\frac{5}{9} + \sqrt{4-1}\cdot\frac{8}{9} + \sqrt{4-\left(1+\sqrt{\frac{3}{5}}\right)^2}\cdot\frac{5}{9}\right] = \frac{1}{4}(3.15607)R^2.$$

The exact integration is $I = \pi R^2 / 4$. To improve the accuracy of numerical integration, we may divide the interval into multiple segments, and apply the linear mapping and the three-point rule in each segment. We can also use more quadrature points in each segment.

### Gaussian Quadrature over a Square and a Hexahedron

For integration over a square in the mapped 2D plane, one may first apply numerical integration rules over the $\xi$-axis, which is followed by integration over the $\eta$-axis such that

$$\int_{-1}^{1} \int_{-1}^{1} f(\xi, \eta) d\xi d\eta = \int_{-1}^{1} \sum_{i=1}^{n} f(a_i, \eta) W_i d\eta = \sum_{j=1}^{n} \sum_{i=1}^{n} f(a_i, a_j) W_i W_j. \qquad (7.5.16)$$

For example, for $n = 2$:

$$a_1 = -a, a_2 = a, a = \frac{1}{\sqrt{3}}, \qquad (7.5.17)$$

and the $2 \times 2$ integration points are shown in Figure 7.18.

**Figure 7.18** Sampling points for the $2 \times 2$ point integration

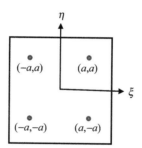

Then, for a quadrilateral element in the 2D domain:

$$\int_{A_e} f_2 dA = \int_{-1}^{1} \int_{-1}^{1} f_2 J_2 d\xi d\eta = \sum_{i=1}^{n} \sum_{j=1}^{n} (f_2 J_2)_{\substack{\xi = a_i \\ \eta = a_j}} W_i W_j, \qquad (7.5.18)$$

where $A_e$ is the element domain in the physical 2D plane.

Similarly, for a hexahedral element:

$$\int_{V_e} f_3 dV = \int_{-1}^{1} \int_{-1}^{1} \int_{-1}^{1} f_3 J_3 d\xi d\eta d\zeta = \sum_{i=1}^{n} \sum_{j=1}^{n} \sum_{k=1}^{n} [f_3 J_3]_{\substack{\xi = a_i \\ \eta = a_j \\ \zeta = a_k}} W_i W_j W_k, \qquad (7.5.19)$$

where $V_e$ is the element domain in the physical 3D space.

**Example 7.2**

Use a numerical integration method to compute the area moment of inertia $I = \int y^2 dA$ of the circular section with a hole, as shown in Figure 7.19.

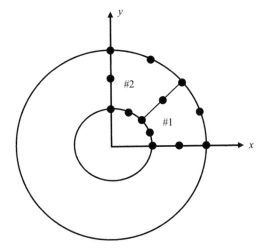

The outer radius of the circle is $R_0 = 6.0$ cm while the inner radius is $R_i = 2.5$ cm. For mapping, use eight-node elements over the section in the first quadrant. Use the $2 \times 2$ point rule $I$ with a mesh of two elements equally spaced in the circumferential direction for the first quadrant. Compare with the exact $I$.

## Solution:

For element #1, we may place element node 1 at the bottom right corner of the outer radius and follow the node numbering scheme as shown in Figure 7.7. The nodal coordinate are then as follows:

**Figure 7.19** Circular cross-section with a hole

$$(x_1, y_1) = (R_0, 0.0), (x_2, y_2) = (R_0 \cos 45°, R_0 \sin 45°),$$
$$(x_3, y_3) = (R_i \cos 45°, R_i \sin 45°), (x_4, y_4) = (R_i, 0.0),$$
$$(x_5, y_5) = (R_0 \cos 22.5°, R_0 \sin 22.5°), (x_6, y_6) = (R_m \cos 45°, R_m \sin 45°),$$
$$(x_7, y_7) = (R_i \cos 22.5°, R_i \sin 22.5°), (x_8, y_8) = (R_m, 0.0),$$

where $R_m = (R_0 + R_i)/2$. According to the mapping:

$$x = \sum_{i=1}^{8} N_i x_i = N_1 x_1 + N_2 x_2 + \cdots + N_8 x_8,$$

$$y = \sum_{i=1}^{8} N_i y_i = N_1 y_1 + N_2 y_2 + \cdots + N_8 y_8.$$

From Eq. (7.4.14):

$$\mathbf{J}_2 = \begin{vmatrix} \dfrac{\partial x}{\partial \xi} & \dfrac{\partial y}{\partial \xi} \\[2ex] \dfrac{\partial x}{\partial \eta} & \dfrac{\partial y}{\partial \eta} \end{vmatrix},$$

where

$$\frac{\partial x}{\partial \xi} = \frac{\partial N_1}{\partial \xi} x_1 + \frac{\partial N_2}{\partial \xi} x_2 + \cdots + \frac{\partial N_8}{\partial \xi} x_8, \quad \frac{\partial x}{\partial \eta} = \frac{\partial N_1}{\partial \eta} x_1 + \frac{\partial N_2}{\partial \eta} x_2 + \cdots + \frac{\partial N_8}{\partial \eta} x_8,$$

$$\frac{\partial y}{\partial \xi} = \frac{\partial N_1}{\partial \xi} y_1 + \frac{\partial N_2}{\partial \xi} y_2 + \cdots + \frac{\partial N_8}{\partial \xi} y_8, \quad \frac{\partial y}{\partial \eta} = \frac{\partial N_1}{\partial \eta} y_1 + \frac{\partial N_2}{\partial \eta} y_2 + \cdots + \frac{\partial N_8}{\partial \eta} y_8.$$

According to Eq. (7.4.15):

$$I_1 = \int y^2 dA = \int_{-1}^{1}\int_{-1}^{1} y^2 J_2 d\xi d\eta = \int_{-1}^{1}\int_{-1}^{1} f d\xi d\eta,$$

where $J_2$ is the determinant of $\mathbf{J}_2$ and $f = y^2 J_2$. According to Eq. (7.5.18):

$$I_1 = \int_{-1}^{1}\int_{-1}^{1} f d\xi d\eta = \sum_{i=1}^{n}\sum_{j=1}^{n} (f)_{\substack{\xi = a_i \\ \eta = a_j}} W_i W_j.$$

For the two-point rule in one dimension:

$$(a_1, a_2) = \left(-\frac{1}{\sqrt{3}}, \frac{1}{\sqrt{3}}\right), (W_1, W_2) = (1.0, 1.0).$$

Applying the $2 \times 2$ two-point rule as shown in Figure 7.18, we can determine that $I_1 = 44.5806$ cm$^4$.

For element # 2, we may place element node 1 at the rightmost corner of the outer radius and follow the node numbering scheme as shown in Figure 7.7. The nodal coordinate are then as follows:

$$(x_1, y_1) = (R_0 \cos 45°, R_0 \sin 45°), (x_2, y_2) = (0.0, R_0),$$
$$(x_3, y_3) = (0.0, R_i), (x_4, y_4) = (R_i \cos 45°, R_i \sin 45°),$$
$$(x_5, y_5) = (R_0 \cos 67.5°, R_0 \sin 67.5°), (x_6, y_6) = (0.0, R_m),$$
$$(x_7, y_7) = (R_i \cos 67.5°, R_i \sin 67.5°), (x_8, y_8) = (R_m \cos 45°, R_m \sin 45°).$$

Following similar procedures as described for element #1, we can determine that $I_2 = 201.7089$ cm$^4$. For the entire section:

$$I = (I_1 + I_2) \times 4 = 985.16 \text{ cm}^4.$$

The above value is close to the exact value of 987.04cm$^4$. We may use more elements or more quadrature points to improve the accuracy.

### 7.5.2  Integration over a Triangle

Once the mapping is established, numerical integration over a right triangle in the mapped plane as shown in Figure 7.20 can be expressed as

$$\int_A f(\xi, \eta) d\xi d\eta = A \sum_{i=1}^{n} f(\xi_i, \eta_i) w_i, \quad A = \frac{1}{2} : \text{area}. \qquad (7.5.20)$$

Setting $f(\xi, \eta) = 1$, we find that for the scaled weights:

$$\sum_{i=1}^{n} w_i = 1. \qquad (7.5.21)$$

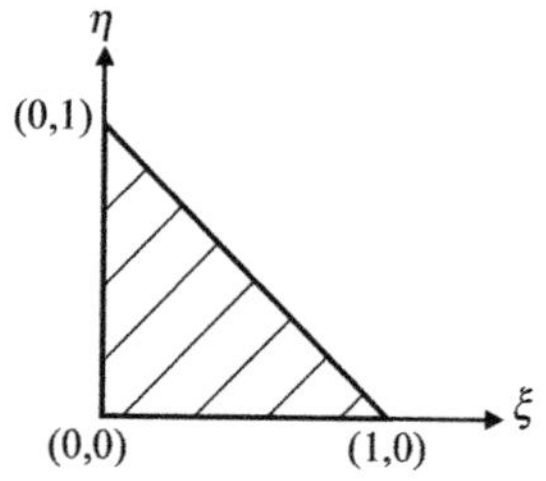

Figure 7.20  Right triangle in the mapped plane

For construction of numerical integration rules over the triangular domain, we may consider a polynomial function expressed as

$$f(\xi, \eta) = c_1 + c_2\xi + c_3\eta + c_4\xi^2 + c_5\xi\eta + c_6\eta^2 + \cdots. \tag{7.5.22}$$

Exact integration of the above polynomial functions over the right triangle in Figure 7.20 is

$$f(\xi, \eta) = \frac{1}{2}c_1 + \frac{1}{6}c_2 + \frac{1}{6}c_3 + \frac{1}{12}c_4 + \frac{1}{24}c_5 + \frac{1}{12}c_6 + \cdots. \tag{7.5.23}$$

Numerical integration rules with various sampling points as shown in Figure 7.21 are given as follows.

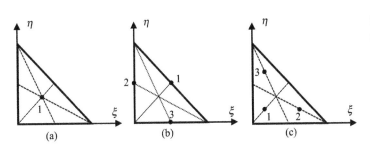

Figure 7.21 Various sampling points for numerical integration over a triangle

(1) One-point rule:

$$\int_A f(\xi, \eta)d\xi d\eta = \frac{1}{2}f(\xi_1, \eta_1)w_1 = \frac{1}{2}f(\xi_1, \eta_1)\cdot 1 \tag{7.5.24}$$

with $w_1 = 1$ according to Eq. (7.5.21). Applying the one-point rule to the polynomial function in Eq. (7.5.2.2):

$$\int_A f(\xi, \eta)d\xi d\eta = \frac{1}{2}f(\xi_1, \eta_1)\cdot 1 = \frac{1}{2}\left(c_1 + c_2\xi_1 + c_3\eta_1 + c_4\xi_1^2 + c_5\xi_1\eta_1 + c_6\eta_1^2\right). \tag{7.5.25}$$

Comparing the right-hand side of the above equation with the exact integration in Eq. (7.5.23):

$$\frac{1}{2}c_2\xi_1 = \frac{1}{6}c_2 \rightarrow \xi_1 = \frac{1}{3}, \quad \frac{1}{2}c_3\eta_1 = \frac{1}{6}c_3 \rightarrow \eta_1 = \frac{1}{3}. \tag{7.5.26}$$

The sampling point is located at $(1/3, 1/3)$ with $w_1 = 1$. The one-point rule can integrate exactly up to a first-order polynomial function.

(2) Three-point rule – option 1:

$$\int_A f(\xi, \eta)d\xi d\eta = \frac{1}{2}[f(\xi_1, \eta_1)w_1 + f(\xi_2, \eta_2)w_2 + f(\xi_3, \eta_3)w_3]. \tag{7.5.27}$$

For simplicity, we may select the sampling points located at the midpoints of the three sides as $(1/2, 1/2), (0, 1/2), (1/2, 0)$ and set $w_1 = w_2 = w_3 = 1/3$. Then

$$\int_A f(\xi, \eta)d\xi d\eta = \frac{1}{6}\left[f\left(\frac{1}{2}, \frac{1}{2}\right) + f\left(0, \frac{1}{2}\right) + f\left(\frac{1}{2}, 0\right)\right]. \tag{7.5.28}$$

Applying the above three-point rule to the polynomial function in Eq. (7.5.22) results in the expression on the right-hand side of Eq. (7.5.23). Accordingly, this three-point rule can integrate exactly up to a quadratic polynomial function. However, it can be shown that this three-point rule cannot integrate exactly a cubic polynomial function.

(3) Three-point rule – option 2: Alternatively, we may consider a three-point rule in which sampling points are located on the lines connecting each of the three vertices through the centroid to the midpoint of the side as shown in Figure 7.20(c). In the figure, the coordinates of points 1, 2, 3 are expressed as $(a, a), (b, a), (a, b)$ and the scaled weights are set as $w_1 = w_2 = w_3 = 1/3$ Then

$$\int_A f(\xi, \eta) d\xi d\eta = \frac{1}{6}[f(a, a) + f(b, a) + f(a, b)]. \tag{7.5.29}$$

Applying the above three-point rule to the polynomial function in Eq. (7.5.22), we can show that

$$a = \frac{1}{6}, b = \frac{2}{3}. \tag{7.5.30}$$

This rule can integrate exactly up to a quadratic polynomial function.

(4) Higher-order integration rules: Following similar procedures, we may construct the seven-point rule and the 13-point rule for integration of higher-order polynomial functions, by solving equations to determine the sampling points and the weights. Higher-order integration rules for a triangle are available in the open literature.

(5) Alternative approaches: We may divide the right triangle into three triangular sub-areas by connecting the centroid of the triangle to each of the three vertices. We can then map each of the sub-areas into the right triangle in another mapped plane and apply the three-point rule to each sub-area for numerical integration. We may also divide the right triangle into three quadrilateral sub-areas by connecting the centroid of the triangle to the midpoint of each side. We can then map each of the sub-areas into a square in another mapped plane and then apply the Gaussian quadrature.

## 7.5.3 Integration over a Tetrahedron

For numerical integration over a tetrahedron in the mapped space as shown in Figure 7.12:

$$\int_V f(\xi, \eta, \zeta) dV = \frac{1}{4} \sum_{i=1}^{n} f(\xi_i, \eta_i, \zeta_i) w_i. \tag{7.5.31}$$

The sampling points and the scaled weights can be determined by considering integration of the polynomial functions. For higher-order integration rules, an iterative method can be used to solve nonlinear equations to determine the sampling points and the weights.

(1) One-point rule: The sampling point is located at $(1/2, 1/2, 1/2)$ and $w_1 = 1$. The one-point rule can integrate exactly up to the first-order polynomial function.

(2) Four-point rule: The sampling points are located at $(a, b, b), (b, a, b), (b, b, a), (b, b, b)$, where:

$$a = 0.58541020, b = 0.13819660,$$

$$w_1 = w_2 = w_3 = w_4 = \frac{1}{4}.$$

Higher-order integration rules are available in the open literature.

(3) Alternative methods: We may divide a tetrahedron into four tetrahedra of smaller size. This can be accomplished by connecting the centroid of the tetrahedron with three vertices of each triangular surface that forms the outer boundary of the tetrahedron. A numerical integration rule such as the four-point rule can then be applied to each of the four smaller tetrahedra after a new mapping is introduced. A tetrahedron can also be divided into four hexahedra. We may then apply the Gaussian quadrature for integration over each hexahedron.

## PROBLEMS

**7.1** For the 16-node element shown in Figure 7.22, all nodes are equally spaced in the mapped plane. Along the $\eta = \pm 1$ boundaries, assumed displacements are cubic in $\xi$ to match four nodes. The assumed displacement can be expressed as

$$u = \sum_{i=1}^{16} N_i u_i, \quad v = \sum_{i=1}^{16} N_i v_i.$$

Construct shape functions $N_1, N_5$, and $N_{16}$.

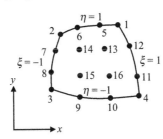

**Figure 7.22** For Problem 7.1

**7.2** For the 10-node element shown in Figure 7.23, construct $N_1, N_3, N_5$, and $N_{10}$. Note that element boundaries can be curved in the physical 2D plane.

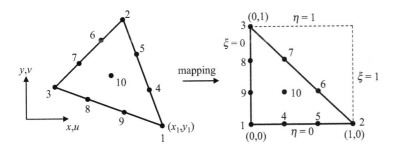

**Figure 7.23** For Problem 7.2

**7.3** Construct by inspection the mapping/shape functions of the six-node element shown in the mapped plane in Figure 7.24.

**Figure 7.24** For Problem 7.3

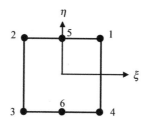

**7.4** Construct the mapping/shape functions for the five-node element shown in the mapped plane (Figure 7.25). You may start from the eight-node element and apply constraints to the midside nodes. This type of element may be placed between a four-node element and an eight-node element for mesh transition.

**Figure 7.25** For Problem 7.4

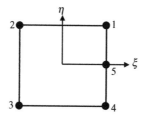

**7.5** For the five-node element shown in the mapped plane (Figure 7.26), construct the shape functions. You may start from the nine-node element and apply constraints.

**Figure 7.26** For Problem 7.5

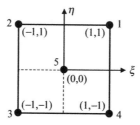

**7.6** For the 27-node element shown in the 3D physical and mapped spaces, confirm that

$$N_1 = \frac{1}{8}(1+\xi)(1+\eta)(1+\zeta)\xi\eta\zeta,$$

$$N_5 = \frac{1}{4}(1+\xi)(1-\xi)(1+\eta)(1+\zeta)\eta\zeta.$$

**7.7** For the 10-node tetrahedral element in the mapped 3D space as shown in Figure 7.28, construct $N_1, N_4, N_6, N_{10}$.

**7.8** Consider numerical integration of the following integral:

$$I = \int_{x=0}^{x=R} \sqrt{R^2 - x^2}dx.$$

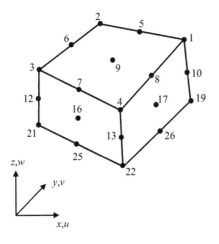

**Figure 7.27** For Problem 7.6

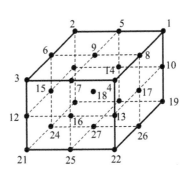

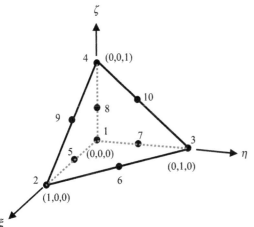

**Figure 7.28** For Problem 7.7

(a) Divide the integration interval into two segments and use the linear mapping and the
three-point rule for each element.

(b) Compare with the exact integration.

**7.9** Use a numerical integration method to compute
the area moment of inertia $I$ of the circular section
with a hole as shown in Example 7.2. For mapping, use the eight-node elements over the section
in the first quadrant. For numerical integration,
use the $3 \times 3$ point rule.

(a) Determine $I$ using two elements equally spaced
in the radial direction and two elements equally
spaced in the circumferential direction for the
first quadrant. Compare with the exact value.

(b) Determine $I$ using two elements equally
spaced in the radial direction and four elements equally spaced in the circumferential direction for the first quadrant. Compare with
the exact value.

# 8      2D and 3D Deformable Solid Bodies

In this chapter we first describe how to construct the element stiffness matrix and load vector in 2D domains. The mapping and shape functions derived in the previous chapter are introduced to express strain components in terms of nodal DOF. Integrations needed for construction of the element stiffness matrix and load vector are carried out in the mapped plane using numerical integration. They are then assembled into the global stiffness matrix and global load vector to form an FE equation for static analysis.

Extension of the FE formulation to 3D domains is demonstrated using the eight-node hexahedron as an example. As in the case of the 2D formulation, mapping and shape functions are used to express six strain components in terms of nodal DOF. Integrations are carried out in the mapped space or plane to generate the element stiffness matrix and element load vector. They are then assembled into the global stiffness matrix and global load vector to form an FE equation for static analysis.

For dynamic problems, the element mass matrix can be formed by treating the inertia effect as a body force applied to the element. The global mass matrix is then assembled to construct the equation of motion for analyses of free vibration and forced vibration.

In the last section, we briefly discuss important aspects of FE modeling and analysis that often arise in 2D and 3D problems where the number of DOF can be large. We discuss issues, such as sparse matrices and mesh generation, which early students of the FE method may find helpful for future reference.

## 8.1    Finite Element Formulation of Plane Stress and Strain Problems

As shown in the previous chapter, the equilibrium for plane stress and plane strain problems is stated as

$$\delta U - \delta W = 0, \tag{8.1.1}$$

where

$$\delta U = \int_A \underset{1\times 3}{\delta \boldsymbol{\varepsilon}^{\mathrm{T}}} \underset{3\times 3}{\mathbf{C}} \underset{3\times 1}{\boldsymbol{\varepsilon}} \, tdA - \int_A \delta \boldsymbol{\varepsilon}^{\mathrm{T}} \mathbf{C} \underset{3\times 1}{\boldsymbol{\varepsilon}^{\circ}} \, tdA, \tag{8.1.2}$$

$$\delta W = \int_A \underset{1\times 2}{\delta \mathbf{u}^{\mathrm{T}}} \underset{2\times 1}{\mathbf{F}_B} \, tdA + \int_{l_\sigma} \delta \mathbf{u}^{\mathrm{T}} \underset{2\times 1}{\bar{\mathbf{T}}} \, tdl. \tag{8.1.3}$$

For FE approximation, the 2D domain is divided into many elements of finite size such that

$$\delta U = \sum \delta U_e, \tag{8.1.4}$$

where

$$\delta U_e = \int_{A_e} \delta\boldsymbol{\varepsilon}^{\mathrm{T}}_{1\times3} \underset{3\times3}{\mathbf{C}} \underset{3\times1}{\boldsymbol{\varepsilon}}\, tdA - \int_{A_e} \delta\boldsymbol{\varepsilon}^{\mathrm{T}} \mathbf{C}\, \underset{3\times1}{\boldsymbol{\varepsilon}^{\circ}}\, tdA. \tag{8.1.5}$$

Likewise:

$$\delta W = \sum \delta W_e, \tag{8.1.6}$$

where

$$\delta W_e = \int_{A_e} \underset{1\times2}{\delta\mathbf{u}^{\mathrm{T}}} \underset{2\times1}{\mathbf{F}_B}\, tdA + \int_{l_{\sigma e}} \delta\mathbf{u}^{\mathrm{T}}\, \underset{2\times1}{\bar{\mathbf{T}}}\, tdl. \tag{8.1.7}$$

Subscript $e$ stands for an element. We can construct the element stiffness matrix and the element load vector, introducing the mapping and the assumed displacement to Eqs (8.1.5) and (8.1.7).

### 8.1.1  Construction of Element Stiffness Matrix

The first step in the construction of the element stiffness matrix is to express strain components in terms of the element nodal displacement vector. Recall that, in the isoparametric formulation for 2D problems, the mapping for the element geometry and the assumed displacement are expressed symbolically as follows.

**Mapping:**

$$x = \sum_{i=1}^{n} N_i(\xi,\eta)x_i = x(\xi,\eta), \quad y = \sum_{i=1}^{n} N_i(\xi,\eta)y_i = y(\xi,\eta). \tag{8.1.8}$$

**Assumed displacement:**

$$u = \sum_{i=1}^{n} N_i(\xi,\eta)u_i = u(\xi,\eta), \quad v = \sum_{i=1}^{n} N_i(\xi,\eta)v_i = v(\xi,\eta), \tag{8.1.9}$$

where $n$ is the total number of nodes in the element. Recall also that strain is related to displacement as

$$\boldsymbol{\varepsilon} = \begin{Bmatrix} \varepsilon_{xx} \\ \varepsilon_{yy} \\ \varepsilon_{xy} \end{Bmatrix} = \begin{Bmatrix} \dfrac{\partial u}{\partial x} \\[2mm] \dfrac{\partial v}{\partial y} \\[2mm] \dfrac{\partial u}{\partial y} + \dfrac{\partial v}{\partial x} \end{Bmatrix}. \tag{8.1.10}$$

In the above equation, the derivatives are taken with respect to the physical $x, y$-coordinates while assumed displacements are functions of the natural $\xi, \eta$-coordinates. Accordingly, we use the chain rule to find the derivatives with respect to $x$ and $y$ such that

$$\frac{\partial u}{\partial \xi} = \frac{\partial u}{\partial x}\frac{\partial x}{\partial \xi} + \frac{\partial u}{\partial y}\frac{\partial y}{\partial \xi},$$
$$\frac{\partial u}{\partial \eta} = \frac{\partial u}{\partial x}\frac{\partial x}{\partial \eta} + \frac{\partial u}{\partial y}\frac{\partial y}{\partial \eta}, \tag{8.1.11}$$

or in matrix form

$$\left\{\begin{array}{c} \dfrac{\partial u}{\partial \xi} \\ \dfrac{\partial u}{\partial \eta} \end{array}\right\} = \begin{bmatrix} \dfrac{\partial x}{\partial \xi} & \dfrac{\partial y}{\partial \xi} \\ \dfrac{\partial x}{\partial \eta} & \dfrac{\partial y}{\partial \eta} \end{bmatrix} \left\{\begin{array}{c} \dfrac{\partial u}{\partial x} \\ \dfrac{\partial u}{\partial y} \end{array}\right\} = \mathbf{J}_2 \left\{\begin{array}{c} \dfrac{\partial u}{\partial x} \\ \dfrac{\partial u}{\partial y} \end{array}\right\}, \tag{8.1.12}$$

where

$$\mathbf{J}_2 = \begin{bmatrix} \dfrac{\partial x}{\partial \xi} & \dfrac{\partial y}{\partial \xi} \\ \dfrac{\partial x}{\partial \eta} & \dfrac{\partial y}{\partial \eta} \end{bmatrix} : \text{ Jacobian matrix in the 2D domain.} \tag{8.1.13}$$

The Jacobian matrix can be evaluated via the mapping in Eq. (8.1.8). For example, the first term in the matrix is

$$\frac{\partial x}{\partial \xi} = \sum_{i=1}^{n} \frac{\partial N_i}{\partial \xi} x_i. \tag{8.1.14}$$

Similar expressions hold for other terms. Inverting Eq. (8.1.12):

$$\left\{\begin{array}{c} \dfrac{\partial u}{\partial x} \\ \dfrac{\partial u}{\partial y} \end{array}\right\} = \mathbf{J}_2^{-1} \left\{\begin{array}{c} \dfrac{\partial u}{\partial \xi} \\ \dfrac{\partial u}{\partial \eta} \end{array}\right\} = \mathbf{Y} \left\{\begin{array}{c} \dfrac{\partial u}{\partial \xi} \\ \dfrac{\partial u}{\partial \eta} \end{array}\right\} = \begin{bmatrix} Y_{11} & Y_{12} \\ Y_{21} & Y_{22} \end{bmatrix} \left\{\begin{array}{c} \dfrac{\partial u}{\partial \xi} \\ \dfrac{\partial u}{\partial \eta} \end{array}\right\}, \tag{8.1.15}$$

where

$$\mathbf{Y} = \begin{bmatrix} Y_{11} & Y_{12} \\ Y_{21} & Y_{22} \end{bmatrix} = \mathbf{J}_2^{-1}. \tag{8.1.16}$$

Similarly

$$\left\{\begin{array}{c} \dfrac{\partial v}{\partial x} \\ \dfrac{\partial v}{\partial y} \end{array}\right\} = \begin{bmatrix} Y_{11} & Y_{12} \\ Y_{21} & Y_{22} \end{bmatrix} \left\{\begin{array}{c} \dfrac{\partial v}{\partial \xi} \\ \dfrac{\partial v}{\partial \eta} \end{array}\right\}. \tag{8.1.17}$$

Using Eqs (8.1.15) and (8.1.17) in Eq. (8.1.10), the strain–displacement relation can then be expressed as

$$\varepsilon = \left\{ \begin{array}{c} \dfrac{\partial u}{\partial x} \\[2mm] \dfrac{\partial v}{\partial y} \\[2mm] \dfrac{\partial u}{\partial y} + \dfrac{\partial v}{\partial x} \end{array} \right\} = \begin{bmatrix} Y_{11} & Y_{12} & 0 & 0 \\ 0 & 0 & Y_{21} & Y_{22} \\ Y_{21} & Y_{22} & Y_{11} & Y_{12} \end{bmatrix} \left\{ \begin{array}{c} \dfrac{\partial u}{\partial \xi} \\[2mm] \dfrac{\partial u}{\partial \eta} \\[2mm] \dfrac{\partial v}{\partial \xi} \\[2mm] \dfrac{\partial v}{\partial \eta} \end{array} \right\} = \mathbf{D} \left\{ \begin{array}{c} \dfrac{\partial u}{\partial \xi} \\[2mm] \dfrac{\partial u}{\partial \eta} \\[2mm] \dfrac{\partial v}{\partial \xi} \\[2mm] \dfrac{\partial v}{\partial \eta} \end{array} \right\},$$    (8.1.18)

where

$$\mathbf{D} = \begin{bmatrix} Y_{11} & Y_{12} & 0 & 0 \\ 0 & 0 & Y_{21} & Y_{22} \\ Y_{21} & Y_{22} & Y_{11} & Y_{12} \end{bmatrix}.$$    (8.1.19)

For simplicity, we now consider two elements, one with three nodes and another with four nodes. Extensions to higher-order elements are straightforward.

### Three-Node Triangular Element

See Figure 8.1.

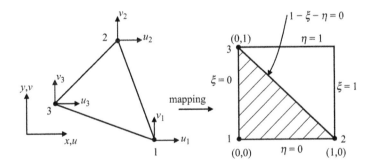

**Figure 8.1** Three-node element in the 2D physical plane and the mapped plane

As shown in Section 7.2.3, the mapping is

$$x = N_1 x_1 + N_2 x_2 + N_3 x_3, \quad y = N_1 y_1 + N_2 y_2 + N_3 y_3$$    (8.1.20)

and the assumed displacement is

$$u = N_1 u_1 + N_2 u_2 + N_3 u_3, \quad v = N_1 v_1 + N_2 v_2 + N_3 v_3,$$    (8.1.21)

where

$$N_1 = 1 - \xi - \eta, \quad N_2 = \xi, \quad N_3 = \eta.$$    (8.1.22)

Then, according to Eq. (8.1.13):

$$\mathbf{J}_2 = \begin{bmatrix} \dfrac{\partial x}{\partial \xi} & \dfrac{\partial y}{\partial \xi} \\ \dfrac{\partial x}{\partial \eta} & \dfrac{\partial y}{\partial \eta} \end{bmatrix} = \begin{bmatrix} x_2 - x_1 & y_2 - y_1 \\ x_3 - x_1 & y_3 - y_1 \end{bmatrix},$$
(8.1.23)

which is constant for the element. Then, according to Eq. (8.1.16):

$$\mathbf{Y} = \mathbf{J}_2^{-1} = \frac{1}{\det \mathbf{J}} \begin{bmatrix} y_3 - y_1 & y_1 - y_2 \\ x_1 - x_3 & x_2 - x_1 \end{bmatrix},$$
(8.1.24)

where

$$\det \mathbf{J}_2 = (x_2 - x_1)(y_3 - y_1) - (x_3 - x_1)(y_2 - y_1).$$
(8.1.25)

Introducing the assumed displacement in Eq. (8.1.21) into Eq. (8.1.18):

$$\varepsilon = \mathbf{D} \left\{ \begin{array}{c} \sum_{i=1}^{3} \dfrac{\partial N_i}{\partial \xi} u_i \\ \sum_{i=1}^{3} \dfrac{\partial N_i}{\partial \eta} u_i \\ \sum_{i=1}^{3} \dfrac{\partial N_i}{\partial \xi} v_i \\ \sum_{i=1}^{3} \dfrac{\partial N_i}{\partial \eta} v_i \end{array} \right\} = \mathbf{D} \left\{ \begin{array}{c} \dfrac{\partial N_1}{\partial \xi} u_1 + \dfrac{\partial N_2}{\partial \xi} u_2 + \dfrac{\partial N_3}{\partial \xi} u_3 \\ \dfrac{\partial N_1}{\partial \eta} u_1 + \dfrac{\partial N_2}{\partial \eta} u_2 + \dfrac{\partial N_3}{\partial \eta} u_3 \\ \dfrac{\partial N_1}{\partial \xi} v_1 + \dfrac{\partial N_2}{\partial \xi} v_2 + \dfrac{\partial N_3}{\partial \xi} v_3 \\ \dfrac{\partial N_1}{\partial \eta} v_1 + \dfrac{\partial N_2}{\partial \eta} v_2 + \dfrac{\partial N_3}{\partial \eta} v_3 \end{array} \right\} = \mathbf{D} \left\{ \begin{array}{c} -u_1 + u_2 \\ -u_1 + u_3 \\ -v_1 + v_2 \\ -v_1 + v_3 \end{array} \right\},$$
(8.1.26)

where $\mathbf{D}$ is determined from Eq. (8.1.19). Placing the nodal DOF into a column vector, the equation above can be expressed as

$$\varepsilon = \mathbf{D} \underbrace{\begin{bmatrix} -1 & 0 & 1 & 0 & 0 & 0 \\ -1 & 0 & 0 & 0 & 1 & 0 \\ 0 & -1 & 0 & 1 & 0 & 0 \\ 0 & -1 & 0 & 0 & 0 & 1 \end{bmatrix}}_{=\mathbf{N}_D:4\times6} \underbrace{\left\{ \begin{array}{c} u_1 \\ v_1 \\ u_2 \\ v_2 \\ u_3 \\ v_3 \end{array} \right\}}_{=\mathbf{d}:6\times1}.$$
(8.1.27)

The above equation can be expressed symbolically as

$$\underset{3\times1}{\varepsilon} = \underset{3\times6}{\mathbf{B}} \underset{6\times1}{\mathbf{d}},$$
(8.1.28)

where

$$\mathbf{B} = \mathbf{D}\mathbf{N}_D$$
(8.1.29)

and

$$\mathbf{d} = \left\{ \begin{matrix} u_1 \\ v_1 \\ u_2 \\ v_2 \\ u_3 \\ v_3 \end{matrix} \right\} : 6 \times 1 \text{ element DOF vector.} \tag{8.1.30}$$

Also, setting $\delta u = \sum N_i \delta u_i$, $\delta v = \sum N_i \delta v_i$:

$$\underset{3 \times 1}{\delta \boldsymbol{\varepsilon}} = \underset{3 \times 6}{\mathbf{B}} \underset{6 \times 1}{\delta \mathbf{d}}, \tag{8.1.31}$$

where

$$\delta \mathbf{d} = \left\{ \begin{matrix} \delta u_1 \\ \delta v_1 \\ \delta u_2 \\ \delta v_2 \\ \delta u_3 \\ \delta v_3 \end{matrix} \right\} : 6 \times 1 \text{ element virtual DOF vector.} \tag{8.1.32}$$

Accordingly, for the first term on the right-hand side of Eq. (8.1.5):

$$\int_{A_e} \delta \boldsymbol{\varepsilon}^{\mathrm{T}} \mathbf{C} \boldsymbol{\varepsilon} t dA = \int_{A_e} \delta \mathbf{d}^{\mathrm{T}} \mathbf{B}^{\mathrm{T}} \mathbf{C} \mathbf{B} \mathbf{d} t dA = \delta \mathbf{d}^{\mathrm{T}} \left( \int_{A_e} \mathbf{B}^{\mathrm{T}} \mathbf{C} \mathbf{B} t dA \right) \mathbf{d} = \delta \mathbf{d}^{\mathrm{T}} \mathbf{k}^e \mathbf{d}, \tag{8.1.33}$$

where

$$\mathbf{k}^e = \int_{A_e} \mathbf{B}^{\mathrm{T}} \mathbf{C} \mathbf{B} t dA : 6 \times 6 \text{ element stiffness matrix.} \tag{8.1.34}$$

For $\mathbf{C}$ and $t$ constant in the element, the integrand in Eq. (8.1.34) is constant and thus

$$\mathbf{k}^e = \mathbf{B}^{\mathrm{T}} \mathbf{C} \mathbf{B} t \int_{A_e} dA = \mathbf{B}^{\mathrm{T}} \mathbf{C} \mathbf{B} t A_e, \tag{8.1.35}$$

where

$$A_e = \int_{A_e} dA : \text{element area.} \tag{8.1.36}$$

To determine the element area, recall from Eqs (7.4.12) and (7.4.13) that

$$dA = J_2 d\xi d\eta. \tag{8.1.37}$$

The area scale $J_2 = \det \mathbf{J}_2$ is determined from Eq. (8.1.25). For the three-node element, $J_2$ is constant. Accordingly:

$$A_e = \int_{A_e} dA = \int J_2 d\xi d\eta = J_2 \int d\xi d\eta = \frac{1}{2} J_2. \tag{8.1.38}$$

## Example 8.1

Consider a three-node plane stress triangular element of uniform thickness with the nodal $(x, y)$ coordinates given as follows:

node 1: $(0, 0)$, node 2: $(a, 0)$, node 3: $(0, a)$.

(a) Determine $\mathbf{J}_2$.
(b) Determine $\mathbf{D}$.
(c) Determine $\mathbf{B}$.
(d) Construct the element stiffness matrix for $v = 1/3$.

### Solution:

(a) Using Eq. (8.1.23):

$$\mathbf{J}_2 = a \begin{bmatrix} 1 & 0 \\ 0 & 1 \end{bmatrix}.$$

(b) According to Eq. (8.1.24):

$$\mathbf{Y} = \mathbf{J}_2^{-1} = \frac{1}{a} \begin{bmatrix} 1 & 0 \\ 0 & 1 \end{bmatrix}.$$

From Eq. (8.1.9):

$$D = \frac{1}{a} \begin{bmatrix} 1 & 0 & 0 & 0 \\ 0 & 0 & 0 & 1 \\ 0 & 1 & 1 & 0 \end{bmatrix}.$$

(c) According to Eq. (8.1.29):

$$\mathbf{B} = \mathbf{DN}_D = \frac{1}{a} \begin{bmatrix} 1 & 0 & 0 & 0 \\ 0 & 0 & 0 & 1 \\ 0 & 1 & 1 & 0 \end{bmatrix} \begin{bmatrix} -1 & 0 & 1 & 0 & 0 & 0 \\ -1 & 0 & 0 & 0 & 1 & 0 \\ 0 & -1 & 0 & 1 & 0 & 0 \\ 0 & -1 & 0 & 0 & 0 & 1 \end{bmatrix},$$

where $\mathbf{N}_D$ is taken from Eq. (8.1.27). Carrying out the multiplication:

$$\mathbf{B} = \frac{1}{a} \begin{bmatrix} -1 & 0 & 1 & 0 & 0 & 0 \\ 0 & -1 & 0 & 0 & 0 & 1 \\ -1 & -1 & 0 & 1 & 1 & 0 \end{bmatrix}.$$

(d) Recall from Section 6.3 of Chapter 6 that for the plane stress case

$$\mathbf{C} = \frac{E}{(1-v^2)} \begin{bmatrix} 1 & v & 0 \\ v & 1 & 0 \\ 0 & 0 & (1-v)/2 \end{bmatrix}.$$

With $v = \frac{1}{3}$ and $A_e = \frac{1}{2}a^2$:

$$\mathbf{k}^e = \mathbf{B}^{\mathrm{T}}\mathbf{C}\mathbf{B}tA_e = \frac{3Et}{16} \begin{bmatrix} 4 & 2 & -3 & -1 & -1 & -1 \\ 2 & 4 & -1 & -1 & -1 & -3 \\ -3 & -1 & 3 & 0 & 0 & 1 \\ -1 & -1 & 0 & 1 & 1 & 0 \\ -1 & -1 & 0 & 1 & 1 & 0 \\ -1 & -3 & 1 & 0 & 0 & 3 \end{bmatrix}.$$

### Four-Node Quadrilateral Element

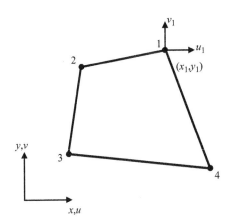

Figure 8.2 Four-node quadrilateral plane element

Consider the four-node quadrilateral element as shown in Figure 8.2. According to Eq. (7.2.6) in the previous chapter:

$$N_1 = \frac{1}{4}(1+\xi)(1+\eta), \quad N_2 = \frac{1}{4}(1-\xi)(1+\eta),$$

$$N_3 = \frac{1}{4}(1-\xi)(1-\eta), \quad N_2 = \frac{1}{4}(1+\xi)(1-\eta).$$

(8.1.39)

According to Eqs (8.1.8) and (8.1.13), the Jacobian matrix is then

$$\mathbf{J}_2 = \begin{bmatrix} \dfrac{\partial x}{\partial \xi} & \dfrac{\partial y}{\partial \xi} \\ \dfrac{\partial x}{\partial \eta} & \dfrac{\partial y}{\partial \eta} \end{bmatrix} = \begin{bmatrix} \displaystyle\sum_{i=1}^{4} \dfrac{\partial N_i}{\partial \xi} x_i & \displaystyle\sum_{i=1}^{4} \dfrac{\partial N_i}{\partial \xi} y_i \\ \displaystyle\sum_{i=1}^{4} \dfrac{\partial N_i}{\partial \eta} x_i & \displaystyle\sum_{i=1}^{4} \dfrac{\partial N_i}{\partial \eta} y_i \end{bmatrix}.$$

(8.1.40)

For example:

$$\frac{\partial N_1}{\partial \xi} = \frac{1}{4}(1+\eta).$$

(8.1.41)

Similar expressions can be written for the other derivatives. The **Y** matrix is then obtained by inverting the $\mathbf{J}_2$ matrix, as shown in Eq. (8.1.16). Introducing the assumed displacement into Eq. (8.1.18):

$$\varepsilon = \mathbf{D}\left\{\begin{array}{c} \displaystyle\sum_{i=1}^{4}\frac{\partial N_i}{\partial \xi}u_i \\[2ex] \displaystyle\sum_{i=1}^{4}\frac{\partial N_i}{\partial \eta}u_i \\[2ex] \displaystyle\sum_{i=1}^{4}\frac{\partial N_i}{\partial \xi}v_i \\[2ex] \displaystyle\sum_{i=1}^{4}\frac{\partial N_i}{\partial \eta}v_i \end{array}\right\} = \mathbf{D}\left\{\begin{array}{c} \dfrac{\partial N_1}{\partial \xi}u_1 + \dfrac{\partial N_2}{\partial \xi}u_2 + \dfrac{\partial N_3}{\partial \xi}u_3 + \dfrac{\partial N_4}{\partial \xi}u_4 \\[2ex] \dfrac{\partial N_1}{\partial \eta}u_1 + \dfrac{\partial N_2}{\partial \eta}u_2 + \dfrac{\partial N_3}{\partial \eta}u_3 + \dfrac{\partial N_4}{\partial \eta}u_4 \\[2ex] \dfrac{\partial N_1}{\partial \xi}v_1 + \dfrac{\partial N_2}{\partial \xi}v_2 + \dfrac{\partial N_3}{\partial \xi}v_3 + \dfrac{\partial N_4}{\partial \xi}v_4 \\[2ex] \dfrac{\partial N_1}{\partial \eta}v_1 + \dfrac{\partial N_2}{\partial \eta}v_2 + \dfrac{\partial N_3}{\partial \eta}v_3 + \dfrac{\partial N_4}{\partial \eta}v_4 \end{array}\right\}, \qquad (8.1.42)$$

in which the $\mathbf{D}$ matrix is determined from Eq. (8.1.19) using the entries in the $\mathbf{Y}$ matrix. The nodal DOF in the above equation can be placed into a column vector such that

$$\varepsilon = \mathbf{D}\underbrace{\left\{\begin{array}{cccccccc} \dfrac{\partial N_1}{\partial \xi} & 0 & \dfrac{\partial N_2}{\partial \xi} & 0 & \dfrac{\partial N_3}{\partial \xi} & 0 & \dfrac{\partial N_4}{\partial \xi} & 0 \\[2ex] \dfrac{\partial N_1}{\partial \eta} & 0 & \dfrac{\partial N_2}{\partial \eta} & 0 & \dfrac{\partial N_3}{\partial \eta} & 0 & \dfrac{\partial N_4}{\partial \eta} & 0 \\[2ex] 0 & \dfrac{\partial N_1}{\partial \xi} & 0 & \dfrac{\partial N_2}{\partial \xi} & 0 & \dfrac{\partial N_3}{\partial \xi} & 0 & \dfrac{\partial N_4}{\partial \xi} \\[2ex] 0 & \dfrac{\partial N_1}{\partial \eta} & 0 & \dfrac{\partial N_2}{\partial \eta} & 0 & \dfrac{\partial N_3}{\partial \eta} & 0 & \dfrac{\partial N_4}{\partial \eta} \end{array}\right\}}_{=\mathbf{N}_D:4\times 8}\underbrace{\left\{\begin{array}{c} u_1 \\ v_1 \\ u_2 \\ v_2 \\ u_3 \\ v_3 \\ u_4 \\ v_4 \end{array}\right\}}_{=\mathbf{d}:8\times 1}. \qquad (8.1.43)$$

The above equation can be expressed symbolically as

$$\underset{3\times 1}{\varepsilon} = \underset{3\times 8}{\mathbf{B}}\ \underset{8\times 1}{\mathbf{d}}, \qquad (8.1.44)$$

$$\text{where } \mathbf{B} = \mathbf{D}\mathbf{N}_D \qquad (8.1.45)$$

and

$$\mathbf{d} = \left\{\begin{array}{c} u_1 \\ v_1 \\ u_2 \\ v_2 \\ u_3 \\ v_3 \\ u_4 \\ v_4 \end{array}\right\} : 8 \times 1 \text{ element DOF vector.} \qquad (8.1.46)$$

Also, setting

$$\delta u = \sum N_i \delta u_i, \quad \delta v = \sum N_i \delta v_i \qquad (8.1.47)$$

results in the following equation:

$$\underset{3\times1}{\delta\boldsymbol{\varepsilon}} = \underset{3\times8}{\mathbf{B}}\ \underset{8\times1}{\delta\mathbf{d}} \rightarrow \delta\boldsymbol{\varepsilon}^{\mathsf{T}} = \delta\mathbf{d}^{\mathsf{T}}\mathbf{B}^{\mathsf{T}}, \tag{8.1.48}$$

where

$$\delta\mathbf{d}^{\mathsf{T}} = \lfloor\, \delta u_1 \quad \delta v_1 \quad \delta u_2 \quad \delta v_2 \quad \delta u_3 \quad \delta v_3 \quad \delta u_4 \quad \delta v_4 \,\rfloor \tag{8.1.49}$$

is the $1 \times 8$ element virtual DOF vector. Accordingly:

$$\int_{A_e} \delta\boldsymbol{\varepsilon}^{\mathsf{T}}\mathbf{C}\boldsymbol{\varepsilon}t dA = \int_{A_e} \delta\mathbf{d}^{\mathsf{T}}\mathbf{B}^{\mathsf{T}}\mathbf{C}\mathbf{B}\mathbf{d}t dA = \delta\mathbf{d}^{\mathsf{T}}\left( \int_{A_e} \mathbf{B}^{\mathsf{T}}\mathbf{C}\mathbf{B}t dA \right)\mathbf{d} = \delta\mathbf{d}^{\mathsf{T}}\mathbf{k}^e\mathbf{d}, \tag{8.1.50}$$

where

$$\mathbf{k}^e = \int_{A_e} \mathbf{B}^{\mathsf{T}}\mathbf{C}\mathbf{B}t dA : 8 \times 8 \text{ element stiffness matrix.} \tag{8.1.51}$$

The element stiffness matrix is symmetric. One can use Eq. (7.5.18) to transform an integration over the physical element to one in the mapped plane. For an element of square and rectangular geometries, the integration in the above equation can be performed analytically to generate the element stiffness matrix. For an element of arbitrary geometry, the element stiffness matrix can be constructed using the quadrature rules in the previous chapter as follows:

$$\mathbf{k}^e = \int_{-1}^{1}\int_{-1}^{1} \mathbf{B}^{\mathsf{T}}\mathbf{C}\mathbf{B}t \det \mathbf{J}_2 d\xi d\eta = \sum_{i=1}^{}\sum_{j=1}^{} \left(\mathbf{B}^{\mathsf{T}}\mathbf{C}\mathbf{B}t \det \mathbf{J}_2\right)_{\substack{\xi = a_i \\ \eta = a_j}} W_i W_j. \tag{8.1.52}$$

It will be shown later in Section 8.3 that the $2 \times 2$ point rule is adequate for the present four-node element.

## Example 8.2

Consider the four-node, 8-DOF square plane stress element shown in Figure 8.3.

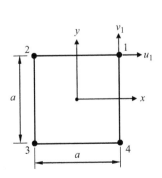

(a) Determine the Jacobian matrix and its inverse.
(b) Determine the **B** matrix.
(c) Determine the element stiffness matrix for $\nu = 1/3$.

**Figure 8.3** Four-node square element

## Solution:

(a) Substituting the following nodal coordinates:

$$x_1 = x_4 = \frac{a}{2}, x_2 = x_3 = -\frac{a}{2}, y_1 = y_2 = \frac{a}{2}, y_3 = y_4 = -\frac{a}{2}$$

into the mapping functions:

$$x = \sum_{i=1}^{4} N_i x_i = \frac{a}{2}\xi, \quad y = \sum_{i=1}^{4} N_i y_i = \frac{a}{2}\eta.$$

According to Eq. (8.1.40):

$$\mathbf{J}_2 = \frac{a}{2}\begin{bmatrix} 1 & 0 \\ 0 & 1 \end{bmatrix}, \mathbf{Y} = \mathbf{J}_2^{-1} = \frac{2}{a}\begin{bmatrix} 1 & 0 \\ 0 & 1 \end{bmatrix}.$$

(b) According to Eqs (8.1.19) and (8.1.43):

$$\mathbf{D} = \frac{2}{a}\begin{bmatrix} 1 & 0 & 0 & 0 \\ 0 & 0 & 0 & 1 \\ 0 & 1 & 1 & 0 \end{bmatrix},$$

$$\mathbf{N}_D = \frac{1}{4}\begin{bmatrix} 1+\eta & 0 & -(1+\eta) & 0 & -(1-\eta) & 0 & (1-\eta) & 0 \\ 1+\xi & 0 & 1-\xi & 0 & -(1-\xi) & 0 & -(1+\xi) & 0 \\ 0 & 1+\eta & 0 & -(1+\eta) & 0 & -(1-\eta) & 0 & 1-\eta \\ 0 & 1+\xi & 0 & 1-\xi & 0 & -(1-\xi) & 0 & -(1+\xi) \end{bmatrix}.$$

Then

$$\mathbf{B} = \mathbf{D}\mathbf{N}_D$$

$$= \frac{1}{2a}\begin{bmatrix} 1+\eta & 0 & -(1+\eta) & 0 & -(1-\eta) & 0 & 1-\eta & 0 \\ 0 & 1+\xi & 0 & 1-\xi & 0 & -(1-\xi) & 0 & -(1+\xi) \\ 1+\xi & 1+\eta & 1-\xi & -(1+\eta) & -(1-\xi) & -(1-\eta) & -(1+\xi) & 1-\eta \end{bmatrix}.$$

(c)

$$\mathbf{k}^e = \int_{-1}^{1}\int_{-1}^{1} \mathbf{B}^{\mathrm{T}}\mathbf{C}\mathbf{B}t(\det \mathbf{J}_2)d\xi d\eta = \frac{Et}{16}\begin{bmatrix} 8 & 3 & -5 & 0 & -4 & -3 & 1 & 0 \\ 3 & 8 & 0 & 1 & -3 & -4 & 0 & -5 \\ -5 & 0 & 8 & -3 & 1 & 0 & -4 & 3 \\ 0 & 1 & -3 & 8 & 0 & -5 & 3 & -4 \\ -4 & -3 & 1 & 0 & 8 & 3 & -5 & 0 \\ -3 & -4 & 0 & -5 & 3 & 8 & 0 & 1 \\ 1 & 0 & -4 & 3 & -5 & 0 & 8 & -3 \\ 0 & -5 & 3 & -4 & 0 & 1 & -3 & 8 \end{bmatrix}.$$

## 8.1.2 Construction of Element Load Vectors

The effect of temperature change in the body, applied body forces, and surface forces contribute to the element load vector. As an illustration, we will consider the four-node element. Extension to other types of elements is straightforward.

### Temperature Change

Introducing Eq. (8.1.48) for the four-node element into the second term on the right-hand side of Eq. (8.1.5):

$$\int_{A_e} \delta\boldsymbol{\varepsilon}^\mathsf{T} \mathbf{C}\boldsymbol{\varepsilon}^o \, t dA = \int_{A_e} \delta\mathbf{d}^\mathsf{T} \mathbf{B}^\mathsf{T} \mathbf{C}\boldsymbol{\varepsilon}^o \, t dA = \delta\mathbf{d}^\mathsf{T} \int_{A_e} \mathbf{B}^\mathsf{T} \mathbf{C}\boldsymbol{\varepsilon}^o \, t dA = \delta\mathbf{d}^\mathsf{T} \mathbf{Q}_T^e, \tag{8.1.53}$$

where

$$\mathbf{Q}_T^e = \int_{A_e} \mathbf{B}^\mathsf{T} \mathbf{C}\boldsymbol{\varepsilon}^o \, t dA : 8 \times 1 \text{ element load vector due to temperature change.} \tag{8.1.54}$$

Once again, the load vector in Eq. (8.1.54) can be constructed by numerical integration in the mapped plane. For the plane stress state case of isotropic materials:

$$\boldsymbol{\varepsilon}^o = \left\{ \begin{array}{c} \varepsilon_{xx}^o \\ \varepsilon_{yy}^o \\ \varepsilon_{xy}^o \end{array} \right\} = \left\{ \begin{array}{c} \alpha\Delta T \\ \alpha\Delta T \\ 0 \end{array} \right\}. \tag{8.1.55}$$

If $\Delta T$ is not constant over the element, it may be interpolated from the nodal values and the shape functions as

$$\Delta T = \sum N_i (\Delta T)_i, \tag{8.1.56}$$

where $(\Delta T)_i$: $\Delta T$ at node $i$.

### Body Force Term

From Eq. (8.1.7):

$$\int_{A_e} \delta\mathbf{u}^\mathsf{T} \mathbf{F}_B \, t dA = \int_{A_e} \lfloor \delta u \quad \delta v \rfloor \left\{ \begin{array}{c} X_B \\ Y_B \end{array} \right\} t dA = \int_{A_e} \delta u X_B \, t dA + \int_{A_e} \delta v Y_B \, t dA. \tag{8.1.57}$$

Introducing Eq. (8.1.47) into the right-hand side of the above equation:

$$\int_{A_e} \delta u X_B t dA = \int_{A_e} (N_1 \delta u_1 + N_2 \delta u_2 + N_3 \delta u_3 + N_4 \delta u_4) X_B t dA$$

$$= \delta u_1 \left( \int_{A_e} N_1 X_B t dA \right) + \cdots + \delta u_4 \left( \int_{A_e} N_4 X_B t dA \right) \tag{8.1.58}$$

and

$$\int_{A_e} \delta v Y_B t dA = \delta v_1 \left( \int_{A_e} N_1 Y_B t dA \right) + \cdots + \delta v_4 \left( \int_{A_e} N_4 Y_B t dA \right). \tag{8.1.59}$$

The above two equations can be combined as

$$\int_{A_e} \delta \mathbf{u}^{\mathsf{T}} \mathbf{F}_B t dA = [\delta u_1 \quad \delta v_1 \quad \cdots \quad \delta u_4 \quad \delta v_4] \begin{Bmatrix} \int_{A_e} N_1 X_B t dA \\ \int_{A_e} N_1 Y_B t dA \\ \vdots \\ \int_{A_e} N_4 X_B t dA \\ \int_{A_e} N_4 Y_B t dA \end{Bmatrix}, \tag{8.1.60}$$

where the column vector in the above equation is the $8 \times 1$ element load vector due to body force. If $X_B$ and $Y_B$ are not constant, they may be interpolated using the shape functions in terms of the nodal values as

$$X_B = \sum_{i=1}^{4} N_i (X_B)_i, \quad Y_B = \sum_{i=1}^{4} N_i (Y_B)_i. \tag{8.1.61}$$

For constant body force of $X_B = C_B$ acting on a square element as shown in Example 8.2, we can show that

$$\int_{A_e} N_1 X_B t dA = \int_{A_e} N_2 X_B t dA = \int_{A_e} N_3 X_B t dA = \int_{A_e} N_4 X_B t dA = \frac{C_B t a^2}{4}. \tag{8.1.62}$$

**Example 8.3**

The nodal coordinates of a four-node plane stress element are given in centimeters as node 1: (2, 2), node 2: (1, 2), node 3: (1, 1), node 4: (2, 1). Material properties are $E = 72\,\text{GPa}$, $\nu = 1/3$, $\alpha = 24 \times 10^{-6}/°\text{C}$.

(a) Construct the element stiffness matrix for thickness $t = 0.02$ m.
(b) Determine the element load vector due to a uniform temperature change of $\Delta T = C_T$.
(c) Determine the element load vector due to a uniform body force of $X_B = C_B$.

**Surface Force Term**

To account for the force applied over the 2D domain boundary, consider an element with one of its sides along the domain boundary. For a curved domain boundary, the four-node element represents an approximation of the actual boundary. From Eq. (8.1.7):

$$\int_{l_{\sigma e}} \delta\mathbf{u}^T\bar{\mathbf{T}}t\,dl = \int_{l_{\sigma e}} \lfloor \delta u \quad \delta v \rfloor \left\{ \begin{array}{c} \bar{T}_x \\ \bar{T}_y \end{array} \right\} t\,dl = \int_{l_{\sigma e}} \delta u \bar{T}_x t\,dl + \int_{l_{\sigma e}} \delta v \bar{T}_y t\,dl. \tag{8.1.63}$$

For an element with surface forces applied over the boundary connecting nodes 1 and 4, as shown in Figure 8.4:

$$\int_{l_4}^{l_1} \delta u \bar{T}_x\, t\,dl = \int \left[ \frac{1}{2}(1+\eta)\delta u_1 + \frac{1}{2}(1-\eta)\delta u_4 \right] \bar{T}_x t\,dl$$

$$= \delta u_1 \underbrace{\left[ \frac{1}{2}\int (1+\eta)\bar{T}_x t\,dl \right]}_{=A} + \delta u_4 \underbrace{\left[ \frac{1}{2}\int (1-\eta)\bar{T}_x t\,dl \right]}_{=B} \tag{8.1.64}$$

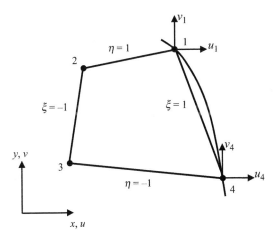

**Figure 8.4** Four-node element with one side along the domain boundary

and

$$\int_{l_4}^{l_1} \delta v \bar{T}_y \, tdl = \int \left[ \frac{1}{2}(1+\eta)\delta v_1 + \frac{1}{2}(1-\eta)\delta v_4 \right] \bar{T}_y tdl$$

$$= \delta v_1 \underbrace{\left[ \frac{1}{2} \int (1+\eta) \bar{T}_y tdl \right]}_{=C} + \delta v_4 \underbrace{\left[ \frac{1}{2} \int (1-\eta) \bar{T}_y tdl \right]}_{=D}. \qquad (8.1.65)$$

Combining Eqs (8.1.64) and (8.1.65) and expressing the result in matrix form:

$$\int \underset{1\times 8}{\delta \mathbf{u}^{\mathrm{T}}} \underset{8\times 1}{\bar{\mathbf{T}}} \, tdl = \begin{bmatrix} \delta u_1 & \delta v_1 & \cdots & \delta u_4 & \delta v_4 \end{bmatrix} \begin{Bmatrix} A \\ C \\ 0 \\ 0 \\ 0 \\ 0 \\ B \\ D \end{Bmatrix}. \qquad (8.1.66)$$

The column vector in the above equation is the $8 \times 1$ element load vector due to surface tractions. As shown in Eq. (7.4.6), line integrals in Eqs (8.1.64) and (8.1.65) can be transformed into those over the mapped line. If $\bar{T}_x$ and $\bar{T}_y$ are not constant along the side 1–4, they may be interpolated using the shape functions as

$$\bar{T}_x = \frac{1}{2}(1+\eta)(\bar{T}_x)_1 + \frac{1}{2}(1-\eta)(\bar{T}_x)_4, \; \bar{T}_y = \frac{1}{2}(1+\eta)(\bar{T}_y)_1 + \frac{1}{2}(1-\eta)(\bar{T}_y)_4. \qquad (8.1.67)$$

Consider the four-node element shown in Example 8.2. For a constant traction of $\bar{T}_x = c$ acting on the surface at $x = a/2$, it can be shown that

$$A = \frac{1}{2}\int (1+\eta)\bar{T}_x \, tdl = \frac{1}{2}\int (1+\eta)ct\frac{a}{2}d\eta = \frac{cat}{4}\int_{-1}^{1}(1+\eta)d\eta = \frac{cat}{2},$$

$$B = \frac{1}{2}\int (1-\eta)\bar{T}_x tdl = \frac{1}{2}\int (1-\eta)ct\frac{a}{2}d\eta = \frac{cat}{4}\int_{-1}^{1}(1-\eta)d\eta = \frac{cat}{2}. \qquad (8.1.68)$$

We note that half of the total distributed load is assigned to each node.

## 8.1.3  Assembly of Global Stiffness Matrix and Global Load Vector

Following the procedure introduced in Chapter 1, element stiffness matrices and load vectors are assembled into the global stiffness matrix and the global load vector, using the connectivity relating the element DOF and the global DOF. For the entire structure:

$$\delta U - \delta W = \sum (\delta U_e - \delta W_e) = \sum \left( \delta \mathbf{d}^T \mathbf{k}^e \mathbf{d} - \delta \mathbf{d}^T \mathbf{f}^e \right)$$

$$= \delta \mathbf{q}^T (\mathbf{Kq} - \mathbf{F}) = 0 \rightarrow \mathbf{Kq} - \mathbf{F} = 0, \tag{8.1.69}$$

in which $\mathbf{K}^e$ is the element stiffness matrix and $\mathbf{f}^e$ is the element load vector due to temperature change, body force, and surface tractions, while $\mathbf{K}$ is the global stiffness matrix, $\mathbf{q}$ is the global DOF vector, and $\mathbf{F}$ is the global load vector. Finally

$$\mathbf{Kq} = \mathbf{F} \tag{8.1.70}$$

is the discretized equilibrium equation. As shown in Chapter 1 and the following chapters, we can apply geometric boundary conditions to construct a reduced version of Eq. (8.1.70), which can then be solved to determine the unknown global DOF vector.

## 8.1.4 Stress Calculation for Each Element

After solving for $\mathbf{q}$, we can determine the stress $\boldsymbol{\sigma}$ in each element from Eqs (6.3.10) and (8.1.44) such that

$$\boldsymbol{\sigma} = \mathbf{C}(\boldsymbol{\varepsilon} - \boldsymbol{\varepsilon}^o) = \mathbf{C}(\mathbf{Bd} - \boldsymbol{\varepsilon}^o), \tag{8.1.71}$$

where $\mathbf{B} = \mathbf{B}(\xi, \eta)$ is determined at a given point in the $\xi, \eta$ domain. In the FE formulation, the displacement is the primary variable and, as shown in Eq. (8.1.18), it is necessary to take derivatives of the assumed displacement to determine strain and thus stress values in each element. Accordingly, in general the stress solutions are not as accurate as the displacement solutions. Starting with the stress values obtained by Eq. (8.1.71), various postprocessing procedures, called the stress recovery methods, have been developed to improve the stress solutions at the nodal points. Some of these methods exploit the existence of the superconvergent stress points within an element, at which highly accurate stress values are obtained. For the four-node plane element, the centroid ($\xi = \eta = 0$) is known as the superconvergent point for stress evaluation. For the eight-node plane element, the $2 \times 2$ sampling points for numerical integration are the superconvergent stress points.

## ► Example 8.4

Consider the $2 \times 2$ mesh of the plane stress elements shown in Figure 8.5. The coordinates of global nodes are given (in meters) as follows:

node 1: (0, 1), node 2: (0, 0.75), node 3: (0, 0),
node 4: (0.7, 1), node 5: (0.6, 0.5), node 6: (0.45, 0),
node 7: (1.2, 1), node 8: (1.2, 0.25), node 9: (1.2, 0).

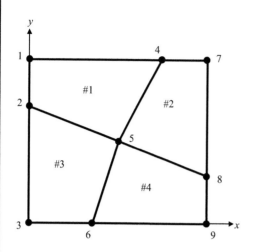

**Figure 8.5** A plane stress problem modeled with a 2 × 2 finite element mesh

The geometric boundary conditions are $u = 0$ along the surface at $x = 0$ and $v = 0$ along the surface at $y = 0$. Tensile stress $\sigma_{xx} = \sigma_0$ (Pa) is applied over the surface at $x = 1.2$ m. The Poisson ratio of the material is $1/3$. The element stiffness matrix of each element is generated using the $2 \times 2$ point integration rule.

(a) Determine the nodal displacements.
(b) Determine the stress at the element centroid of each element.

For this simple problem we can compare the FE solution with the exact displacements and stresses, which are

$$u = \frac{\sigma_0}{E} x, \quad v = -\frac{\sigma_0}{3E} y, \quad \sigma_{xx} = \sigma_0, \quad \sigma_{yy} = 0, \quad \sigma_{xy} = 0.$$

## Solution:

(a) For the FE model, the global stiffness matrix is initially an $18 \times 18$ matrix and, using Eq. (8.1.64), the global load due to the applied traction is

$$F_{13} = 0.375\sigma_0 t, \quad F_{15} = 0.5\sigma_0 t, \quad F_{17} = 0.125\sigma_0 t.$$

Accordingly:

$$F_1 = R_1, F_2 = 0, F_3 = R_3, F_4 = 0, F_5 = R_5, F_6 = R_6,$$
$$F_7 = F_8 = F_9 = F_{10} = F_{11} = 0, F_{12} = R_{12}, F_{13} = 0.375\sigma_0 t, F_{14} = 0,$$
$$F_{15} = 0.5\sigma_0 t, F_{16} = 0, F_{17} = 0.125\sigma_0 t, F_{18} = R_{18},$$

where $R$ stands for reaction force. The geometric boundary conditions are $q_1 = q_3 = q_5 = q_6 = q_{12} = q_{18} = 0$. We can show that the FE solution for displacement is identical to the exact solution everywhere, as well as at the nodal points.

(b) The stress values within each element are calculated using Eq. (8.1.71) with zero thermally induced strain. It turns out that $\sigma_{xx} = \sigma_0, \sigma_{yy} = 0, \sigma_{xy} = 0$ everywhere in the element, which is identical to the exact solution.

## 8.2    Finite Element Modeling of 3D Solids and Structures

Recall from Chapter 6 that for a 3D solid or structural element:

$$\delta U_e = \int_{V_e} \delta\boldsymbol{\varepsilon}^T \boldsymbol{\sigma} dV = \int_{V_e} \delta\boldsymbol{\varepsilon}^T \mathbf{C}(\boldsymbol{\varepsilon} - \boldsymbol{\varepsilon}^o) dV$$

$$= \int_{V_e} \delta\boldsymbol{\varepsilon}^T \mathbf{C}\boldsymbol{\varepsilon} dV - \int_{V_e} \delta\boldsymbol{\varepsilon}^T \mathbf{C}\boldsymbol{\varepsilon}^o dV \qquad (8.2.1)$$

and

$$\delta W_e = \int_{V_e} \delta\mathbf{u}^T \mathbf{F}_B dV + \int_{S_{\sigma e}} \delta\mathbf{u}^T \bar{\mathbf{T}} dS. \qquad (8.2.2)$$

### 8.2.1    3D Element Stiffness Matrix

Recall from Appendix 1 that the strain–displacement relation for 3D solids is

$$\varepsilon_{xx} = \frac{\partial u}{\partial x}, \varepsilon_{yy} = \frac{\partial v}{\partial y}, \varepsilon_{zz} = \frac{\partial w}{\partial z},$$

$$\varepsilon_{xy} = \frac{\partial u}{\partial y} + \frac{\partial v}{\partial x}, \varepsilon_{yz} = \frac{\partial v}{\partial z} + \frac{\partial w}{\partial y}, \varepsilon_{xz} = \frac{\partial u}{\partial z} + \frac{\partial w}{\partial x}. \qquad (8.2.3)$$

Assumed displacements are expressed as functions of the mapped coordinates. Accordingly, for the derivatives of assumed displacements with respect to the physical Cartesian coordinate system, we invoke the chain rule for differentiation. Namely

$$\begin{Bmatrix} \dfrac{\partial u}{\partial \xi} \\[2mm] \dfrac{\partial u}{\partial \eta} \\[2mm] \dfrac{\partial u}{\partial \zeta} \end{Bmatrix} = \begin{bmatrix} \dfrac{\partial x}{\partial \xi} & \dfrac{\partial y}{\partial \xi} & \dfrac{\partial z}{\partial \xi} \\[2mm] \dfrac{\partial x}{\partial \eta} & \dfrac{\partial y}{\partial \eta} & \dfrac{\partial z}{\partial \eta} \\[2mm] \dfrac{\partial x}{\partial \zeta} & \dfrac{\partial y}{\partial \zeta} & \dfrac{\partial z}{\partial \zeta} \end{bmatrix} \begin{Bmatrix} \dfrac{\partial u}{\partial x} \\[2mm] \dfrac{\partial u}{\partial y} \\[2mm] \dfrac{\partial u}{\partial z} \end{Bmatrix} = \mathbf{J}_3 \begin{Bmatrix} \dfrac{\partial u}{\partial x} \\[2mm] \dfrac{\partial u}{\partial y} \\[2mm] \dfrac{\partial u}{\partial z} \end{Bmatrix}, \qquad (8.2.4)$$

where

$$\mathbf{J}_3 = \begin{bmatrix} \dfrac{\partial x}{\partial \xi} & \dfrac{\partial y}{\partial \xi} & \dfrac{\partial z}{\partial \xi} \\[2mm] \dfrac{\partial x}{\partial \eta} & \dfrac{\partial y}{\partial \eta} & \dfrac{\partial z}{\partial \eta} \\[2mm] \dfrac{\partial x}{\partial \zeta} & \dfrac{\partial y}{\partial \zeta} & \dfrac{\partial z}{\partial \zeta} \end{bmatrix} : \text{Jacobian matrix in the 3D space.} \qquad (8.2.5)$$

Inverting Eq. (8.2.4):

$$
\left\{
\begin{array}{c}
\dfrac{\partial u}{\partial x} \\[2mm]
\dfrac{\partial u}{\partial y} \\[2mm]
\dfrac{\partial u}{\partial z}
\end{array}
\right\}
= \mathbf{J}_3^{-1}
\left\{
\begin{array}{c}
\dfrac{\partial u}{\partial \xi} \\[2mm]
\dfrac{\partial u}{\partial \eta} \\[2mm]
\dfrac{\partial u}{\partial \zeta}
\end{array}
\right\}
= \mathbf{Y}
\left\{
\begin{array}{c}
\dfrac{\partial u}{\partial \xi} \\[2mm]
\dfrac{\partial u}{\partial \eta} \\[2mm]
\dfrac{\partial u}{\partial \zeta}
\end{array}
\right\}
=
\begin{bmatrix}
Y_{11} & Y_{12} & Y_{13} \\
Y_{21} & Y_{22} & Y_{23} \\
Y_{31} & Y_{32} & Y_{33}
\end{bmatrix}
\left\{
\begin{array}{c}
\dfrac{\partial u}{\partial \xi} \\[2mm]
\dfrac{\partial u}{\partial \eta} \\[2mm]
\dfrac{\partial u}{\partial \zeta}
\end{array}
\right\},
\tag{8.2.6}
$$

where $\mathbf{Y} = \mathbf{J}_3^{-1}$. Similarly

$$
\left\{
\begin{array}{c}
\dfrac{\partial v}{\partial x} \\[2mm]
\dfrac{\partial v}{\partial y} \\[2mm]
\dfrac{\partial v}{\partial z}
\end{array}
\right\}
=
\begin{bmatrix}
Y_{11} & Y_{12} & Y_{13} \\
Y_{21} & Y_{22} & Y_{23} \\
Y_{31} & Y_{32} & Y_{33}
\end{bmatrix}
\left\{
\begin{array}{c}
\dfrac{\partial v}{\partial \xi} \\[2mm]
\dfrac{\partial v}{\partial \eta} \\[2mm]
\dfrac{\partial v}{\partial \zeta}
\end{array}
\right\},
\quad
\left\{
\begin{array}{c}
\dfrac{\partial w}{\partial x} \\[2mm]
\dfrac{\partial w}{\partial y} \\[2mm]
\dfrac{\partial w}{\partial z}
\end{array}
\right\}
=
\begin{bmatrix}
Y_{11} & Y_{12} & Y_{13} \\
Y_{21} & Y_{22} & Y_{23} \\
Y_{31} & Y_{32} & Y_{33}
\end{bmatrix}
\left\{
\begin{array}{c}
\dfrac{\partial w}{\partial \xi} \\[2mm]
\dfrac{\partial w}{\partial \eta} \\[2mm]
\dfrac{\partial w}{\partial \zeta}
\end{array}
\right\}.
\tag{8.2.7}
$$

Then

$$
\varepsilon_{xx} = \frac{\partial u}{\partial x} = \begin{bmatrix} Y_{11} & Y_{12} & Y_{13} \end{bmatrix}
\left\{
\begin{array}{c}
\dfrac{\partial u}{\partial \xi} \\[2mm]
\dfrac{\partial u}{\partial \eta} \\[2mm]
\dfrac{\partial u}{\partial \zeta}
\end{array}
\right\}.
\tag{8.2.8}
$$

For the eight-node element:

$$
\varepsilon_{xx} = \begin{bmatrix} Y_{11} & Y_{12} & Y_{13} \end{bmatrix}
\left\{
\begin{array}{c}
\displaystyle\sum_{i=1}^{8} \dfrac{\partial N_i}{\partial \xi} u_i \\[4mm]
\displaystyle\sum_{i=1}^{8} \dfrac{\partial N_i}{\partial \eta} u_i \\[4mm]
\displaystyle\sum_{i=1}^{8} \dfrac{\partial N_i}{\partial \zeta} u_i
\end{array}
\right\}
\tag{8.2.9}
$$

or

$$
\varepsilon_{xx} = \begin{bmatrix} Y_{11} & Y_{12} & Y_{13} \end{bmatrix}
\underbrace{
\begin{bmatrix}
\dfrac{\partial N_1}{\partial \xi} & 0 & 0 & \dfrac{\partial N_2}{\partial \xi} & 0 & 0 & \cdots & \dfrac{\partial N_8}{\partial \xi} & 0 & 0 \\[3mm]
\dfrac{\partial N_1}{\partial \eta} & 0 & 0 & \dfrac{\partial N_2}{\partial \eta} & 0 & 0 & \cdots & \dfrac{\partial N_8}{\partial \eta} & 0 & 0 \\[3mm]
\dfrac{\partial N_1}{\partial \zeta} & 0 & 0 & \dfrac{\partial N_2}{\partial \zeta} & 0 & 0 & \cdots & \dfrac{\partial N_8}{\partial \zeta} & 0 & 0
\end{bmatrix}
}_{3 \times 24}
\underbrace{
\left\{
\begin{array}{c}
u_1 \\ v_1 \\ w_1 \\ u_2 \\ v_2 \\ w_2 \\ \vdots \\ u_8 \\ v_8 \\ w_8
\end{array}
\right\}
}_{\substack{\mathbf{d}: \\ 24 \times 1}}.
\tag{8.2.10}
$$

Written symbolically:

$$\varepsilon_{xx} = \mathbf{B}_{xx}\mathbf{d}, \tag{8.2.11}$$

where $\mathbf{d}$ is a $24 \times 1$ column vector of nodal DOF for the eight-node element. Similar expressions hold for other strain components. Then, the strain vector can be expressed symbolically as

$$\underset{6\times 1}{\boldsymbol{\varepsilon}} = \begin{Bmatrix} \varepsilon_{xx} \\ \varepsilon_{yy} \\ \varepsilon_{zz} \\ \varepsilon_{xy} \\ \varepsilon_{yz} \\ \varepsilon_{zx} \end{Bmatrix} = \underbrace{\begin{Bmatrix} \mathbf{B}_{xx} \\ \mathbf{B}_{yy} \\ \mathbf{B}_{zz} \\ \mathbf{B}_{xy} \\ \mathbf{B}_{yz} \\ \mathbf{B}_{zx} \end{Bmatrix}}_{\substack{=\,\mathbf{B}\,: \\ 6\,\times\,24}} \underset{24\times 1}{\mathbf{d}} = \mathbf{B}\mathbf{d} \tag{8.2.12}$$

and

$$\delta\boldsymbol{\varepsilon} = \mathbf{B}\delta\mathbf{d}. \tag{8.2.13}$$

Introducing Eqs (8.2.12) and (8.2.13) into the first term in Eq. (8.2.1):

$$\int_{V_e} \delta\boldsymbol{\varepsilon}^{\mathrm{T}}\mathbf{C}\boldsymbol{\varepsilon}dV = \int_{V_e} \delta\mathbf{d}^{\mathrm{T}}\mathbf{B}^{\mathrm{T}}\mathbf{C}\mathbf{B}\mathbf{d}dV = \delta\mathbf{d}^{\mathrm{T}}\int_{V_e} \mathbf{B}^{\mathrm{T}}\mathbf{C}\mathbf{B}dV\mathbf{d} = \delta\mathbf{d}^{\mathrm{T}}\mathbf{k}^e\mathbf{d}, \tag{8.2.14}$$

where

$$\mathbf{k}^e = \int_{V_e} \mathbf{B}^{\mathrm{T}}\mathbf{C}\mathbf{B}dV \tag{8.2.15}$$

is the $24 \times 24$ element stiffness matrix for the eight-node element. We can use Eq. (7.5.19) to transform the integration over the element volume in the physical 3D space into that in the mapped 3D space and then apply the numerical integration rule. For the eight-node 3D element:

$$\mathbf{k}^e = \int_{-1}^{1}\int_{-1}^{1}\int_{-1}^{1} \mathbf{B}^{\mathrm{T}}\mathbf{C}\mathbf{B}J_3 d\xi d\eta d\zeta = \sum_{i=1}\sum_{j=1}\sum_{k=1} \left[\mathbf{B}^{\mathrm{T}}\mathbf{C}\mathbf{B}J_3\right]_{\substack{\xi = a_i \\ \eta = a_j \\ \zeta = a_k}} W_i W_j W_k. \tag{8.2.16}$$

## 8.2.2  3D Element Load Vectors

### Element Load Vector Due to Temperature Change

Introducing Eq. (8.2.13) into the second term on the right-hand side of Eq. (8.2.1):

$$\int_{V_e} \delta\boldsymbol{\varepsilon}^{\mathrm{T}}\mathbf{C}\boldsymbol{\varepsilon}^o dV = \delta\mathbf{d}^{\mathrm{T}}\int_{V_e} \mathbf{B}^{\mathrm{T}}\mathbf{C}\boldsymbol{\varepsilon}^o dV = \delta\mathbf{d}^{\mathrm{T}}\mathbf{Q}_T^e, \tag{8.2.17}$$

where

$$\mathbf{Q}_T^e = \int_{V_e} \mathbf{B}^T \mathbf{C}\boldsymbol{\varepsilon}^o dV. \tag{8.2.18}$$

For the eight-node element, $\mathbf{Q}_T^e$ is the 24 × 1 element load vector due to temperature change.

**Element Load Vector Due to Body Force**

Introducing the virtual displacement field into the first term on the right-hand side of Eq. (8.2.2):

$$\int_{V_e} \delta\mathbf{u}^T \mathbf{F}_B dV = \delta\mathbf{d}^T \int_{V_e} \mathbf{N}^T \mathbf{F}_B dV = \delta\mathbf{d}^T \mathbf{Q}_B^e, \tag{8.2.19}$$

where

$$\mathbf{Q}_B^e = \int_{V_e} \mathbf{N}^T \mathbf{F}_B dV. \tag{8.2.20}$$

For the eight-node element, $\mathbf{Q}_B^e$ is the 24 × 1 element load vector due to body force.

**Element Load Vector Due to Surface Force**

Introducing the virtual displacement field into the second term on the right-hand side of Eq. (8.2.2):

$$\int_{S_{\sigma e}} \delta\mathbf{u}^T \bar{\mathbf{T}} dS = \delta\mathbf{d}^T \int_{S_{\sigma e}} \mathbf{N}^T \bar{\mathbf{T}} dS = \delta\mathbf{d}^T \mathbf{Q}_S^e, \tag{8.2.21}$$

where

$$\mathbf{Q}_S^e = \int_{V_S} \mathbf{N}^T \bar{\mathbf{T}} dS. \tag{8.2.22}$$

For the eight-node element, $\mathbf{Q}_S^e$ is the 24 × 1 element load vector due to applied surface tractions. Integration over an element surface in the physical 3D space can be performed in the mapped surface following the procedures shown in Section 7.4.4 of Chapter 7.

## 8.3    Number of Sampling Points for Numerical Integration

Numerical integration is used for the construction of element stiffness matrices and load vectors. To minimize computing effort, it is desirable to use as small a number of sampling or integration points as possible. We will discuss how a proper number of sampling points can be determined first using the four-node element in the 2D domain as an example. Extension to 3D elements is straightforward.

**Four-Node Plane Element**

For simplicity, consider a four-node element of rectangular geometry as shown in Figure 8.6.

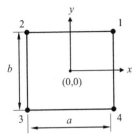

**Figure 8.6** Four-node element of rectangular geometry in the 2D domain

For this element, the mapping is simplified to

$$x = \frac{a}{2}\xi, \quad y = \frac{b}{2}\eta \tag{8.3.1}$$

and the Jacobian matrix and its determinant are constant. For the four-node element, the assumed displacement can be expressed as

$$u = a_1 + a_2\xi + a_3\eta + a_4\xi\eta,$$
$$v = b_1 + b_2\xi + b_3\eta + b_4\xi\eta. \tag{8.3.2}$$

Accordingly, strain components

$$\varepsilon_{xx} = \frac{\partial u}{\partial x}, \quad \varepsilon_{yy} = \frac{\partial v}{\partial y}, \quad \varepsilon_{xy} = \frac{\partial u}{\partial y} + \frac{\partial v}{\partial x} \tag{8.3.3}$$

are linear in $\xi$ and $\eta$. Since $\varepsilon = \mathbf{B}\mathbf{d}$, $\mathbf{B}$ is linear in $\xi$ and $\eta$. We may then determine the number of sampling points by considering the element stiffness matrix, expressed as

$$\mathbf{k}^e = \int_{-1}^{1}\int_{-1}^{1} \mathbf{B}^{\mathsf{T}}\mathbf{C}\mathbf{B} \det \mathbf{J} t d\xi d\eta. \tag{8.3.4}$$

For an element of rectangular shape and constant thickness, the integrand in the above equation is at most quadratic in $\xi$ and $\eta$. Accordingly, two sampling points in each direction is adequate for exact integration.

A simpler approach is to note that strain components are linear in $\xi$ and $\eta$. Accordingly, two points in the $\xi$ direction and two points in the $\eta$ direction are adequate to fit linear strain distributions. For non-rectangular elements, $\mathbf{J}$ is not constant. However, we may still use the $2 \times 2$ point rule.

**Eight-Node Plane Element**

For the eight-node element, strain is quadratic and thus the $3 \times 3$ point rule is needed to exactly integrate the element stiffness matrix when the element is rectangular. However, the

$2 \times 2$ point integration is not only acceptable but also preferable in certain cases. This will be explained later in Chapter 10.

### Nine-Node Plane Element

For the nine-node element, strain is quadratic and thus the $3 \times 3$ point rule is needed to exactly integrate the element stiffness matrix when the element is rectangular.

## 8.4  Dynamic Problems

Following the procedure described in Chapter 4, the element mass matrix can be constructed by treating the inertia force as a body force. For a solid in equilibrium:

$$\int_V \delta\varepsilon^T \sigma dV - \int_V \delta u^T F_B dV - \int_{S_\sigma} \delta u^T \bar{T} dS = 0. \tag{8.4.1}$$

Treating inertia force as a body force:

$$F_B dV = -(\rho dV)\ddot{u}, \tag{8.4.2}$$

where $\rho$ is the mass density per volume. Introducing the above equation into the second term in Eq. (8.4.1):

$$-\int_V \delta u^T F_B dV = \int_V \delta u^T \ddot{u} \rho dV. \tag{8.4.3}$$

### Element Mass Matrix

For FE modeling, the assumed displacement vector can be expressed as

$$u = Nd. \tag{8.4.4}$$

For example, for the three-node element in the 2D domain as shown in Figure 8.1:

$$u = \begin{Bmatrix} u \\ v \end{Bmatrix} = \begin{bmatrix} N_1 & 0 & N_2 & 0 & N_3 & 0 \\ 0 & N_1 & 0 & N_2 & 0 & N_3 \end{bmatrix} \begin{Bmatrix} u_1 \\ v_1 \\ u_2 \\ v_2 \\ u_3 \\ v_3 \end{Bmatrix} = Nd, \tag{8.4.5}$$

where

$$N = \begin{bmatrix} N_1 & 0 & N_2 & 0 & N_3 & 0 & N_4 & 0 \\ 0 & N_1 & 0 & N_2 & 0 & N_3 & 0 & N_4 \end{bmatrix}. \tag{8.4.6}$$

Then

$$\delta\mathbf{u} = \mathbf{N}\delta\mathbf{d}, \quad \ddot{\mathbf{u}} = \mathbf{N}\ddot{\mathbf{d}}. \tag{8.4.7}$$

Introducing the above equation into Eq. (8.4.3) corresponding to an element:

$$-\int_{V_e} \delta\mathbf{u}^T \mathbf{F}_B dV = \int_{V_e} \delta\mathbf{u}^T \ddot{\mathbf{u}}\rho dV = \int_{V} \delta\mathbf{d}^T \mathbf{N}^T \mathbf{N}\ddot{\mathbf{d}}\rho dV = \delta\mathbf{d}^T \left( \int_{V_e} \rho\mathbf{N}^T \mathbf{N}dV \right)\ddot{\mathbf{d}} \tag{8.4.8}$$

$$= \delta\mathbf{d}^T \mathbf{m}^e \ddot{\mathbf{d}},$$

where

$$\mathbf{m}^e = \int_{V_e} \rho\mathbf{N}^T \mathbf{N}dV \tag{8.4.9}$$

is the element mass matrix which can be evaluated using numerical integration.

## Example 8.5

For the three-node plane element shown in Example 8.1, construct the element mass matrix.

## Solution:

$$\mathbf{m}^e = \int_{V_e} \rho\mathbf{N}^T \mathbf{N}dV = \int_{A_e} \rho\mathbf{N}^T \mathbf{N}t dA = \rho t \int_{A_e} \mathbf{N}^T \mathbf{N}dA = \rho t \int \mathbf{N}^T \mathbf{N}J_2 d\xi d\eta$$

$$= \rho t J_2 \int \mathbf{N}^T \mathbf{N}d\xi d\eta.$$

Substituting Eq. (8.4.6) into the above equation and carrying out the integration, it can be shown that

$$\mathbf{m}^e = \frac{\rho A_e t}{12}\begin{bmatrix} 2 & 0 & 1 & 0 & 1 & 0 \\ 0 & 2 & 0 & 1 & 0 & 1 \\ 1 & 0 & 2 & 0 & 1 & 0 \\ 0 & 1 & 0 & 2 & 0 & 1 \\ 1 & 0 & 1 & 0 & 2 & 0 \\ 0 & 1 & 0 & 1 & 0 & 2 \end{bmatrix}, A_e = \frac{1}{2}a^2.$$

### Equation of Motion

For the entire body:

$$-\int_{V_e} \delta\mathbf{u}^T \mathbf{F}_B dV = \int_{V_e} \delta\mathbf{u}^T \rho\ddot{\mathbf{u}}dV = \delta\mathbf{q}^T \mathbf{M}\ddot{\mathbf{q}}, \tag{8.4.10}$$

where $\mathbf{M}$ is the global mass matrix and $\ddot{\mathbf{q}}$ is the global acceleration vector.

The global mass matrix is assembled from the element mass matrices using the identical procedure for the assembly of the global stiffness matrix such that

$$\delta \mathbf{q}^T (\mathbf{Kq} + \mathbf{M\ddot{q}} - \mathbf{F}) = \mathbf{0} \rightarrow \mathbf{Kq} + \mathbf{M\ddot{q}} - \mathbf{F} = \mathbf{0} \qquad (8.4.11)$$

or

$$\mathbf{M\ddot{q}} + \mathbf{Kq} = \mathbf{F}, \qquad (8.4.12)$$

which is the equation of motion. Initial conditions at time $t = 0$ are prescribed as

$$\mathbf{q}(0) = \mathbf{q}_0, \quad \dot{\mathbf{q}}(0) = \dot{\mathbf{q}}_0. \qquad (8.4.13)$$

For dynamic response under applied loads, the equation of motion can be solved using various numerical integration methods as described in Chapter 4. In most practical structural dynamics problems, only a few lowest modes are of interest. Accordingly, there is no need to find all natural frequencies and natural modes in free vibration analysis, even if the FE model has a large number of DOF. Eigenvalue analysis methods, such as the subspace iteration or the Lanczos method, provide efficient means to compute a few lowest natural frequencies and modes as needed.

## Diagonal Mass Matrix

Numerical integration rules as discussed in Chapter 7 can be used to construct an element matrix, called a "consistent" mass matrix. Alternatively, one may use numerical integration rules in which sampling points coincide with the nodes of an element. This will result in a diagonal or lumped element mass matrix. A diagonal mass matrix can be useful when a method such as the CDS is used for numerical integration in time of an equation of motion under the assumption of negligible damping.

As an example, consider the two-node uniaxial element with mapping and assumed displacement as follows:

$$x = N_1 x_1 + N_2 x_2, \quad u = N_1 u_1 + N_2 u_2, \qquad (8.4.14)$$

where

$$N_1(\xi) = \frac{1}{2}(1 - \xi), \quad N_2(\xi) = \frac{1}{2}(1 + \xi), \qquad (8.4.15)$$

$$\mathbf{m}^e = \frac{1}{2} l \int_{\xi=-1}^{\xi=1} \begin{bmatrix} N_1^2 & N_1 N_2 \\ N_2 N_1 & N_2^2 \end{bmatrix} m \, d\xi, \qquad (8.4.16)$$

where $m$ is the mass per length. Assuming $m$ is uniform over the element, the exact integration results in

$$\mathbf{m}^e = \frac{ml}{6}\begin{bmatrix} 2 & 1 \\ 1 & 2 \end{bmatrix}, \tag{8.4.17}$$

which is the consistent mass matrix. However, consider the following two-point integration rule with the sampling points located at the nodal points:

$$\int_{-1}^{1} f(\xi)d\xi = [f(-1) + f(1)] \cdot 1. \tag{8.4.18}$$

The weight is equal to 1. Applying the above rule to Eq. (8.4.16) results in the lumped mass matrix as follows:

$$\mathbf{m}^e = \frac{ml}{2}\begin{bmatrix} 1 & 0 \\ 0 & 0 \end{bmatrix} + \frac{ml}{2}\begin{bmatrix} 0 & 0 \\ 0 & 1 \end{bmatrix} = \frac{ml}{2}\begin{bmatrix} 1 & 0 \\ 0 & 1 \end{bmatrix}. \tag{8.4.19}$$

The mass matrix is now a diagonal matrix with half of the total element mass shared equally per node. Extension to the four-node plane element and the eight-node hexahedron element is straightforward.

### Kinetic Energy

Kinetic energy of solids is expressed as

$$T = \int_V \left( \frac{1}{2}dm\dot{\mathbf{u}} \cdot \dot{\mathbf{u}} \right) = \frac{1}{2}\int_V (\dot{\mathbf{u}} \cdot \dot{\mathbf{u}}\rho dV). \tag{8.4.20}$$

For an element:

$$\dot{\mathbf{u}} = \mathbf{N}\dot{\mathbf{d}} \tag{8.4.21}$$

and

$$T_e = \frac{1}{2}\int_{V_e} (\dot{\mathbf{u}}^T\dot{\mathbf{u}}\rho dV) = \frac{1}{2}\dot{\mathbf{d}}^T\int_{V_e} (\mathbf{N}^T\mathbf{N}\rho dV)\dot{\mathbf{d}} = \frac{1}{2}\dot{\mathbf{d}}^T\mathbf{m}^e\dot{\mathbf{d}}, \tag{8.4.22}$$

where the element mass matrix $\mathbf{m}^e$ is identical to that in Eq. (8.4.9). For the entire body:

$$T = \sum_e T_e = \frac{1}{2}\dot{\mathbf{q}}^T\mathbf{M}\dot{\mathbf{q}}, \tag{8.4.23}$$

where the global mass matrix is assembled from the element mass matrices.

## 8.5 Solution of Finite Element Equations

Consider a system of linear simultaneous equations expressed in matrix form as

$$\mathbf{Kq} = \mathbf{F}, \tag{8.5.1}$$

where $\mathbf{K}$ is symmetric and positive definite. The above equation can be solved via the Gaussian elimination method, which involves triangularization processes. Alternatively, one may attempt to solve the above equation via an iterative method such as the conjugate gradient method. A description of the Gaussian elimination method with triple factorization is given in Appendix 2.

## Dynamic Problems

For structural dynamic problems in which a numerical time-marching method such as the trapezoidal rule or the CDS is used, the resulting equation is a system of linear simultaneous equations as shown in Section 4.3 of Chapter 4. Accordingly, these equations can be solved using the same solver as used for the static problems.

## Sparse Matrix

In the FE method, an element shares nodes only with adjacent elements with common boundaries. Accordingly, $K_{mn} = 0$ if DOF $m$ and DOF $n$ do not belong to the same element and thus the global stiffness matrix is sparse. A good solution scheme must account for the sparseness of the global stiffness matrix. In Appendix 2 we introduce the skyline method, which is one of the simplest ways of storing sparse matrices. The same observation is made on the global mass matrix, which is non-diagonal.

## Mesh Generation

Construction of an FE model for a body of given geometry involves dividing the domain into many elements of identical or different types. The union of these individual elements constitutes a finite element mesh. Generation of a finite mesh can be tedious and time-consuming for 3D bodies of complicated shape, often encountered in practical applications. Accordingly, automation of mesh generation has been an integral part of any commercial software for FE analysis. However, an in-depth discussion of this subject – including error estimates of FE solutions – is beyond the scope of this book.

## Eigenvalue Analysis

Mathematically, static buckling analysis and free vibration analysis are exercises in eigenvalue analysis. For static buckling analysis, only the smallest eigenvalue is adequate to determine the buckling load and buckling mode. In structural vibration, only the few modes with lower frequencies are excited. Accordingly, it is not necessary to determine all of the natural frequencies and modes of an FE model. The subspace iteration method and the Lanczos method provide computationally efficient approaches to determine only the lowest frequencies and modes as needed for FE models with very large DOF.

## PROBLEMS

**8.1** For the eight-node square element of side length $a$ as shown in Figure 8.7, determine the element load vector due to constant applied traction $\bar{T}_x = c$ as shown.

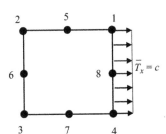

**Figure 8.7** For Problem 8.1

**8.2** Consider the four-node square element of side length $a$ as shown in Figure 8.8. The element is subjected to an applied traction vector $\bar{T}_x$ which is linearly distributed along the 1–4 side of the element.
The applied traction can be expressed in terms of the nodal values as

$$\bar{T}_x = \frac{1}{2}(1 + \eta)(\bar{T}_x)_1 + \frac{1}{2}(1 - \eta)(\bar{T}_x)_4,$$

where the nodal values are given. Determine the element load vector corresponding to the applied traction.

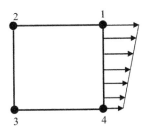

**Figure 8.8** For Problem 8.2

**8.3** Consider the six-node triangular element with straight sides as shown in Figure 8.9. For midside node 2, $x_4 = (x_1 + x_3)/2, y_4 = (y_1 + y_3)/2$. Similar relations hold for nodes 5 and 6.

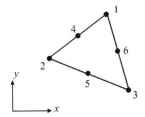

**Figure 8.9** For Problem 8.3

(a) Is the Jacobian matrix constant?

(b) How does strain in the element vary with $\xi$ and $\eta$? Linear or quadratic?

(c) How does strain in the element vary with $x$ and $y$? Linear or quadratic?

(d) Determine the minimum number of sampling points needed to exactly integrate the element stiffness matrix.

**8.4** Figure 8.10 shows the five-node element in the mapped plane.

(a) Construct the shape functions. You may start from the nine-node element and apply constraints.

(b) Is the Jacobian matrix constant for an element of rectangular geometry?

(c) How does the strain vary in $\xi$ and $\eta$ for an element of rectangular geometry?

(d) Determine the minimum number of sampling points needed to exactly integrate the stiffness matrix of an element with rectangular shape.

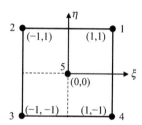

**Figure 8.10** For Problem 8.4

**8.5** Consider the same geometry, material, and FE model as described in Example 8.4. The bottom surface at $y = 0$ is fixed, while other surfaces are subject to shear stress of $\sigma_{xy} = \sigma_{yx} = \tau_0$, where $\tau_0$ (Pa) is a given value.

(a) Determine the unknown nodal displacements.

(b) Determine the stress at the element centroid as well as the integration points of each element.

**8.6** Consider a plane stress problem with the same geometry and FE model as described in Example 8.4 and shown in Figure 8.5. The geometric boundary conditions are such that the displacement in the $y$-direction is zero along the $y = 0$ side and the displacement in the $x$-direction is zero along the $x = 0$ and $x = 1.2$ sides. The material properties are $E = 72$ GPa, $v = 1/3$, and $\alpha = 24 \times 10^{-6}/°C$. The body, initially stress-free, is subjected to a temperature change of uniform $\Delta T = 100°C$.

(a) Determine the unknown nodal displacements.

(b) Determine the stress at the element centroid as well as at the integration points of each element.

**8.7** The nodal coordinates of a four-node plane stress element are given in meters as node 1: (2, 2), node 2: (1, 2), node 3: (1, 1), node 4: (2, 1). Material properties are $E = 72$ GPa, $v = 1/3$, mass density $\rho = 2,780$ kg/m$^3$.

(a) Use the $2 \times 2$ point integration to construct the element mass matrix for thickness $t = 0.02$ m.

(b) Construct the diagonal mass matrix using the nodal points as the integration points.

*Note:* Do the following problems using any software available to you.

**8.8** Consider a cantilevered solid body described as follows:

Dimensions: length 100 cm,
width 1 cm, height 1 cm.
Material: $E = 200$ GPa and $v = 0.3$.
   The solid body is fixed at the $x = 0$ surface and a bending load of 1 N is applied at the tip ($x = 100$ cm).
(a) Model the body with solid elements as follows:

   Element type: 20-node 3D brick elements.
   Mesh size: $10 \times 2 \times 2$ (i.e. 10 elements in the length direction, 2 elements in the thickness direction, 2 elements in the width direction).
   Compare the tip displacement with the solution obtained by the Bernoulli–Euler (B–E) beam bending theory to appreciate the validity of the B–E theory. Report the displacement ratio $R$ defined as

$$R = \frac{\text{max tip displacement\_3D finite element analysis}}{\text{max tip displacement\_B} - \text{E beam bending theory}}.$$

(b) Repeat with a length of 25 cm.
(c) Repeat with a length of 10 cm.
(d) Repeat with a length of 5 cm.

**8.9** Consider a cantilevered solid body described as follows:

Dimensions: length 50 cm,
width 1 cm, height 1 cm.
Material: $E = 200$ GPa and $v = 0.3$.
The solid body is fixed at the $x = 0$ surface.

Model the body with solid elements as follows:

Element type: 20-node brick elements.
Mesh size: $10 \times 2 \times 2$ (i.e. 10 elements in the length direction, 2 elements in the thickness direction, 2 elements in the width direction).

$q_1$: transverse displacement at the midpoint ($x = 25$ cm).
$q_2$: transverse displacement at the tip ($x = 50$ cm).

(a) Determine $q_1$ and $q_2$ for force $F_1 = 1$ N applied at the midpoint.
(b) Determine $q_1$ and $q_2$ for force $F_2 = 1$ N applied at the tip.
(c) Determine the terms in the **C** matrix as shown below:

$$\left\{ \begin{array}{c} q_1 \\ q_2 \end{array} \right\} = \left[ \begin{array}{cc} C_{11} & C_{12} \\ C_{21} & C_{22} \end{array} \right] \left\{ \begin{array}{c} F_1 \\ F_2 \end{array} \right\}.$$

(d) Determine the stiffness matrix for the 2-DOF model of the body such that

$$\begin{Bmatrix} F_1 \\ F_2 \end{Bmatrix} = \begin{bmatrix} K_{11} & K_{12} \\ K_{21} & K_{22} \end{bmatrix} \begin{Bmatrix} q_1 \\ q_2 \end{Bmatrix}.$$

**8.10** Repeat Problem 8.9 for a tapered body of identical mass and length. The body is tapered such that the height of the body at the fixed end is equal to 1.4 cm, while the width is identical to that in Problem 8.9.

**8.11** Consider a cylindrical shell with the following dimensions and material properties:
Length: $L = 20$ mm.
Radius to the midwall: $a = 10.5$ mm.
Wall thickness: $t = 0.4$ mm.
$E = 200$ GPa, $v = 0.3$, mass density $\rho = 8,000$ kg/m$^3$.

The cylindrical shell is fixed at one end and free at the other end. Use a model with the 20-node solid elements.
(a) Determine the natural frequencies of the first eight modes, counting double frequencies separately.
(b) Determine the natural modes of the first eight modes. Sketch the modes.

**8.12** Consider a hemispherical shell with a circular cutout at the crown. The cutout area is attached to a rigid rod of circular cross-section. The rod itself is fixed to a base. Shell dimensions and material properties are given as follows:
Radius to the shell midwall: $a = 10$ mm.
Radius of the rigid rod: $b = 2$ mm.
Wall thickness: $t = 0.125$ mm.
$E = 200$ GPa, $v = 0.3$, mass density $\rho = 8,000$ kg/m$^3$.

Use a model with the 20-node solid elements to do the following:
(a) Determine the natural frequencies of the first eight modes, counting double frequencies separately.
(b) Determine the natural modes of the first eight modes. Sketch the modes.

**8.13** Repeat Problem 8.12 with $t = 0.25$ mm.

**8.14** For the solid body described in Problem 8.9, determine the first 10 natural frequencies and identify the bending and torsional modes. Compare with the frequencies obtained by the beam bending theory and the torsion theory.

**8.15** Consider a solid body identical in geometry and material to that described in Problem 8.9. The body is fixed at $x = 0$ and subjected to a force couple at $x = L$ corresponding to a torque of 1 N-m.
(a) Determine the twist angle of the cross-section and the derivative of twist angle along the body length.
(b) According to the elementary theory of torsion:

$$T = GJ \frac{\partial \phi}{\partial x},$$

where $T$ is the torque acting on the cross-section, $G$ is the shear modulus, $J$ is the torsion constant of the cross-section, and $\phi$ is the twist angle of the cross-section. Determine the torsion constant using the results from part (a).

(c) Compare the constant obtained in part (b) with that from St. Venant theory, given as $J = 0.1406a^4$, where $a$ is the side length of the square cross-section.

**8.16** A square composite panel of a $[0/90/90/0]$ lay-up is clamped at all four edges and subjected to a uniformly distributed unit pressure load. For the panel shown in Figure 8.11, the side length is $L = 1.3$ m and the thickness is 1.3 cm in the $z$ direction. The material properties of a ply are given as follows:

$$E_1 = 163 \text{ GPa}, E_2 = 10.5 \text{ GPa}, G_{12} = G_{13} = 5.93 \text{ GPa},$$
$$G_{23} = 5.15 \text{ GPa}, v_{12} = v_{13} = 0.36, v_{23} = 0.3,$$

where 1, 2, 3 are the fiber direction, the in-plane direction normal to the fiber, and the out-of-plane direction normal to the fiber. The ply angles are defined with respect to the $x$-axis.

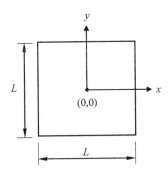

**Figure 8.11** For Problem 8.16

The panel is modeled with the 20-node brick elements. Due to symmetry in the geometry, applied load, and geometric constraints, only one-quarter of the panel in the $xy$-plane is modeled with a uniform mesh of $15 \times 15$ elements. For the modeling in the thickness direction, each ply is modeled with three elements of equal thickness.

(a) Determine the transverse displacement along the $x = 0$ line.

(b) Determine the in-plane stresses through the thickness at the center of the modeled quadrant.

(c) Determine the transverse stresses through the thickness at the center of the modeled quadrant.

# 9 Plates and Shells

This chapter deals with the FE formulation for thin plate and shell structures. All structures, including thin plates and shells, are geometrically 3D bodies. Accordingly, they can be modeled using 3D solid elements. However, for plates and shells, certain assumptions on the kinematics of deformation have been introduced to simplify the mathematical description of their behavior under applied loads. These assumptions provide the foundation of classical theories for plates and shells.

As a prelude to the formulation of classical plate bending theories, we will first look at the assumptions on the kinematics of slender bodies undergoing bending deformation in which the cross-section normal to the body axis is assumed to be a rigid plane that translates and rotates into a corresponding plane in the deformed configuration. This assumption is based on the recognition that contribution to strain energy is mostly from axial or longitudinal strain and thus contributions from other strain components can be neglected. Accordingly, within the context of classical beam bending theories, the material is treated as anisotropic even if the actual material is isotropic – such as aluminum or steel.

In the classical plate bending theory, it is assumed that a line normal to the plate midsurface is a rigid line that translates and rotates without change in length as the plate undergoes bending deformation. Accordingly, the contribution of thickness change on strain energy is assumed negligible. Under this assumption on the kinematics of plate bending deformation, material is being treated as anisotropic even if actual material is isotropic.

Subsequently, it will be shown that the assumption on the kinematics of deformation for plate bending can be extended to the FE formulation of curved shell structures, in accordance with the concept of isoparametric formulation. This approach will result in shell elements with 5 DOF (three displacements and two rotational angles) per node.

For 3D solid elements that can be used for plates and shell analysis, we will first look at solid elements with three nodes through the thickness. We will then show how solid elements with two nodes through the thickness can be constructed for analysis of plate and shell structures.

## 9.1 Slender Body in Bending

For slender bodies in bending, we will first examine the assumptions on the kinematics of deformation. This will be followed by a discussion on the strain–stress relation and the strain–displacement relation.

## Assumptions on the Kinematics of Deformation

Figure 9.1 shows a portion of a slender body in the undeformed configuration and the same portion in the deformed configuration in 3D space.

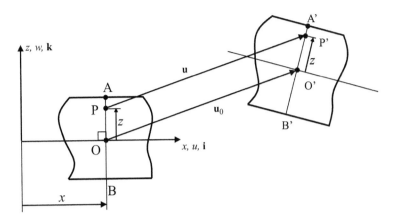

**Figure 9.1** Cross-section of a slender body (or line vector normal to the plate midsurface) in the undeformed configuration and deformed configuration

Note that line AB is a projection of a flat cross-sectional surface normal to the body axis (i.e. the $x$-axis). According to the beam bending theory, the flat cross-section that contains line AB is assumed rigid and translates and rotates as a rigid plane to the surface that contains line A′B′. We will now examine where the assumptions regarding the kinematics of deformation lead.

## Stress–Strain Relationship

Assuming no temperature change, for 3D solids

$$\varepsilon = \hat{\mathbf{C}}\sigma \tag{9.1.1}$$

or

$$\begin{Bmatrix} \varepsilon_{xx} \\ \varepsilon_{yy} \\ \varepsilon_{zz} \\ \varepsilon_{yz} \\ \varepsilon_{zx} \\ \varepsilon_{xy} \end{Bmatrix} = \begin{bmatrix} \hat{C}_{11} & \hat{C}_{12} & \hat{C}_{13} & \hat{C}_{14} & \hat{C}_{15} & \hat{C}_{16} \\ \hat{C}_{21} & \hat{C}_{22} & \hat{C}_{23} & \hat{C}_{24} & \hat{C}_{25} & \hat{C}_{26} \\ \hat{C}_{31} & \hat{C}_{32} & \hat{C}_{33} & \hat{C}_{34} & \hat{C}_{35} & \hat{C}_{36} \\ \hat{C}_{41} & \hat{C}_{42} & \hat{C}_{43} & \hat{C}_{44} & \hat{C}_{45} & \hat{C}_{46} \\ \hat{C}_{51} & \hat{C}_{52} & \hat{C}_{53} & \hat{C}_{54} & \hat{C}_{55} & \hat{C}_{56} \\ \hat{C}_{61} & \hat{C}_{62} & \hat{C}_{63} & \hat{C}_{64} & \hat{C}_{65} & \hat{C}_{66} \end{bmatrix} \begin{Bmatrix} \sigma_{xx} \\ \sigma_{yy} \\ \sigma_{zz} \\ \sigma_{yz} \\ \sigma_{zx} \\ \sigma_{xy} \end{Bmatrix}, \tag{9.1.2}$$

where $\hat{\mathbf{C}}$ is a symmetric matrix of elastic compliance coefficients. Under the assumption of the cross-section that translates and rotates as a rigid plane:

$$\varepsilon_{yy} = \varepsilon_{zz} = \varepsilon_{yz} = 0. \tag{9.1.3}$$

Since $\varepsilon_{yy} = 0$ for any combination of stress states, all entries in the second row of $\hat{C}$ must be equal to zero. Then, all entries in the second column of $\hat{C}$ are equal to zero due to the symmetry. Similarly, since $\varepsilon_{zz} = \varepsilon_{yz} = 0$ for any combinations of stress states, all entries in the third and fourth rows of $\hat{C}$ are equal to zero, and then all entries in the third and fourth columns of $\hat{C}$ are equal to zero due to the symmetry.

For a slender body in bending, the cross-sections normal to the body axis in the undeformed configuration remain almost normal to the body axis in the deformed configuration. Accordingly, the contribution of transverse shear strains $\varepsilon_{zx}$ and $\varepsilon_{xy}$ to strain energy is very small compared with that of strain $\varepsilon_{xx}$. Accordingly, in the B–E beam bending theory, it is further assumed that $\varepsilon_{zx} = \varepsilon_{xy} = 0$, while the Timoshenko beam bending theory does not adopt this assumption. For the B–E theory, all entries in the fifth and sixth rows of $\hat{C}$ are now equal to zero, and then all entries in the fifth and sixth columns of $\hat{C}$ are equal to zero due to the symmetry.

From Eq. (A1.4.12) for isotropic materials such as aluminum and steel, the strain–stress relation for the Timoshenko theory is

$$\varepsilon_{xx} = \frac{1}{E}\sigma_{xx}, \quad \varepsilon_{zx} = \frac{1}{G}\sigma_{zx}, \quad \varepsilon_{xy} = \frac{1}{G}\sigma_{xy}, \tag{9.1.4}$$

while for the B–E theory,

$$\varepsilon_{xx} = \frac{1}{E}\sigma_{xx}. \tag{9.1.5}$$

Note that Poisson's ratio $\nu$ is equal to zero and shear modulus $G \to \infty$ in the B–E beam bending theory.

### Kinematic Variables and Strain–Displacement Relation

The displacement vector $\mathbf{u}$ of a generic point $P(x, y, z)$ on the cross-section in the undeformed configuration can be expressed as

$$\mathbf{u} = \mathbf{u}_0 + z\mathbf{k}' - z\mathbf{k}. \tag{9.1.6}$$

As shown in Figure 9.1, $\mathbf{u}_0$ represents the displacement vector due to the translation of the cross-section, $z\mathbf{k}$ is the vector drawn from point O to point P, and $z\mathbf{k}'$ is the vector drawn from point O′ to point P′. Since the cross-section is assumed rigid, line vector $z\mathbf{k}$ translates and rotates into line vector $z\mathbf{k}'$ in the deformed configuration with no change in length.

### Bending in the 2D Plane

If we further assume that bending occurs only in the $xz$-plane, rotation of line vector $z\mathbf{k}$ to $z\mathbf{k}'$ can be expressed in terms of angle $\theta_y$, introduced as shown in Figure 9.2.

In the figure, angle $\theta_y$ is defined positive clockwise following the right-hand rule. Under the assumption of the rigid cross-section:

$$\begin{aligned} z\mathbf{k}' &= z(\sin\theta_y\mathbf{i} + \cos\theta_y\mathbf{k}), \\ z\mathbf{k}' - z\mathbf{k} &= z\sin\theta_y\mathbf{i} + z(\cos\theta_y - 1)\mathbf{k}. \end{aligned} \tag{9.1.7}$$

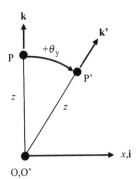

**Figure 9.2** Rotational angle of cross-section around the $y$-axis

The rotation angle is small in many practical situations. Then

$$\sin \theta_y \cong \theta_y, \quad \cos \theta_y \cong 1, \quad z\mathbf{k}' - z\mathbf{k} \simeq z\theta_y\mathbf{i}. \tag{9.1.8}$$

The displacement vector $\mathbf{u}$ can now be written as

$$\mathbf{u} = \mathbf{u}_0 + z\theta_y\mathbf{i}. \tag{9.1.9}$$

Displacement vectors $\mathbf{u}$ and $\mathbf{u}_0$ can be expressed in component form as

$$\mathbf{u} = u\mathbf{i} + w\mathbf{k}, \quad \mathbf{u}_0 = u_0\mathbf{i} + w_0\mathbf{k}. \tag{9.1.10}$$

Comparing the components:

$$\begin{aligned} u(x, z) &= u_0(x) + z\theta_y(x), \\ w(x, z) &= w_0(x). \end{aligned} \tag{9.1.11}$$

So, the displacement of a generic point P on a cross-section can be expressed in terms of translation and rotation of the cross-section as a rigid surface.

### Bending in 3D Space

The cross-section translates and rotates as a rigid plane in 3D space. Accordingly, the displacement vector $\mathbf{u}$ of point $P(x, y, z)$ can now be expressed as

$$\mathbf{u} = \mathbf{u}_0 + \left(z\theta_y - y\theta_z\right)\mathbf{i}, \tag{9.1.12}$$

where $\theta_z$ is the rotation of the cross-section around the $z$-axis. Displacement vectors $\mathbf{u}$ and $\mathbf{u}_0$ can be expressed in component form as

$$\mathbf{u} = u\mathbf{i} + v\mathbf{j} + w\mathbf{k}, \quad \mathbf{u}_0 = u_0\mathbf{i} + v_0\mathbf{j} + w_0\mathbf{k}. \tag{9.1.13}$$

Comparing the components:

$$\begin{aligned} u(x, z) &= u_0(x) + z\theta_y(x) - y\theta_z(x), \\ v(x, z) &= v_0(x), \\ w(x, z) &= w_0(x). \end{aligned} \tag{9.1.14}$$

**Strain–Displacement Relation**

Under the assumption of rigid cross-sections, $\varepsilon_{yy} = \varepsilon_{zz} = \varepsilon_{yz} = 0$. Introducing Eq. (9.1.14) to the strain–displacement relation in Appendix 1, the remaining strain components for the Timoshenko theory can be expressed in terms of the two translations and one rotation:

$$\varepsilon_{xx} = \frac{\partial u}{\partial x} = \frac{\partial u_0}{\partial x} + z\frac{\partial \theta_y}{\partial x} - y\frac{\partial \theta_z}{\partial x},$$

$$\varepsilon_{xz} = \frac{\partial u}{\partial z} + \frac{\partial w}{\partial x} = \theta_y + \frac{\partial w_0}{\partial x}, \quad \varepsilon_{xy} = \frac{\partial u}{\partial y} + \frac{\partial v}{\partial x} = -\theta_z + \frac{\partial v_0}{\partial x}. \tag{9.1.15}$$

For a slender body in bending, transverse shear strains $\varepsilon_{xz}, \varepsilon_{xy}$ are small. Accordingly, in the B–E beam theory, they are set equal to zero. Therefore

$$\theta_y = -\frac{\partial w_0}{\partial x}, \theta_z = \frac{\partial v_0}{\partial x} \tag{9.1.16}$$

and $\theta_y, \theta_z$ are no longer independent kinematic variables.

## 9.2    Plate Bending

With this background in mechanics of beam bending in mind, let us now consider bending of a flat plate in which the $z$-axis is normal to the plate midsurface and the $x$ and $y$-axes are embedded in the midsurface. Referring to Figure 9.1, $z\mathbf{k}$, the vector drawn from point O to point P, is normal to the midsurface in the undeformed configuration, and $z\mathbf{k}'$ is the vector drawn from point O′ to point P′. Line vector $z\mathbf{k}$ is assumed rigid and translates and rotates into line vector $z\mathbf{k}'$ in the deformed configuration with no change in length.

**Strain–Stress Relationship**

Consider the strain–stress relation for 3D solids in Eq. (9.1.2). Since $\varepsilon_{zz} = 0$ for any stress state, all entries in the third row of $\hat{\mathbf{C}}$ are equal to zero and, due to symmetry of $\hat{\mathbf{C}}$, all entries in the third column of $\hat{\mathbf{C}}$ are also equal to zero. Accordingly, in the Reissner–Mindlin (R–M) plate bending theory, the strain–stress relation reduces to

$$\boldsymbol{\varepsilon} = \hat{\mathbf{C}}\boldsymbol{\sigma}, \tag{9.2.1}$$

where $\boldsymbol{\varepsilon}$ and $\boldsymbol{\sigma}$ are $5\times1$ column vectors and $\hat{\mathbf{C}}$ is a $5\times5$ matrix. For isotropic materials such as aluminum and steel:

$$\begin{Bmatrix} \varepsilon_{xx} \\ \varepsilon_{yy} \\ \varepsilon_{yz} \\ \varepsilon_{zx} \\ \varepsilon_{xy} \end{Bmatrix} = \frac{1}{E}\begin{bmatrix} 1 & -v & 0 & 0 & 0 \\ -v & 1 & 0 & 0 & 0 \\ 0 & 0 & 2(1+v) & 0 & 0 \\ 0 & 0 & 0 & 2(1+v) & 0 \\ 0 & 0 & 0 & 0 & 2(1+v) \end{bmatrix}\begin{Bmatrix} \sigma_{xx} \\ \sigma_{yy} \\ \sigma_{yz} \\ \sigma_{zx} \\ \sigma_{xy} \end{Bmatrix}. \tag{9.2.2}$$

Solving for $\boldsymbol{\sigma}$:

$$\boldsymbol{\sigma} = \mathbf{C}\boldsymbol{\varepsilon}, \quad \mathbf{C} = \hat{\mathbf{C}}^{-1}. \tag{9.2.3}$$

In the Kirchhoff plate bending theory, two transverse shear strains $\varepsilon_{xz}, \varepsilon_{yz}$ are assumed to be zero. Accordingly, all entries in the fourth and fifth rows of the $6 \times 6$ $\hat{\mathbf{C}}$ matrix are equal to zero and all entries in the fourth and fifth columns are equal to zero. The strain–stress relation then reduces to an expression symbolically identical to Eq. (9.2.1). However, $\boldsymbol{\varepsilon}$ and $\boldsymbol{\sigma}$ are now $3 \times 1$ column vectors and $\hat{\mathbf{C}}$ is a $3 \times 3$ matrix. For isotropic materials:

$$\begin{Bmatrix} \varepsilon_{xx} \\ \varepsilon_{yy} \\ \varepsilon_{xy} \end{Bmatrix} = \frac{1}{E} \begin{bmatrix} 1 & -\nu & 0 \\ -\nu & 1 & 0 \\ 0 & 0 & 2(1+\nu) \end{bmatrix} \begin{Bmatrix} \sigma_{xx} \\ \sigma_{yy} \\ \sigma_{xy} \end{Bmatrix}. \tag{9.2.4}$$

### Kinematic Variables

Referring to Figure 9.1, the displacement vector $\mathbf{u}$ of point $\mathrm{P}(x, y, z)$ can be expressed as

$$\mathbf{u}(x, y, z) = \mathbf{u}_0(x, y) + z\mathbf{k}'(x, y) - z\mathbf{k}(x, y). \tag{9.2.5}$$

For small rotation, the rotation of $z\mathbf{k}$ to $z\mathbf{k}'$ can be described by rotation $\theta_y$ about the $y$-axis and rotation $\theta_x$ about the $x$-axis, as shown in Figure 9.3, such that

$$z\mathbf{k}' - z\mathbf{k} = z(\theta_y \mathbf{i} - \theta_x \mathbf{j}). \tag{9.2.6}$$

**Figure 9.3** Two rotational angles for description of the kinematics of plate bending

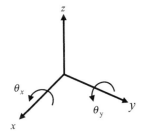

Then

$$\mathbf{u}(x, y, z) = \mathbf{u}_0(x, y) + z[\theta_y(x, y)\mathbf{i} - \theta_x(x, y)\mathbf{j}]. \tag{9.2.7}$$

In component form:

$$\mathbf{u}(x, y, z) = u(x, y, z)\mathbf{i} + v(x, y, z)\mathbf{j} + w(x, y, z)\mathbf{k} \tag{9.2.8}$$

and

$$\mathbf{u}_0(x, y) = u_0(x, y)\mathbf{i} + v_0(x, y)\mathbf{j} + w_0(x, y)\mathbf{k}. \tag{9.2.9}$$

Comparing both sides of $\mathbf{u}$:

$$u(x, y, z) = u_0(x, y) + z\theta_y(x, y),$$
$$v(x, y, z) = v_0(x, y) - z\theta_x(x, y), \qquad (9.2.10)$$
$$w(x, y, z) = w_0(x, y).$$

There are five kinematic variables with three translations $(u_0, v_0, w_0)$ and two rotational angles $(\theta_x, \theta_y)$.

### Strain–Displacement Relation

Introducing Eqs (9.2.10) to the strain–displacement relation in Appendix 1 yields

$$\varepsilon_{xx} = \frac{\partial u}{\partial x} = \frac{\partial u_0}{\partial x} + z\frac{\partial\theta_y}{\partial x}, \quad \varepsilon_{yy} = \frac{\partial v}{\partial y} = \frac{\partial v_0}{\partial y} - z\frac{\partial\theta_x}{\partial y}, \quad \varepsilon_{zz} = \frac{\partial w}{\partial z} = 0,$$

$$\varepsilon_{xy} = \frac{\partial u}{\partial y} + \frac{\partial v}{\partial x} = \frac{\partial u_0}{\partial y} + \frac{\partial v_0}{\partial x} + z\left(\frac{\partial\theta_y}{\partial y} - \frac{\partial\theta_x}{\partial x}\right), \qquad (9.2.11)$$

$$\varepsilon_{yz} = \frac{\partial v}{\partial z} + \frac{\partial w}{\partial y} = -\theta_x + \frac{\partial w_0}{\partial y}, \quad \varepsilon_{xz} = \frac{\partial u}{\partial z} + \frac{\partial w}{\partial x} = \theta_y + \frac{\partial w_0}{\partial x}.$$

For thin plates in bending, the two transverse shear strains $(\varepsilon_{xz}, \varepsilon_{yz})$ are very small relative to the other strain components. Accordingly, the Kirchhoff plate bending theory assumes that these two strains are equal to zero and thus

$$\theta_y = -\frac{\partial w_0}{\partial x}, \quad \theta_x = \frac{\partial w_0}{\partial y}. \qquad (9.2.12)$$

The two rotational angles are thus no longer independent, and $u_0, v_0, w_0$ are the three remaining kinematic variables.

### Equilibrium

For solids and structures in equilibrium:

$$\delta U - \delta W = 0. \qquad (9.2.13)$$

For the internal virtual work:

$$\delta U = \int_V \delta\boldsymbol{\varepsilon}^{\mathrm{T}}\boldsymbol{\sigma}dV = \int_V \delta\boldsymbol{\varepsilon}^{\mathrm{T}}\mathbf{C}(\boldsymbol{\varepsilon} - \boldsymbol{\varepsilon}^o)dV. \qquad (9.2.14)$$

Since strain is linear through the thickness, integration through the plate thickness can be carried out analytically.

**Finite Element Formulation Based on the Reissner–Mindlin Theory**

According to the isoparametric formulation, mapping and assumed displacements are expressed in terms of nodal values as follows.

Mapping:

$$x = \sum N_i x_i, \quad y = \sum N_i y_i. \tag{9.2.15}$$

Assumed displacements:

$$u_0 = \sum N_i u_{0i}, \quad v_0 = \sum N_i v_{0i}, \quad w_0 = \sum N_i w_{0i},$$
$$\theta_x = \sum N_i \theta_{xi}, \quad \theta_y = \sum N_i \theta_{yi}. \tag{9.2.16}$$

Nodes are placed on the midsurface of the plate and there are 5 DOF per node, as shown in Figure 9.4 for the four-node element.

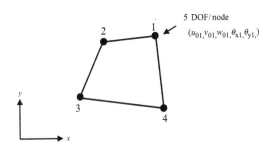

5 DOF/node

$(u_{01}, v_{01}, w_{01}, \theta_{x1}, \theta_{y1})$

**Figure 9.4** Four-node element with 5 DOF per node

A plate element based on the R–M theory may suffer from transverse shear locking. This will be discussed in Chapter 10.

**Finite Element Formulation Based on the Kirchhoff Theory**

Rotational angles are related to the derivatives of $w_0$ as shown in Eq. (9.2.12). Accordingly, the two derivatives of $w_0$ must be continuous across element boundaries, in addition to the continuity of $u_0$, $v_0$, $w_0$. It turns out that it is not a straightforward task to construct an assumed field for $w_0$ that satisfies this requirement, as will be shown in a problem at the end of this chapter.

## 9.3 Shell Element Formulation

Thin structures of curved geometry are called shell structures in which kinematics of in-plane deformation and bending deformation are coupled to each other. For thin shells, the assumptions on kinematics of deformation introduced in the R–M plate bending theory can be extended to formulation of shell elements, within the context of the isoparametric formulation. For this, we will look at the description of shell geometry and kinematics of deformation.

## 9.3.1    Geometry and Kinematics of Deformation

Consider a generic point P in the shell structure in the original undeformed configuration as shown in Figure 9.5, in which $x^G, y^G, z^G$ are three axes of a global Cartesian coordinate system for description of the geometry and the kinematics of deformation.

**Figure 9.5** Portion of shell structure in the undeformed configuration and deformed configuration

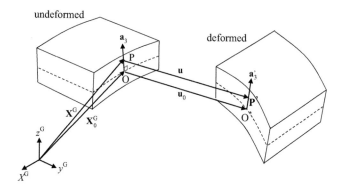

### Description of Geometry

The position vector $\mathbf{x}^G$ of point P can be expressed as

$$\mathbf{x}^G = \mathbf{x}_0^G + z\mathbf{a_3}, \tag{9.3.1}$$

where

$\mathbf{x}_0^G$: position vector of point O on the midsurface
$z\mathbf{a_3}$: line vector drawn from point O to point P
$\mathbf{a_3}$: unit vector normal to the midsurface located at point O
$z$: local coordinate normal to the midsurface located at point O.

We may introduce a non-dimensional coordinate $\zeta$ such that

$$z = \zeta\frac{t}{2} \quad (-1 \leq \zeta \leq 1), \tag{9.3.2}$$

where $t$ is the shell thickness at point O and $\zeta = 0$ is on the midsurface. Then

$$\mathbf{x}^G = \mathbf{x}_0^G + \zeta\frac{t}{2}\mathbf{a_3}. \tag{9.3.3}$$

### Kinematics of Deformation

We assume that the line vector $z\mathbf{a_3}$ translates and rotates into the line vector $z\mathbf{a}_3'$ with no change in length. The displacement vector of point P can then be expressed as

$$\mathbf{u} = \mathbf{u_0} + z\mathbf{a}_3' - z\mathbf{a_3}, \tag{9.3.4}$$

where

$\mathbf{u}$: displacement vector of point P
$\mathbf{u_0}$: displacement vector of point O

$\mathbf{a}_3'$: unit vector $\mathbf{a}_3$ after rotation

$z\mathbf{a}_3'$: line vector drawn from point $O'$ to point $P'$.

At a point on the midsurface, one may define a local coordinate system as shown in Figure 9.6, in which $\mathbf{a}_1, \mathbf{a}_2$ are the unit vectors tangent to the shell midsurface while $\mathbf{a}_3$ is the unit vector normal to the midsurface.

In Figure 9.6, angle $\theta_1$ is the rotation of $\mathbf{a}_3$ around the $\mathbf{a}_1$-axis while angle $\theta_2$ is the rotation of $\mathbf{a}_3$ around the $\mathbf{a}_2$-axis, following the right-hand rule. For small rotation:

$$z\left(\mathbf{a}_3' - \mathbf{a}_3\right) = z(\theta_2\mathbf{a}_1 - \theta_1\mathbf{a}_2). \tag{9.3.5}$$

Thus $\mathbf{u}$ can be expressed in terms of five kinematic variables $(\mathbf{u}_0, \theta_1, \theta_2)$, as

$$\mathbf{u} = \mathbf{u}_0 + z(\theta_2\mathbf{a}_1 - \theta_1\mathbf{a}_2) = \mathbf{u}_0 + \zeta\frac{t}{2}(-\mathbf{a}_2\theta_1) + \zeta\frac{t}{2}(\mathbf{a}_1\theta_2). \tag{9.3.6}$$

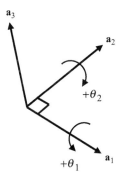

**Figure 9.6** Local coordinate system defined at a point on the shell midsurface and two rotational angles

## 9.3.2 Finite Element Formulation

As an example, consider the eight-node shell element as shown in Figure 9.7. The element has 5 DOF per node. Introducing mapping functions, the position vector can be expressed in terms of nodal values as

$$\mathbf{x}^G = \sum N_i(\xi, \eta)\left(\mathbf{x}_0^G\right)_i + \zeta\sum N_i(\xi, \eta)\left(\frac{t}{2}\mathbf{a}_3\right)_i. \tag{9.3.7}$$

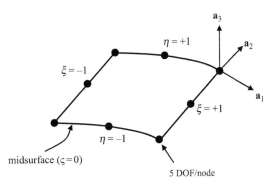

**Figure 9.7** Eight-node 40-DOF shell element with the nodes placed on the midsurface

Introducing shape functions, the assumed displacement can be expressed as

$$\mathbf{u} = \sum N_i(\xi, \eta)(\mathbf{u}_0)_i + \zeta \sum N_i(\xi, \eta)\left(-\frac{t}{2}\mathbf{a}_2\theta_1\right)_i + \zeta \sum N_i(\xi, \eta)\left(\frac{t}{2}\mathbf{a}_1\theta_2\right)_i. \qquad (9.3.8)$$

For 3D solids, the internal virtual work is expressed as

$$\delta U = \int_V \delta\boldsymbol{\varepsilon}^T \boldsymbol{\sigma} dV = \int_V \underset{1\times6}{\delta\boldsymbol{\varepsilon}^T} \underset{6\times6}{\mathbf{C}} \underset{6\times1}{\boldsymbol{\varepsilon}} \, dV, \qquad (9.3.9)$$

assuming no temperature change. In the local coordinate system with unit vectors $\mathbf{a}_1, \mathbf{a}_2, \mathbf{a}_3$, $\varepsilon_{zz}^l = 0$:

$$\underset{5\times1}{\boldsymbol{\varepsilon}^l} = \underset{5\times5}{\hat{\mathbf{C}}^l} \, \underset{5\times1}{\boldsymbol{\sigma}^l}. \qquad (9.3.10)$$

From the above equation:

$$\underset{5\times1}{\boldsymbol{\sigma}^l} = \underset{5\times5}{\mathbf{C}^l} \, \underset{5\times1}{\boldsymbol{\varepsilon}^l}, \qquad (9.3.11)$$

where $\mathbf{C}^l$ is the inverse of $\hat{\mathbf{C}}^l$. In the global $X, Y, Z$ system:

$$\underset{6\times1}{\boldsymbol{\sigma}} = \underset{6\times6}{\mathbf{C}} \, \underset{6\times1}{\boldsymbol{\varepsilon}}. \qquad (9.3.12)$$

The $\mathbf{C}$ matrix for the global coordinate system can be obtained by

$$\underset{6\times6}{\mathbf{C}} = \underset{6\times5}{\mathbf{T}^T} \, \underset{5\times5}{\mathbf{C}^l} \, \underset{5\times6}{\mathbf{T}}, \qquad (9.3.13)$$

where $\mathbf{T}$ is constructed at integration points using direction cosines. The element stiffness matrix for the eight-node, 40-DOF element can be constructed using $2 \times 2 \times 2$ point integration. The six-node curved triangular shell element can also be formulated following a similar procedure.

## 9.4    Plates and Shells Modeled using 3D Solid Elements

Although plates and shells are thin, they are geometrically 3D structures. Accordingly, we may model them with 3D solid elements. Consider a simple example in which a cantilevered slender body of isotropic material is under a tip force acting downward, as shown in Figure 9.8. At and near the top surface, strain $\varepsilon_{xx}$ is positive while $\varepsilon_{yy}, \varepsilon_{zz}$ are negative due to the Poisson's ratio effect. In contrast, at and near the bottom surface, $\varepsilon_{xx}$ is negative while $\varepsilon_{yy}, \varepsilon_{zz}$ are positive. Accordingly, $\varepsilon_{xx}, \varepsilon_{yy}, \varepsilon_{zz}$ experience a change in sign through the thickness. For thin plates and shells, it is reasonable to assume that these strains vary linearly. With this in mind we will now consider solid elements with two or three nodes through the thickness.

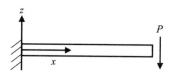

**Figure 9.8** Cantilevered slender body of rectangular cross-section under tip force

### 9.4.1  Solid Elements with Three Nodes through the Thickness

Elements under this category are: (1) the 20-node brick element with eight nodes on the top and bottom surfaces and four nodes on the midsurface; (2) the 24-node brick element with eight nodes each on the top, mid, and bottom surfaces; (3) the 15-node triangular prism element with six nodes on the top and bottom surface and three nodes on the midsurface; and (4) the 18-node triangular prism element with six nodes each on the top, mid, and bottom surfaces.

For an element of flat rectangular geometry, strain $\varepsilon_{zz}$ is linear through the thickness. Accordingly, one may employ models with a single element through the thickness for plate and shell analyses. Table 9.1 shows the results of plate analysis using FE models with 20-node elements and 24-node elements. The square plate is clamped along all boundaries and is under uniform pressure. According to the exact solution from the Kirchhoff plate bending theory, the maximum displacement occurring at the plate centroid is

$$w_{\max} = 0.0012645 \frac{pL^4}{D}, \quad D = \frac{Et^3}{12(1-v^2)}, \tag{9.4.1}$$

where $p$ is the uniform pressure and $L$ is the plate side length. Table 9.1 lists the maximum displacement values normalized with respect to the exact solution given in the above equation. It is observed that the FE solutions are in good agreement with the exact solution, especially for the models with 24-node elements.

**Table 9.1 Non-dimensional maximum displacement for a clamped plate under uniform pressure ($L/t = 100$)**

| Element types | 4×4 mesh | 8×8 mesh | 16×16 mesh |
|---|---|---|---|
| 20-node element ($2 \times 2 \times 3$ point integration) | 0.92696 | 0.96705 | 0.98465 |
| 24-node element ($2 \times 2 \times 3$ point integration) | 0.99301 | 1.00085 | 1.00085 |
| 16-node element ($2 \times 2 \times 2$ point integration) | 0.75480 | 0.75960 | 0.75968 |
| 16-node element ($2 \times 2 \times 2$ point integration) with $v_{xz} = v_{yz} = 0$ | 0.99064 | 1.00188 | 1.00196 |

### 9.4.2  Solid Elements with Two Nodes through the Thickness

Plate and shell structures can be modeled using solid shell elements with two nodes through the thickness and the modified 3D constitutive equation. For example, we may consider the 16-node solid element and the 18-node solid element with two nodes in the thickness direction, as shown in Figure 9.9. Since the assumed displacement for the element is linear in the thickness direction, strain $\varepsilon_{zz}$ is constant through the thickness. Accordingly, the element cannot represent linearly varying $\varepsilon_{zz}$ and $\varepsilon_{zz} \to 0$ as the thickness becomes smaller. For isotropic material:

$$\varepsilon_{zz} = \frac{1}{E}\left(\sigma_{zz} - v\sigma_{xx} - v\sigma_{yy}\right) = 0 \to \sigma_{zz} = v\left(\sigma_{xx} + \sigma_{yy}\right). \tag{9.4.2}$$

As an example, consider the cantilevered beam under tip force $P$ applied at the tip, as shown in Figure 9.8. On the top and bottom surfaces:

**Figure 9.9** Elements of rectangular geometry with two nodes through the thickness

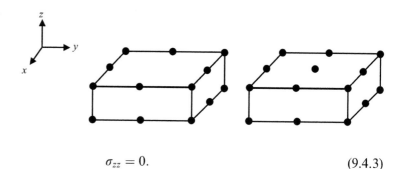

$$\sigma_{zz} = 0. \tag{9.4.3}$$

However, according to Eq. (9.4.2), with $\sigma_{yy} = 0$:

$$\sigma_{zz} = \nu\sigma_{xx}, \tag{9.4.4}$$

which is erroneous. To remedy this situation, one may revisit the strain–stress relation for 3D solids expressed as

$$\boldsymbol{\varepsilon} = \hat{\mathbf{C}}\boldsymbol{\sigma}. \tag{9.4.5}$$

We may then treat the shell material as orthotropic with $\nu_{xz} = \nu_{yz} = 0$, such that

$$
\begin{Bmatrix}
\varepsilon_{xx} \\
\varepsilon_{yy} \\
\varepsilon_{zz} \\
\varepsilon_{yz} \\
\varepsilon_{zx} \\
\varepsilon_{xy}
\end{Bmatrix}
=
\begin{bmatrix}
\dfrac{1}{E_x} & \dfrac{-\nu_{yx}}{E_y} & 0 & 0 & 0 & 0 \\[2mm]
\dfrac{-\nu_{xy}}{E_x} & \dfrac{1}{E_y} & 0 & 0 & 0 & 0 \\[2mm]
0 & 0 & \dfrac{1}{E_z} & 0 & 0 & 0 \\[2mm]
0 & 0 & 0 & \dfrac{1}{G_{yz}} & 0 & 0 \\[2mm]
0 & 0 & 0 & 0 & \dfrac{1}{G_{zx}} & 0 \\[2mm]
0 & 0 & 0 & 0 & 0 & \dfrac{1}{G_{xy}}
\end{bmatrix}
\begin{Bmatrix}
\sigma_{xx} \\
\sigma_{yy} \\
\sigma_{zz} \\
\sigma_{yz} \\
\sigma_{zx} \\
\sigma_{xy}
\end{Bmatrix}. \tag{9.4.6}
$$

For curved shell structures, the above relationship between strain and stress must be established at the local coordinate system defined at integration points. For isotropic materials such as aluminum and steel, we further set

$$E_x = E_y = E_z = E, \quad \nu_{xy} = \nu_{yx} = \nu, \quad G_{yz} = G_{zx} = G_{xy} = G. \tag{9.4.7}$$

The results of FE analysis using the 16-node elements are shown in Table 9.1, demonstrating the effectiveness of the present approach.

Alternatively, one may replace two nodes through the thickness with a single node with 6 DOF placed on the midsurface. The displacement can be expressed as

$$\mathbf{u} = \mathbf{u}_0 + z\mathbf{u}_1, \tag{9.4.8}$$

where $\mathbf{u}_0$ is the displacement at the midplane. Then, at the top surface with $z = t/2$:

$$\mathbf{u}_{\text{top}} = \mathbf{u}_0 + \frac{t}{2}\mathbf{u}_1. \tag{9.4.9}$$

While at the bottom surface with $z = -t/2$:

$$\mathbf{u}_{\text{bottom}} = \mathbf{u}_0 - \frac{t}{2}\mathbf{u}_1. \qquad (9.4.10)$$

For example, the 16-node element with 3 DOF per node can be transformed into the eight-node element with 6 DOF per node. Using the relationship described in the preceding equations:

$$\mathbf{d}_{16} = \mathbf{T}\mathbf{d}_8, \qquad (9.4.11)$$

where $\mathbf{d}_{16}$ is the element DOF vector of the 16-node element, $\mathbf{d}_8$ is that of the eight-node element, and $\mathbf{T}$ is the transformation matrix. Then

$$\mathbf{k}_8^e = \mathbf{T}^T\mathbf{k}_{16}^e\mathbf{T}, \qquad (9.4.12)$$

where $\mathbf{k}_8^e$ is the element stiffness matrix of the eight-node element and $\mathbf{k}_{16}^e$ is that of the 16-node element.

## PROBLEMS

**9.1** Consider a slender body under the Timoshenko theory of the kinematics of deformation. For simplicity, further assume that the body displaces only in the $xz$-plane. Recall that

$$u(x, y, z) = u_0(x) + z\theta_y(x), \quad w(x, y, z) = w_0(x)$$

and

$$\varepsilon_{xx} = \frac{\partial u}{\partial x} = \frac{\partial u_0}{\partial x} + z\frac{\partial \theta_y}{\partial x}, \quad \varepsilon_{xz} = \frac{\partial u}{\partial z} + \frac{\partial w}{\partial x} = \theta_y + \frac{\partial w_0}{\partial x},$$

$$\sigma_{xx} = E\varepsilon_{xx}, \quad \sigma_{xz} = G\varepsilon_{xz}.$$

All other strains are equal to zero.

(a) Show that

$$\delta U = \int \delta\varepsilon_{xx}\sigma_{xx}dV + \int \delta\varepsilon_{xz}\sigma_{xz}dV = \delta U_A + \delta U_B + \delta U_S,$$

where

$$\delta U_A = \int_{x=0}^{x=L} EA\frac{\partial \delta u_0}{\partial x}\frac{\partial u_0}{\partial x}dx, \quad \delta U_B = \int_{x=0}^{x=L} EI_y\frac{\partial \delta\theta_y}{\partial x}\frac{\partial \theta_y}{\partial x}dx,$$

$$\delta U_S = \int_{x=0}^{x=L} GA\left(\delta\theta_y + \frac{\partial \delta w_0}{\partial x}\right)\left(\theta_y + \frac{\partial w_0}{\partial x}\right)dx,$$

where $A$ is the cross-sectional area and

$$\int_A zdydz = 0, \quad I_y = \int_A z^2dydz.$$

(b) Assume $u_0, w_0, \theta$ to be linear for an element, such that

$$u_0 = \frac{1}{2}(1-\xi)u_{01} + \frac{1}{2}(1+\xi)u_{02}, \text{etc.}$$

Then for the element

$$\delta U_e = \delta \mathbf{d}^\mathrm{T} \mathbf{k}^e \mathbf{d},$$

where

$$\mathbf{d} = \left\{ \begin{array}{c} u_{01} \\ w_{01} \\ \hat{\theta}_1 \\ u_{02} \\ w_{02} \\ \hat{\theta}_2 \end{array} \right\} : \text{element DOF vector}$$

with $\hat{\theta} = l\theta$ and $l$ the element length. Construct the element stiffness matrix using the one-point rule for numerical integration.

**9.2** A cantilever beam of length $L = 100\,\mathrm{cm}$, thickness $t = 1\,\mathrm{cm}$, and width $b = 1\,\mathrm{cm}$ is subjected to a tip load of $P$.
(a) Determine the tip displacement using a model with five elements of equal size as developed in Problem 9.1. Compare with the exact solution.
(b) Repeat using a model with 10 elements. Compare with the exact solution.

**9.3** A circular ring of square cross-section is subjected to a pair of point loads applied radially outward at the diametrical position as shown in Figure 9.10. A quarter of the ring is modeled using the element developed in Problem 9.1. Geometric and material data are given as follows:

Radius to the midwall of the ring: $R = 100\,\mathrm{cm}$.
Thickness: $t = 1\,\mathrm{cm}$.
Width: $b = 1\,\mathrm{cm}$.
$E = 72\,\mathrm{GPa}, v = 0.3$.

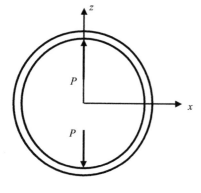

**Figure 9.10** For Problem 9.3

(a) Construct the element stiffness matrix with DOF corresponding to displacement in the Cartesian coordinate system shown.
(b) For $P = 1\mathrm{N}$, determine the outward radial displacement at the load point using a model of 10 elements of equal size.
(c) Repeat (b) using a model with 15 elements of equal size.
(d) Compare with the solution obtained by the eight-node shell elements.

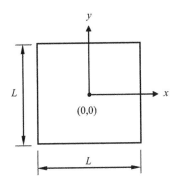

**Figure 9.11** For Problem 9.4

**9.4** A square plate shown in Figure 9.11 is clamped along all four edges and subjected to a uniformly distributed pressure load. For the plate, the side length is $L = 1.3$ m and the thickness is 1.3 cm in the $z$ direction. The material is isotropic with the properties given as $E = 72$ GPa, $v = 0.3$, and $\rho = 2{,}780$ kg/m$^3$.

The plate is modeled with the six-node or eight-node shell elements. Due to symmetry in the geometry, applied load, and geometric constraints, only one-quarter of the plate in the $xy$-plane is modeled with a uniform $7 \times 7$ mesh of plate elements.

(a) For a unit pressure load, determine the transverse displacement along the $x = 0$ line.
(b) Determine the in-plane stresses through the thickness at the center of the modeled quadrant.
(c) Determine the first eight natural frequencies and modes of the plate. Sketch the modes.

**9.5** Consider a cylindrical shell with dimensions and material properties given as follows:

Length: $L = 20$ mm.
Radius to the midwall: $a = 10.5$ mm.
Wall thickness: $t = 0.4$ mm.
$E = 200$ GPa, $v = 0.3$, and mass density $\rho = 8{,}000$ kg/m$^3$.

The cylindrical shell is fixed at one end and free at the other end.
(a) Use a model with the six-node or eight-node shell elements to determine the natural frequencies of the first eight modes, counting double frequencies separately.
(b) Use a model with the 20-node solid elements to determine the natural frequencies of the first eight modes. Compare with the results obtained in part (a).
(c) Determine the natural modes of the first eight modes. Sketch the modes.

**9.6** Repeat Problem 9.5 with $t = 0.2$ mm.

**9.7** Consider a hemispherical shell with a circular cutout at the crown. The cutout area is attached to a rigid rod of circular cross-section. The rod itself is fixed to a base. Shell dimensions and material properties are given as follows:

Radius to the shell midwall: $a = 10$ mm.
Radius of the rigid rod: $b = 2$ mm.
Wall thickness: $t = 0.125$ mm.
$E = 200$ GPa, $v = 0.3$, and mass density $\rho = 8{,}000$ kg/m$^3$.

(a) Use a model with the six-node or eight-node shell elements to determine the natural frequencies of the first eight modes, counting double frequencies separately.
(b) Use a model with the 20-node solid elements to determine the natural frequencies of the first eight modes. Compare with the results obtained in part (a).
(c) Sketch the natural modes of the first eight modes.

**9.8** Repeat Problem 9.7 with $t = 0.0625$ mm.

**Figure 9.12** For Problem 9.9

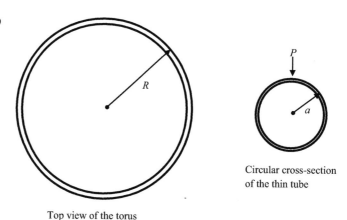

Top view of the torus

Circular cross-section
of the thin tube

**9.9** Consider a torus structure shown in Figure 9.12. The geometrical and material data are given as follows:

Outer radius of the torus: $R = 1.90$ m.
Outer radius of the circular tube: $a = 0.09$ m.
Skin thickness of the circular tube: $t = 4.2 \times 10^{-4}$ m
$E = 72$ GPa, $v = 0.3$, and mass density $\rho = 2,780$ kg/m$^3$.

The torus is fixed at three points (120° apart), located at the bottom of the tubular cross-section. Use a model with the six or eight-node shell elements to do the following.
(a) A vertical downward static force of $P = 30$ N is applied at a location 60° away from a fixed point. Determine and plot the displacement and stress along the torus crown.
(b) Plot the deformed shape of the torus.
(c) Determine the first eight natural frequencies of the torus. Plot the corresponding natural modes.

**9.10** Repeat Problem 9.9 with the skin thickness of the circular tube equal to $9 \times 10^{-4}$ m. Compare with the frequencies found in Problem 9.9 and comment on what you observe.

**9.11** Try Problem 9.9 using a model with 20-node solid elements.

**9.12** Consider the formulation of a triangular plate bending element based on the Kirchhoff theory, as shown in Figure 9.13(a). Each corner node has 3 DOF $(w_0, \theta_y, \theta_x)$, while the center node has 1 DOF $(w_0)$. The element has 10 DOF to match a complete cubic polynomial function. So, one may try the assumed displacement with the following 10 polynomial terms:

$$w_0 \sim 1, x, y, x^2, xy, y^2, x^2y, xy^2, x^3, y^3.$$

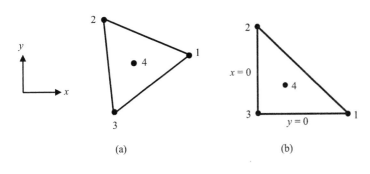

**Figure 9.13** For Problem 9.12

(a)                                          (b)

For simplicity, consider side 1–3 of a right triangle as shown in Figure 9.13(b), and show that $w_0$ is cubic in $x$ and $\theta_x$ and $\theta_y$ are quadric in $x$ along the side. Accordingly, similar to the beam bending element based on the B–E theory, one can match cubic $w_0$ with $(w_0)_3, (w_0)_1, (\theta_y)_3, (\theta_y)_1$ along the side 1–3.

(a) Is it possible to match quadratic variation of $\theta_x$ along the side 1–3 with $(\theta_x)_3, (\theta_x)_1$?

(b) Can you conclude that it is not possible to find an assumed $w_0$ that satisfies all continuity requirements?

# 10    Element Locking

Under certain conditions, a finite element may lose its ability to deform and become excessively stiff. This phenomenon is called "element locking." In this chapter, we will consider the following three types of element locking:

1. transverse shear locking
2. membrane locking
3. incompressibility locking.

Transverse shear locking occurs when the FE approximation is introduced to model bending behavior of slender bodies or thin plates or shells based on the kinematics of deformation that allows transverse shear straining. One can show that the condition of vanishing transverse shear strain can place excessive constraints on the deformation modes of an FE model. This will limit the ability of the element to represent the bending behavior, resulting in an excessively stiff element, especially for an element with lower-order polynomials in the assumed displacement field.

Membrane locking occurs when the FE approximation is introduced to model structural components such as thin curved arches and shells. As an arch gets slender, the arch becomes inextensible and the primary mode of deformation becomes bending dominant. The condition of inextensibility can place excessive constraint on the assumed displacement of an arch element, limiting its ability to properly represent bending behavior. Membrane locking can occur in a shell element as the shell thickness gets smaller and the shell becomes more flexible in bending.

Incompressibility locking occurs when the FE approximation is used to model a solid body of incompressible material such as rubber, which experiences no volume change when deforming under applied loads. The condition of zero volume change places constraints on the assumed displacement of a finite element. Especially for an element with the assumed displacement of a lower-order polynomial function, the constraining condition of zero volume change can be extremely severe, rendering the element almost rigid.

Element locking can be alleviated or avoided via a reduced-order or lower-order integration, with the number of sampling points smaller than is necessary for exact integration of the element stiffness matrix. However, a lower-order integration can trigger spurious kinematic modes which render the element kinematically unstable. Alternatively, as described at the end of this chapter, a formulation with an independently assumed strain field or stress field, in addition to the assumed displacement, can be used to alleviate the detrimental effect of locking.

## 10.1 Transverse Shear Locking

Transverse shear locking occurs when the FE approximation is introduced to model bending behavior of slender bodies, thin plates, or shells based on the kinematics of deformation that allows transverse shear straining. To appreciate this, consider pure bending of a beam displacing in the $xz$-plane. The cross-section located at position $x$ is assumed to translate by $w$ in the $z$ direction and rotate by angle $\theta$ as a rigid plane.

Accordingly, the horizontal displacement of point P on the cross-section as shown in Figure 10.1 can be expressed as

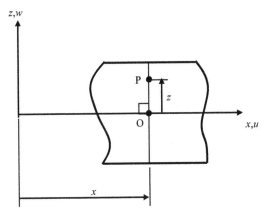

$$u = z\theta. \tag{10.1.1}$$

The strain–displacement relations are then expressed as

**Figure 10.1** A portion of a beam in the original undeformed configuration

$$\varepsilon_{xx} = \frac{\partial u}{\partial x} = z\frac{\partial \theta}{\partial x},$$
$$\varepsilon_{xz} = \frac{\partial u}{\partial z} + \frac{\partial w}{\partial x} = \theta + \frac{\partial w}{\partial x}. \tag{10.1.2}$$

For thin beams, $\varepsilon_{xz}$ is very small. In the B–E beam theory, the transverse shear strain $\varepsilon_{xz}$ is assumed zero and

$$\theta = -\frac{\partial w}{\partial x}. \tag{10.1.3}$$

In contrast, the Timoshenko beam theory allows small but nonzero transverse shear strain. Accordingly, $\theta$ and $w$ are independent. Consider now an FE model based on the Timoshenko beam bending theory.

### 10.1.1 Two-Node Timoshenko Theory Beam Bending Element

For a two-node element as shown in Figure 10.2, assumed rotation and displacement can be expressed as

$$\theta = a_1 + a_2 x, \quad w = b_1 + b_2 x. \tag{10.1.4}$$

Note that, for convenience, the assumed rotation and displacement are expressed in the physical $x$-coordinate rather than a mapped non-dimensional coordinate system.

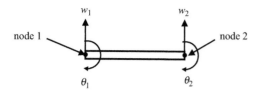

Figure 10.2 Two-node element based on Timoshenko beam theory

As the beam gets thinner, the transverse shear strain $\varepsilon_{xz}$ approaches zero. Then, in the limiting case of $\varepsilon_{xz} = 0$:

$$\theta + \frac{\partial w}{\partial x} = 0 \tag{10.1.5}$$

or

$$a_1 + a_2 x + b_2 = 0 \tag{10.1.6}$$

and thus

$$\begin{aligned} a_1 + b_2 &= 0, \\ a_2 &= 0. \end{aligned} \tag{10.1.7}$$

Equation (10.1.7) represents two constraints among the 4 DOF. The original assumed displacement in Eq. (10.1.4) then reduces to

$$\begin{aligned} \theta &= a_1, \\ w &= b_1 - a_1 x. \end{aligned} \tag{10.1.8}$$

Accordingly, only two of the coefficients remain free. To appreciate the effect of these constraints, consider now a cantilever beam of length $L$ and thickness $t$ modeled by two-node elements as shown in Figure 10.3.

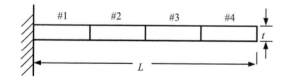

Figure 10.3 Cantilevered beam modeled with four two-node elements

For element #1 to satisfy the geometric boundary conditions at the clamped end, the following must hold:

$$\begin{aligned} \theta &= a_1 = 0, \\ w &= b_1 = 0. \end{aligned} \tag{10.1.9}$$

The element therefore does not deform as $L/t \to \infty$. Similarly, elements #2, #3, and so on do not deform. This type of locking is called "transverse shear locking."

 **Example 10.1**

In order to appreciate the effect of locking as the $L/t$ ratio increases, consider a cantilever beam as shown in Figure 10.4.

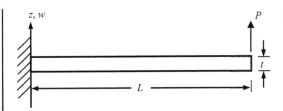

**Figure 10.4** Cantilevered beam under tip force

Table 10.1 lists the tip displacement, non-dimensionalized by the exact solution obtained according to the B–E beam bending theory, such that

$$\bar{w}_{tip} = \frac{w_{tip}}{\left(PL^3/3EI_y\right)}.$$

**Table 10.1 Effect of transverse shear locking on tip displacement of cantilevered beam**

|  | $L/t = 10$ | $L/t = 100$ |
|---|---|---|
| Exact (Timoshenko theory) | 1.006 | 1.000 |
| FE model (20 two-node elements) | 0.9182 | 0.0942 |

The results clearly show that the effect of locking becomes more pronounced as the beam gets thinner relative to the length.

## 10.1.2 Effect of Numerical Integration

Let's see now what happens when numerical integration is used to construct a beam bending element based on the Timoshenko theory. The following strains are evaluated at the integration or sampling points:

$$\varepsilon_{xx} = z\frac{\partial\theta}{\partial x},$$

$$\varepsilon_{xz} = \theta + \frac{\partial w}{\partial x}. \qquad (10.1.10)$$

Consider again the two-node element with

$$\theta = a_1 + a_2x, \quad w = b_1 + b_2x. \qquad (10.1.11)$$

Substituting Eq. (10.1.2) into Eq. (10.1.10) gives

$$\varepsilon_{xx} = z\frac{\partial\theta}{\partial x} = za_2,$$

$$\varepsilon_{xz} = \theta + \frac{\partial w}{\partial x} = a_1 + a_2x + b_2. \qquad (10.1.12)$$

We observe that the transverse shear strain is linear, and so can be fitted with two points. As $L/t$ increases, $\varepsilon_{xz}$ gets smaller. In the limiting case

$$\varepsilon_{xz} = 0 \qquad\qquad (10.1.13)$$

at the integration points, which imposes constraints on the coefficients in Eq. (10.1.11). So, the number of constraint equations is identical to the number of integration points.

node 1

node 2

$x_1 = -1$

$x_2 = +1$

**Figure 10.5** Two-node element with nodes located at $x_1 = -1$ and $x_2 = +1$

For simplicity, let's consider an element with $x_1 = -1$ and $x_2 = +1$, as shown in Figure 10.5.

### Two-Point Rule

For the two-point rule, $\varepsilon_{xz}$ is evaluated at $x = \pm 1/\sqrt{3}$ (see Table 7.1). Then, from Eqs (10.1.12) and (10.1.13), the constraint equations are

$$a_1 - \frac{1}{\sqrt{3}}a_2 + b_2 = 0,$$
$$a_1 + \frac{1}{\sqrt{3}}a_2 + b_2 = 0, \qquad\qquad (10.1.14)$$

from which one finds that

$$a_1 + b_2 = 0,$$
$$a_2 = 0. \qquad\qquad (10.1.15)$$

We note that Eq. (10.1.15) is identical to the constraint equations we observed previously in Eq. (10.1.7).

### Reduced-Order Integration

We may alleviate the problem of over-constraining the element if we use the one-point integration instead of the two-point integration. In this case, $\varepsilon_{xz} = 0$ at the integration point $(x = 0)$, and thus

$$a_1 + b_2 = 0. \qquad\qquad (10.1.16)$$

So, there is only one constraint, giving the element more ability to represent bending behavior.

### Example 10.2

In order to appreciate the effect of the reduced-order integration with the one-point rule, let's consider the non-dimensional tip displacement of the cantilever beam (Figure 10.4). One observes from Table 10.2 that reduced-order integration with one sampling point results in a dramatic improvement, compared with the results shown in Table 10.1.

**Table 10.2** Effect of the one-point integration on alleviating transverse shear locking

|                                      | $L/t = 10$ | $L/t = 100$ |
| ------------------------------------ | ---------- | ----------- |
| Exact (Timoshenko theory)            | 1.006      | 1.000       |
| Five elements (one-point integration)| 0.9965     | 0.9901      |

### 10.1.3 Spurious Kinematic Modes

We have observed that reduced integration can alleviate locking in the case of the two-node element. However, excessively lower-order integration or reduced integration may trigger "spurious" kinematic modes. A kinematic mode is a displacement mode that does not produce strain. Examples are the rigid-body modes that are "legitimate" kinematic modes. A spurious kinematic mode is any mode, other than the rigid-body modes, that does not produce strain. As an example, consider the three-node bending element based on the Timoshenko beam theory, as shown in Figure 10.6.

$x = -c \qquad x = 0 \qquad x = c$    **Figure 10.6** Three-node bending element with nodes at $x = -c, 0, c$

   The assumed rotational angle and displacement are quadratic and, for convenience, one may express them as

$$\theta = a_1 + a_2 x + a_3 x^2,$$
$$w = b_1 + b_2 x + b_3 x^2. \tag{10.1.17}$$

Then

$$\varepsilon_{xx} = z \frac{\partial \theta}{\partial x} = z(a_2 + 2a_3 x) \tag{10.1.18}$$

is linear, which requires two sampling points to fit. The transverse shear strain is

$$\varepsilon_{xz} = \theta + \frac{\partial w}{\partial x} = a_1 + a_2 x + a_3 x^2 + b_2 + 2b_3 x, \tag{10.1.19}$$

which is quadratic and therefore requires three sampling points to fit. In the limiting case, $\varepsilon_{xz} = 0$ at the integration points. Accordingly, the three-point rule will result in the three constraint equations. However, if the one-point (at $x = 0$) rule is used to sample $\varepsilon_{xz}$ in Eq. (10.1.19):

$$\varepsilon_{xz} = a_1 + b_2. \tag{10.1.20}$$

We observe that the $a_2, a_3,$ and $b_3$ terms do not contribute to $\varepsilon_{xz}$. Among these, the $a_2$ and $a_3$ terms contribute to $\varepsilon_{xx}$. However, the $b_3$ term does not produce any strain.

Therefore, $w = b_3 x^2$ is a spurious kinematic mode. This mode is called "compatible" in that the kinematic mode of one element is compatible with that of an adjacent element. If the two-point rule is used for $\varepsilon_{xz}$, it can fit a linear function and thus the $b_3$ term contributes to $\varepsilon_{xz}$. Accordingly, there is no spurious kinematic mode for the two-point rule.

## 10.1.4 Reissner–Mindlin Theory Plate Bending Element

Consider pure plate bending with no in-plane displacement such that $u_0 = 0$ and $v_0 = 0$. For simplicity, consider a four-node element of rectangular geometry of side lengths $a$ and $b$. The assumed displacement and rotational angles can be expressed as

$$w = a_1 + a_2\xi + a_3\eta + a_4\xi\eta,$$
$$\theta_x = b_1 + b_2\xi + b_3\eta + b_4\xi\eta, \tag{10.1.21}$$
$$\theta_y = c_1 + c_2\xi + c_3\eta + c_4\xi\eta.$$

For the element domain given by $-a/2 \le x \le a/2$ and $-b/2 \le y \le b/2$, the mapping can be expressed as

$$x = \frac{a}{2}\xi, \quad y = \frac{b}{2}\eta. \tag{10.1.22}$$

The strain–displacement relations can then be expressed as

$$\varepsilon_{xx} = z\frac{\partial\theta_y}{\partial x} = z\frac{2}{a}(c_2 + c_4\eta),$$

$$\varepsilon_{yy} = \frac{\partial v}{\partial y} = -z\frac{\partial\theta_x}{\partial y} = -z\frac{2}{b}(b_3 + b_4\xi),$$

$$\varepsilon_{zz} = \frac{\partial w}{\partial z} = 0,$$

$$\varepsilon_{xy} = \frac{\partial u}{\partial y} + \frac{\partial v}{\partial x} = z\left(\frac{\partial\theta_y}{\partial y} - \frac{\partial\theta_x}{\partial x}\right) = z\left[\frac{2}{b}(c_3 + c_4\xi) - \frac{2}{a}(b_2 + b_4\eta)\right], \tag{10.1.23}$$

$$\varepsilon_{yz} = \frac{\partial v}{\partial z} + \frac{\partial w}{\partial y} = -\theta_x + \frac{\partial w_0}{\partial y} = -(b_1 + b_2\xi + b_3\eta + b_4\xi\eta) + \frac{2}{b}(a_3 + a_4\xi),$$

$$\varepsilon_{xz} = \frac{\partial u}{\partial z} + \frac{\partial w}{\partial x} = \theta_y + \frac{\partial w_0}{\partial x} = -(c_1 + c_2\xi + c_3\eta + c_4\xi\eta) + \frac{2}{a}(a_2 + a_4\eta).$$

In the limiting case of zero transverse shear strains:

$$b_1 + \frac{2}{b}a_3 = 0, \quad b_2 + \frac{2}{b}a_4 = 0, \quad b_3 = 0, \quad b_4 = 0,$$
$$-c_1 + \frac{2}{a}a_3 = 0, \quad -c_3 + \frac{2}{a}a_4 = 0, \quad c_2 = 0, \quad c_4 = 0. \tag{10.1.24}$$

Accordingly, there are eight constraints among 12 DOF for the element. If numerical integration is used to construct the element stiffness matrix, $2 \times 2$ points are required to sample linearly distributed strains. However, the resulting element exhibits the effect of transverse shear locking.

### Selective Integration

If the one-point integration rule is selectively used for the transverse shear strain components while keeping the $2 \times 2$ point rule for other strain components, the $b_2, b_3, b_4, c_2, c_3, c_4, a_4$ terms do not contribute to the transverse shear strains. Among these, the $b_2, b_3, b_4, c_2, c_3, c_4$ terms contribute to $\varepsilon_{xx}, \varepsilon_{yy}, \varepsilon_{xy}$ while the $a_4$ term does not contribute to any strain. Accordingly

$$w = a_4 \xi \eta \qquad (10.1.25)$$

is a spurious kinematic mode. One can observe that this mode is compatible or communicative with the kinematic mode of an adjacent element. Accordingly, the element using the one-point integration rule for transverse shear strains is not suitable for the four-node element.

## 10.2 Membrane Locking

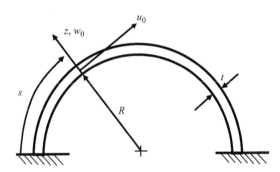

**Figure 10.7** Thin circular arch with clamped ends

Membrane locking may occur when the FE approximation is introduced to model structural components such as thin curved arches and shells. To illustrate this, consider a simple circular arch of radius $R$ as shown in Figure 10.7.

In the figure, $s$ is the coordinate along the arch axis, $u_0$ is the tangential or in-plane displacement along $s$, and $w_0$ is the displacement normal to the body axis. It can be shown that axial strain $\varepsilon$ is expressed as

$$\varepsilon = \varepsilon_0 + z\kappa, \qquad (10.2.1)$$

where $\varepsilon_0$ is the membrane, in-plane, or stretching strain and $\kappa$ is the out-of-plane or bending strain. Assuming transverse shear strain to be equal to zero, it can be shown that

$$\varepsilon_0 = \frac{du_0}{ds} + \frac{w_0}{R},$$
$$\kappa = \frac{1}{R}\frac{du_0}{ds} - \frac{d^2 w_0}{ds^2}. \qquad (10.2.2)$$

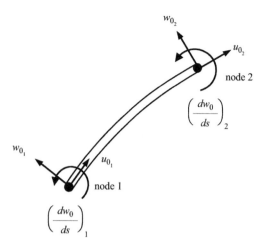

Figure 10.8 Two-node, 6-DOF arch element

For thin arches, $\varepsilon_0 \to 0$ as $R/t \to \infty$ and in the limiting case:

$$\varepsilon_0 = \frac{du_0}{ds} + \frac{w_0}{R} = 0. \qquad (10.2.3)$$

The resulting constraints on the assumed displacement field may cause element locking called "membrane locking." To appreciate the effect of constraints, let's consider the following example of the two-node, 6-DOF arch element.

Assume that $u_0$ is linear and $w_0$ is cubic such that

$$\begin{aligned} u_0 &= a_1 + a_2 s, \\ w_0 &= b_1 + b_2 s + b_3 s^2 + b_4 s^3. \end{aligned} \qquad (10.2.4)$$

Then

$$\varepsilon_0 = \frac{du_0}{ds} + \frac{w_0}{R} = a_2 + \frac{1}{R}\left(b_1 + b_2 s + b_3 s^2 + b_4 s^3\right). \qquad (10.2.5)$$

In the limiting case $(R/t \to \infty)$, $\varepsilon_0 = 0$ which results in

$$\begin{aligned} a_2 + \frac{b_1}{R} &= 0, \\ b_2 = b_3 = b_4 &= 0. \end{aligned} \qquad (10.2.6)$$

Accordingly, there are four constraint conditions that are imposed on the six coefficients in Eq. (10.2.4). Upon applying these constraint conditions, the assumed displacement field in Eq. (10.2.4) reduces to

$$\begin{aligned} u_0 &= a_1 + a_2 s, \\ w_0 &= -R a_2. \end{aligned} \qquad (10.2.7)$$

So, the element has only two remaining coefficients. To appreciate the consequence of having only two free coefficients in an element, consider the arch in Figure 10.9.

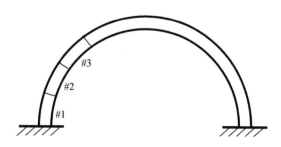

Figure 10.9 Clamped arch modeled with two-node elements

First consider element number 1. The element is fixed at $s = 0$. Applying the boundary conditions to Eq. (10.2.7):

$$u_0 = a_1 = 0,$$
$$w_0 = -Ra_2 = 0 \rightarrow a_2 = 0, \tag{10.2.8}$$

which therefore leaves the displacement as

$$u_0 = 0, \quad w_0 = 0 \tag{10.2.9}$$

everywhere in the element. Element #1 is therefore rigid and does not deform. Similarly, elements #2, #3, and so on are also rigid and do not deform, and the FE model locks. This type of locking is called membrane locking. We note that locking is dependent on the geometric boundary conditions.

**Numerical Integration**

Consider now applying numerical integration. In the limiting case $(R/t \rightarrow \infty)$

$$\varepsilon_0 = 0 \tag{10.2.10}$$

at the integration points, imposing constraint conditions on the coefficients of the assumed displacement field. For the two-node, 6-DOF element, assumed displacement $u_0$ is linear while $w_0$ is cubic. Then, from Eq. (10.2.2), we observe that $\varepsilon_0$ is cubic in $s$ while $\kappa$ is linear in $s$. One may use the three-point rule, the two-point rule, and the one-point rule to evaluate $\varepsilon_0$. Using fewer integration points means that fewer constraint conditions are imposed on the assumed displacement field. However, the reduced-order integration to alleviate element locking may also trigger spurious kinematic modes, which requires a careful examination.

## 10.3 Incompressibility Locking

Incompressibility locking occurs when the FE approximation is used to model a solid body of incompressible material – such as rubber with no volume change while undergoing deformation under applied loads. Referring to Eq. (A1.1.24) in Appendix 1, relative volume change (RVC), a non-dimensional measure of volume change, is

$$\text{RVC} = \varepsilon_{xx} + \varepsilon_{yy} + \varepsilon_{zz}. \tag{10.3.1}$$

For isotropic materials:

$$\varepsilon_{xx} = \left(\sigma_{xx} - v\sigma_{yy} - v\sigma_{zz}\right)/E,$$
$$\varepsilon_{yy} = \left(\sigma_{yy} - v\sigma_{xx} - v\sigma_{zz}\right)/E, \tag{10.3.2}$$
$$\varepsilon_{zz} = \left(\sigma_{zz} - v\sigma_{xx} - v\sigma_{yy}\right)/E.$$

Then

$$\varepsilon_{xx} + \varepsilon_{yy} + \varepsilon_{zz} = \left(\sigma_{xx} + \sigma_{yy} + \sigma_{zz}\right)(1 - 2v)/E. \tag{10.3.3}$$

For rubber-like materials:

$$RVC = \varepsilon_{xx} + \varepsilon_{yy} + \varepsilon_{zz} \to 0. \tag{10.3.4}$$

Therefore $v \to 1/2$. For a finite element, the condition of zero relative volume change may place constraints on the assumed displacement field and prevent the element from deforming. This phenomenon is called "incompressibility locking." To appreciate this, we may consider a four-node plane strain element as shown in Figure 10.10. For an element of rectangular shape, the assumed displacement can be expressed as

$$\begin{aligned} u &= a_1 + a_2 x + a_3 y + a_4 xy, \\ v &= b_1 + b_2 x + b_3 y + b_4 xy. \end{aligned} \tag{10.3.5}$$

For the plane strain problem, $\varepsilon_{zz} = 0$ and thus

$$\varepsilon_{xx} + \varepsilon_{yy} + \cancel{\varepsilon_{zz}} = \frac{\partial u}{\partial x} + \frac{\partial v}{\partial y} = a_2 + a_4 y + b_3 + b_4 x. \tag{10.3.6}$$

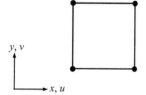

**Figure 10.10** Four-node plane strain element

$y, v$

$x, u$

As $v \to 1/2$, $\varepsilon_{xx} + \varepsilon_{yy} \to 0$. In the limiting case:

$$\varepsilon_{xx} + \varepsilon_{yy} = 0. \tag{10.3.7}$$

Therefore, from Eq. (10.3.6), we observe that the incompressibility condition imposes three constraints on the assumed displacement field, as follows:

$$\begin{aligned} a_2 + b_3 &= 0, \\ a_4 &= 0, \\ b_4 &= 0. \end{aligned} \tag{10.3.8}$$

The assumed displacement field in Eq. (10.3.5) then reduces to

$$\begin{aligned} u &= a_1 + a_2 x + a_3 y, \\ v &= b_1 + b_2 x - a_2 y. \end{aligned} \tag{10.3.9}$$

To appreciate the effect of the constraint conditions in Eq. (10.3.8), let's consider the case shown in Figure 10.11.

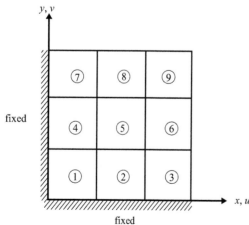

**Figure 10.11** Plane strain problem modeled with the four-node elements

For element #1:

Along the $y = 0$ line: $u = a_1 + a_2 x$.
Since $u = 0$ along the $y = 0$ line: $a_1 = a_2 = 0$.
Along the $x = 0$ line: $u = a_3 y$.
Since $u = 0$ along the $x = 0$ line: $a_3 = 0$.

Accordingly, $u = 0$ everywhere in element #1. Similarly, $v = 0$ everywhere in element #1. So, element #1 cannot deform. We can then observe that elements #2 and #4 cannot deform. This reasoning can be extended to all elements. So, the FE model is rigid and cannot deform, regardless of the applied loads.

Suppose now numerical integration is used to construct the element stiffness matrix. For incompressible materials, $\varepsilon_{xx} + \varepsilon_{yy} = 0$ at the integration points. If the $2 \times 2$ point rule is used to integrate the four-node element, it appears that four constraint conditions are imposed. However, as shown in Eq. (10.3.8), we know that there are three constraints. This means that one of the four constraint equations from the $2 \times 2$ point rule is linearly dependent. To reduce locking, one may try reduced-order integration with the one-point rule. However, once again, this will trigger spurious kinematic modes, as shown in the following section.

## 10.4 Spurious Kinematic Modes of 2D and 3D Elements

Reduced-order integration may trigger spurious kinematic modes which render individual elements kinematically unstable.

### Four-Node Rectangular Element

As an example, consider a four-node element of rectangular shape as shown in Figure 10.10, with the assumed displacement field expressed in Eq. (10.3.5). Then

$$\varepsilon_{xx} = \frac{\partial u}{\partial x} = a_2 + a_4 y,$$

$$\varepsilon_{yy} = \frac{\partial v}{\partial y} = b_3 + b_4 x, \tag{10.4.1}$$

$$\varepsilon_{xy} = \frac{\partial u}{\partial y} + \frac{\partial v}{\partial x} = a_3 + b_2 + a_4 x + b_4 y.$$

For simplicity, assume that the origin of the coordinates is located at the element centroid. If the one-point rule is used for reduced-order integration:

$$\varepsilon_{xx} = a_2,$$
$$\varepsilon_{yy} = b_3, \tag{10.4.2}$$
$$\varepsilon_{xy} = a_3 + b_2.$$

We observe that the $a_4$ and $b_4$ terms do not produce strain. The corresponding displacement modes are the spurious kinematic modes, called the "hour glass" modes. They are

$$u = a_4 xy, \quad v = b_4 xy. \tag{10.4.3}$$

Consider a $2 \times 2$ mesh of square elements of identical geometry, as shown in Figure 10.12. Set $a_4 = c$, where $c$ is a positive constant, then $u = cxy$ is the spurious kinematic mode. The figure also shows that the kinematic mode of individual elements is compatible or communicative with the adjacent elements. We can draw a similar sketch for the $v = b_4 xy$ mode.

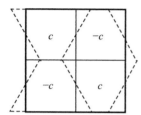

**Figure 10.12** Spurious hour-glass modes (with $c > 0$) in the $2 \times 2$ element mesh

In general, for identification of the spurious modes of an element, we note that

$$\mathbf{k}^e \mathbf{d}_s = \mathbf{0}, \tag{10.4.4}$$

where $\mathbf{k}^e$ is the element stiffness matrix and $\mathbf{d}_s$ is the DOF vector corresponding to the kinematic modes. Accordingly, we may carry out eigenvalue analysis of the element stiffness matrix as follows:

$$\mathbf{k}^e \mathbf{d} = \lambda \mathbf{Id}, \tag{10.4.5}$$

where $\mathbf{I}$ is an identity matrix and $\mathbf{d}$ is the element DOF vector. The above equation is in the standard form for eigenvalue analysis, which will reveal zero eigenvalues and eigenvectors corresponding to the rigid-body modes as well as nonzero eigenvalues and corresponding elastic modes. Accordingly, for a 2D element for plane stress and plane strain analysis, there exist three zero eigenvalues corresponding to two rigid-body translations and one rotation, while for a 3D element, there are six zero eigenvalues corresponding to three rigid-body translations and three rigid-body rotations. Any spurious mode existing in the element will manifest as a mode with additional zero eigenvalue.

**Eight-Node Plane Element**

For an eight-node element of square shape as shown in Figure 10.13, $3 \times 3$ point integration is needed for exact integration of the element stiffness matrix. However, eigenvalue analysis will show that $2 \times 2$ point integration triggers a spurious kinematic mode. For an element with side length of $a = 2$, the spurious mode is as follows:

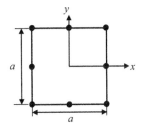

**Figure 10.13** Eight-node element of square geometry

$$u = c\xi(1 - 3\eta^2),$$
$$v = -c\eta(1 - 3\xi^2). \qquad (10.4.6)$$

We can also sketch the above mode on a $2 \times 2$ mesh to show that it is incompatible and gets suppressed. Accordingly, one can use the eight-node element with $2 \times 2$ reduced-order integration as long as the mesh size is equal to or greater than $2 \times 2$. The details are left as a problem at the end of this chapter.

### 20-Node Solid Element

For the 20-node element of rectangular shape, $3 \times 3 \times 3$ point integration is needed for exact integration of the element stiffness matrix. $2 \times 2 \times 2$ point integration triggers spurious kinematic modes, which can be identified by conducting eigenvalue analysis. However, one can show that these kinematic modes are incompatible and disappear if more than one element is used in each of the three directions. The reader may try one of the problems at the end of this chapter to appreciate the kinematic stability of the 20-node element with $2 \times 2 \times 2$ point integration.

## 10.5 Formulations with the Assumed Strain or Stress Field

In order to alleviate the detrimental effect of locking on an element, an assumed strain formulation may be used in which the strain field is assumed independent of the assumed displacement field. The coefficients of the assumed strain field are eliminated at the element level. For example, the independently assumed strain can be symbolically expressed as

$$\hat{\varepsilon} = \mathbf{P}\alpha, \qquad (10.5.1)$$

while the strain derived from the assumed displacement can be expressed as

$$\varepsilon = \mathbf{B}\mathbf{d}, \qquad (10.5.2)$$

where $\mathbf{d}$ is the column vector of nodal DOF. For an example of the assumed strain field, check Problem 10.6 at the end of this chapter. Considering the difference between the two strain vectors, one may construct a weighted least-square function $L_\alpha$ such that

$$L_\alpha = \int_{V_e} (\hat{\varepsilon} - \varepsilon)^{\mathrm{T}} \mathbf{C}(\hat{\varepsilon} - \varepsilon)dV. \qquad (10.5.3)$$

In the above equation, the $\mathbf{C}$ matrix serves as a weight and the integration is over the element. Substituting Eqs (10.5.1) and (10.5.2) into Eq. (10.5.3):

$$L_\alpha = \int_{V_e} (\mathbf{P}\alpha - \mathbf{B}\mathbf{d})^T \mathbf{C}(\mathbf{P}\alpha - \mathbf{B}\mathbf{d})dV = \alpha^T\mathbf{H}\alpha - 2\alpha^T\mathbf{G}\mathbf{d} + \cdots, \quad (10.5.4)$$

where

$$\mathbf{H} = \int_{V_e} \mathbf{P}^T\mathbf{C}\mathbf{P}dV, \mathbf{G} = \int_{V_e} \mathbf{P}^T\mathbf{C}\mathbf{B}dV. \quad (10.5.5)$$

Taking the derivative $L_\alpha$ in Eq. (10.5.4) with respect to $\alpha$ and setting it to zero results in

$$\mathbf{H}\alpha = \mathbf{G}\mathbf{d} \rightarrow \alpha = \mathbf{H}^{-1}\mathbf{G}\mathbf{d}. \quad (10.5.6)$$

The above equation relates $\alpha$ to the element DOF vector. Then the assumed strain in Eq. (10.5.1) can be expressed as

$$\hat{\varepsilon} = \mathbf{P}\mathbf{H}^{-1}\mathbf{G}\mathbf{d}, \quad (10.5.7)$$

which will be used to substitute the displacement-dependent strain in the FE formulation. In terms of the assumed strain, the element strain energy can be expressed as

$$U_e = \frac{1}{2}\int_{V_e} \hat{\varepsilon}^T\mathbf{C}\hat{\varepsilon}dV. \quad (10.5.8)$$

Substituting Eq. (10.5.7) into Eq. (10.5.8):

$$U_e = \frac{1}{2}\mathbf{d}^T\mathbf{G}^T\mathbf{H}^{-1}\mathbf{G}\mathbf{d} = \frac{1}{2}\mathbf{d}^T\mathbf{k}^e\mathbf{d}, \quad (10.5.9)$$

where

$$\mathbf{k}^e = \mathbf{G}^T\mathbf{H}^{-1}\mathbf{G}: \text{ element stiffness matrix.} \quad (10.5.10)$$

The assumed strain field must be judiciously selected to alleviate locking without triggering any spurious kinematic modes when the elements are assembled. There exist other approaches to introducing the assumed strain field. Detailed information on these approaches can be found in the open literature.

Alternatively, we may introduce an independently assumed stress field in addition to the assumed displacement. The coefficient of the assumed stress field can then be expressed in terms of the nodal DOF, which can be used for construction of the element stiffness matrix. We may also consider the FE formulation in which both an assumed stress field and an assumed strain field are introduced, in addition to the assumed displacement, to construct the element stiffness matrix.

## PROBLEMS

**10.1** The cantilever beam shown in Figure 10.14 is modeled with the four-node plane stress elements of rectangular shape. The beam is modeled using a $2 \times 10$ mesh with two elements in the thickness direction and 10 elements in the length direction. Do the following and compare with the B–E beam theory solution for the vertical displacement at the tip.

(a) The solution obtained using the FE model exhibits the effect of transverse shear locking when the $2 \times 2$ point rule is used. What are the constraint equations on the four-node element that cause element locking? What are the element displacement fields when the constraints are active?

(b) A dramatic improvement is observed when the $1 \times 2$ point rule (with the one-point rule in the horizontal direction and the two-point rule in the vertical direction) is used. What are the constraint equations corresponding to the $1 \times 2$ point rule?

(c) Are there any spurious kinematic modes when the $1 \times 2$ point rule is used?

$$p_y = A\left(1 - \frac{x}{L}\right) \text{ lb/in}, E = 10^7 \text{psi}, \quad \nu = 0.3,$$

$$L = 10 \text{ in}, \quad h = 1.0 \text{ in},$$

$$\text{width (in the } z \text{ direction)} = 0.4 \text{ in}.$$

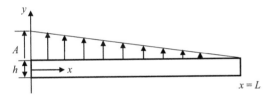

**Figure 10.14** For Problem 10.2

**10.2** Consider an eight-node plane stress element of square shape as shown in Figure 10.13.

(a) Show that the mapping reduces to

$$x = \frac{a}{2}\xi, \quad y = \frac{a}{2}\eta.$$

(b) The $2 \times 2$ point rule triggers one spurious kinematic mode. Confirm via an eigen value analysis that for $a = 2$ the displacement field given below is indeed the spurious kinematic mode:

$$u = c\xi\left(1 - 3\eta^2\right), \quad v = -c\eta\left(1 - 3\xi^2\right).$$

(c) Sketch the kinematic mode for the element.

(d) Show that, for a $2 \times 2$ mesh of square elements, the kinematic mode identified in (b) is suppressed.

**10.3** For the 20-node solid element, rectangular in shape and bounded by $x = \pm 1$, $y = \pm 1$, and $z = \pm 1$, the assumed displacement can be expressed as

$$\begin{Bmatrix} u \\ v \\ w \end{Bmatrix} = \begin{bmatrix} a_1 & a_2 & \cdots & \cdots & a_{20} \\ b_1 & b_2 & \cdots & \cdots & b_{20} \\ c_1 & c_2 & \cdots & \cdots & c_{20} \end{bmatrix} \mathbf{P},$$

where $\mathbf{P}$ is a $20 \times 1$ column vector of polynomial terms placed as follows:

$$1, x, y, z, x^2, y^2, z^2, xy, yz, zx, x^2y, x^2z, xy^2, y^2z, xz^2, yz^2, xyz, x^2yz, xy^2z, xyz^2.$$

(a) For the element with $2 \times 2 \times 2$ point integration, carry out eigenvalue analysis to identify the spurious kinematic modes as follows:

$$
\begin{aligned}
u &= a_2 x(1 - 3y^2), & v &= -a_2 y(1 - 3x^2), \\
u &= a_{10} zx(1 - 3y^2), & v &= -a_{10} zy(1 - 3x^2), \\
u &= a_2 x(1 - 3z^2), & w &= -a_2 z(1 - 3x^2), \\
u &= a_8 yx(1 - 3z^2), & w &= -a_8 yz(1 - 3x^2), \\
v &= b_3 y(1 - 3z^2), & w &= -b_3 z(1 - 3y^2), \\
v &= b_8 xy(1 - 3z^2), & w &= -b_8 xz(1 - 3y^2),
\end{aligned}
$$

where $a_2, a_8, b_3, b_8, a_{10}$ are undetermined constants.

(b) Show that these modes are incompatible and disappear if more than one elements is used in the $x, y,$ and $z$ directions.

**10.4** For the 20-node element of rectangular shape bounded by $x = \pm 1$, $y = \pm 1$, and $z = \pm 1$, $2 \times 2 \times 3$ point integration is used with three points in the $z$ direction.

(a) Show that the spurious kinematic modes are

$$
\begin{aligned}
u &= a_2 x(1 - 3y^2), & v &= -a_2 y(1 - 3x^2), \\
u &= a_{10} zx(1 - 3y^2), & v &= -a_{10} zy(1 - 3x^2).
\end{aligned}
$$

(b) Show that these modes are incompatible and disappear if more than one elements is used in the $x$ and $y$ directions.

**10.5** For construction of the element stiffness matrix via the assumed strain formulation, consider the four-node plane stress element as shown in Figure 10.10. The nodal coordinates (in meters) are node 1: (2, 2), node 2: (1, 2), node 3: (1, 1), node 4: (2, 1). Material properties are $E = 72$ GPa and $\nu = 0.33$. The independent strain is assumed as

$$
\begin{aligned}
\varepsilon_{xx} &= \alpha_1 + \alpha_2 y, \\
\varepsilon_{yy} &= \alpha_3 + \alpha_4 x, \\
\varepsilon_{xy} &= \alpha_5.
\end{aligned}
$$

In matrix form, the assumed strain can be expressed as

$$
\left\{ \begin{array}{c} \varepsilon_{xx} \\ \varepsilon_{yy} \\ \varepsilon_{xy} \end{array} \right\} = \begin{bmatrix} 1 & y & 0 & 0 & 0 \\ 0 & 0 & 1 & x & 0 \\ 0 & 0 & 0 & 0 & 1 \end{bmatrix} \left\{ \begin{array}{c} \alpha_1 \\ \alpha_2 \\ \alpha_3 \\ \alpha_4 \\ \alpha_5 \end{array} \right\} \rightarrow \boldsymbol{\varepsilon} = \mathbf{P}\boldsymbol{\alpha}.
$$

Construct the element stiffness matrix using $2 \times 2$ point integration.

# 11 Heat Transfer

In Chapters 2 and 8, it was shown that temperature changes can cause thermally induced strain and stress in solids and structures. In this chapter, we present the FE formulation of heat transfer problems which can be used to determine temperature distributions in solid bodies, starting with heat conduction in the 1D domain. For 1D problems, heat energy balance under the steady-state condition is considered for a segment of infinitesimal length to derive the governing equation involving temperature and a heat source. Similar to the notion of virtual displacement in earlier chapters, a virtual temperature or an arbitrary weight function is introduced to derive an integral equivalent of the governing equation to which the FE formulation is applied. A two-node element is used to illustrate construction of the heat transfer matrix and load vector, including the effect of heat convection. This is followed by discussions of the unsteady, time-dependent heat conduction in the 1D domain. The FE formulation results in a set of first-order equations in time involving the capacitance matrix, which can be solved to determine nodal temperatures as a function of time.

Subsequently, we consider heat transfer in 2D and 3D domains. First, the governing equation is derived for the steady-state heat conduction in the 3D domain by considering heat energy balance of an infinitesimal volume. A virtual temperature or an arbitrary weight function is then introduced to derive an integral equivalent of the governing equation in a scalar form. The equation for 2D heat conduction is then derived from that for 3D heat conduction under simplifying assumptions. For the FE formulation, steady heat conduction in the 2D domain is introduced to illustrate construction of the conduction matrix using the four-node element as an example. This is followed by the FE formulation of the steady state and unsteady, time-dependent heat conduction in the 3D domain.

We may note that mathematical equivalences exist between the heat conduction equation and scalar field equations for other physical phenomena. Examples include diffusion, incompressible potential flow, fluid flow through porous bodies, and electric currents. Accordingly, the FE formulation described in this chapter can be extended to modeling and analysis of physical problems beyond the example of heat transfer covered in this chapter.

## 11.1 Steady-State Heat Transfer in the 1D Domain

Consider heat conduction through a solid wall of thickness $L$ with one surface located at $x = 0$ and another surface at $x = L$. The dimension of the wall normal to the $x$-axis is very large or small compared with the wall thickness. Boundary conditions and heat source and temperature distributions are independent of the $y$ and $z$-coordinates and thus heat conduction is considered 1D. As shown in Figure 11.1, heat conduction may also be considered 1D in thin fins of small thickness (in the $z$ direction) and large width (in the $y$ direction) in which $f_s(x)$ is the heat source per unit length, dependent only on the $x$-coordinate.

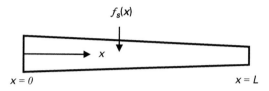

**Figure 11.1** Thin body undergoing 1D heat conduction

## 11.1.1 Heat Conduction Equation

To construct the governing equation for heat conduction in the 1D domain, consider heat energy balance for a body of infinitesimal length $dx$, as shown in Figure 11.2.

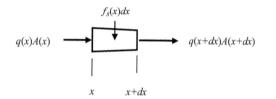

**Figure 11.2** Energy balance for a body of infinitesimal length

In the figure

$q(x)$: heat flux $(W/m^2)$
$A(x)$: cross-sectional area
$f_s(x)$: heat source per unit length $(W/m)$.

Introducing the heat flow $F(x)$,

$$F(x) = q(x)A(x),\tag{11.1.1}$$

the heat energy balance under steady-state condition is

$$F(x) + f_s(x)dx - F(x + dx) = 0$$

$$\rightarrow F + f_s dx - \left(F + \frac{\partial F}{\partial x}dx\right) = 0 \rightarrow \left(\frac{\partial F}{\partial x} - f_s\right)dx = 0.\tag{11.1.2}$$

From the above equation, we conclude that

$$\frac{\partial F}{\partial x} - f_s = 0.\tag{11.1.3}$$

According to Fourier's heat conduction law, heat flux is related to the temperature gradient as

$$q = -k\frac{\partial T}{\partial x},\tag{11.1.4}$$

where $T$ is the temperature and $k$ is the thermal conductivity (W/m°C) of the material. Substituting Eqs (11.1.1) and (11.1.4) into Eq. (11.1.3):

$$\frac{\partial}{\partial x}\left(kA\frac{\partial T}{\partial x}\right) + f_s = 0. \tag{11.1.5}$$

Introducing a virtual temperature or a weight function $\delta T$ which is arbitrary, and integrating over the entire domain, we can construct the following integral equation from Eq. (11.1.3):

$$\int_{x=0}^{x=L} \delta T\left(\frac{\partial F}{\partial x} - f_s\right) dx = 0. \tag{11.1.6}$$

For the first integral in the above equation, we apply integration by parts such that

$$\int_{x=0}^{x=L} \delta T \frac{\partial F}{\partial x} dx = \int_{x=0}^{x=L} \delta T dF = (F\delta T)_{x=0}^{x=L} - \int_{x=0}^{x=L} F d(\delta T) = (F\delta T)_{x=0}^{x=L} - \int_{x=0}^{x=L} F \frac{\partial \delta T}{\partial x} dx. \tag{11.1.7}$$

Substituting the last of the above equation into Eq. (11.1.6):

$$-\int_{x=0}^{x=L} F\frac{\partial \delta T}{\partial x} dx - \int_{x=0}^{x=L} \delta T f_s dx + (F\delta T)_{x=0}^{x=L} = 0. \tag{11.1.8}$$

Introducing Eqs (11.1.1) and (11.1.4) into the above equation:

$$\int_{x=0}^{x=L} k\frac{\partial \delta T}{\partial x}\frac{\partial T}{\partial x} A dx - \int_{x=0}^{x=L} \delta T f_s dx - \delta W_B = 0, \tag{11.1.9}$$

where

$$\delta W_B = -(F\delta T)_{x=L} + (F\delta T)_{x=0} \tag{11.1.10}$$

is the boundary term. For example, we may consider a boundary condition in which temperature and heat flux are prescribed as follows: $T = T_0$ at $x = 0$ and $q = q_L$ at $x = L$. In this case, the boundary term becomes

$$\delta W_B = -(q_L A\delta T)_{x=L} + (qA\delta T)_{x=0}. \tag{11.1.11}$$

Note that $\delta T = 0$ at $x = 0$, where the temperature is prescribed. Therefore, the second term on the right-hand side of the above equation is equal to zero. However, it is kept in the above equation to show the heat flux corresponding to the prescribed temperature. Note also that a mathematical equivalence exists between heat conduction in the 1D domain and uniaxial deformation of a slender body as described in Chapter 1.

## 11.1.2 Finite Element Formulation for 1D Heat Transfer

For the FE formulation, we can use Eq. (11.1.9) along with Eq. (11.1.10).

### Element Conduction Matrix

The domain is divided into elements such that the first term in Eq. (11.1.9) can be expressed as

$$\int_{x=0}^{x=L} kA \frac{\partial \delta T}{\partial x} \frac{\partial T}{\partial x} dx = \sum_e \int_{x=x_1}^{x=x_2} kA \frac{\partial \delta T}{\partial x} \frac{\partial T}{\partial x} dx, \tag{11.1.12}$$

where "$e$" stands for element number. For two-node elements, the mapping is expressed as

$$x = (1 - s)x_1 + sx_2 = x_1 + ls, \tag{11.1.13}$$

where $s = 0$ at node 1, $s = 1$ at node 2, and $l = x_2 - x_1$ is the element length. The temperature is assumed linear as

$$T = (1 - s)T_1 + sT_2, \tag{11.1.14}$$

where $T_1$ and $T_2$ are the nodal temperatures. Accordingly:

$$\frac{\partial T}{\partial x} = \frac{\partial T}{\partial s} \frac{ds}{dx} = \frac{1}{l}(T_2 - T_1) = \frac{1}{l} \lfloor -1 \ \ 1 \rfloor \begin{Bmatrix} T_1 \\ T_2 \end{Bmatrix}. \tag{11.1.15}$$

$$\text{Setting} \quad \delta T = (1 - s)\delta T_1 + s\delta T_2, \tag{11.1.16}$$

$$\frac{\partial \delta T}{\partial x} = \frac{1}{l}(\delta T_2 - \delta T_2) = \frac{1}{l} \lfloor \delta T_1 \ \ \delta T_2 \rfloor \begin{Bmatrix} -1 \\ 1 \end{Bmatrix}. \tag{11.1.17}$$

Then

$$\frac{\partial \delta T}{\partial x} \frac{\partial T}{\partial x} = \frac{1}{l^2} \lfloor \delta T_1 \ \ \delta T_2 \rfloor \begin{Bmatrix} -1 \\ 1 \end{Bmatrix} \lfloor -1 \ \ 1 \rfloor \begin{Bmatrix} T_1 \\ T_2 \end{Bmatrix} = \frac{1}{l^2} \lfloor \delta T_1 \ \ \delta T_2 \rfloor \begin{bmatrix} 1 & -1 \\ -1 & 1 \end{bmatrix} \begin{Bmatrix} T_1 \\ T_2 \end{Bmatrix}. \tag{11.1.18}$$

Substituting the above equation into an element in Eq. (11.1.12):

$$\int_{x=x_1}^{x=x_2} kA \frac{\partial \delta T}{\partial x} \frac{\partial T}{\partial x} dx = \int_{s=0}^{s=1} kA \frac{1}{l^2} \lfloor \delta T_1 \ \ \delta T_2 \rfloor \begin{bmatrix} 1 & -1 \\ -1 & 1 \end{bmatrix} \begin{Bmatrix} T_1 \\ T_2 \end{Bmatrix} l ds$$

$$= \left( \frac{1}{l} \int_{s=0}^{s=1} kA ds \right) \lfloor \delta T_1 \ \ \delta T_2 \rfloor \begin{bmatrix} 1 & -1 \\ -1 & 1 \end{bmatrix} \begin{Bmatrix} T_1 \\ T_2 \end{Bmatrix} = \lfloor \delta T_1 \ \ \delta T_2 \rfloor \begin{bmatrix} k_{11}^e & k_{12}^e \\ k_{21}^e & k_{22}^e \end{bmatrix} \begin{Bmatrix} T_1 \\ T_2 \end{Bmatrix}, \tag{11.1.19}$$

where

$$\begin{bmatrix} k_{11}^e & k_{12}^e \\ k_{21}^e & k_{22}^e \end{bmatrix} = \left( \frac{1}{l} \int_{s=0}^{s=1} kA \, ds \right) \begin{bmatrix} 1 & -1 \\ -1 & 1 \end{bmatrix} : \text{ element heat conduction matrix.} \quad (11.1.20)$$

For constant $kA$:

$$\begin{bmatrix} k_{11}^e & k_{12}^e \\ k_{21}^e & k_{22}^e \end{bmatrix} = \frac{kA}{l} \begin{bmatrix} 1 & -1 \\ -1 & 1 \end{bmatrix}. \quad (11.1.21)$$

The element conduction matrix bears a strong resemblance to the element stiffness matrix in Chapter 1. It is also symmetric and singular.

**Element Load Vector due to Heat Source**

For the second term in Eq. (11.1.9):

$$\int_{x=0}^{x=L} \delta T f_s \, dx = \sum_e \int_{x=x_1}^{x=x_2} \delta T f_s \, dx. \quad (11.1.22)$$

Accordingly

$$\int_{x=x_1}^{x=x_2} \delta T f_s \, dx = \int_{s=0}^{s=1} [(1-s)\delta T_1 + s\delta T_2] f_s l \, ds = \delta T_1 \int_{s=0}^{s=1} (1-s) f_s l \, ds + \delta T_2 \int_{s=0}^{s=1} s f_s l \, ds. \quad (11.1.23)$$

The above equation can be expressed as

$$\int_{x=x_1}^{x=x_2} \delta T f_s \, dx = \delta T_1 Q_1^e + \delta T_2 Q_2^e = \lfloor \delta T_1 \quad \delta T_2 \rfloor \begin{Bmatrix} Q_1^e \\ Q_2^e \end{Bmatrix}, \quad (11.1.24)$$

where

$$Q_1^e = l \int_{s=0}^{s=1} (1-s) f_s \, ds, \quad Q_2^e = l \int_{s=0}^{s=1} s f_s \, ds \quad (11.1.25)$$

and

$$\begin{Bmatrix} Q_1^e \\ Q_2^e \end{Bmatrix} : \text{ element heat load vector due to } f_s.$$

The FE formulation based on Eq. (11.1.9) results in the following equation:

$$\delta \mathbf{q}^T (\mathbf{Kq} - \mathbf{F}) = 0 \rightarrow \mathbf{Kq} = \mathbf{F}, \quad (11.1.26)$$

where **q** is the column vector of global nodal temperatures, **K** is the global conduction matrix assembled from element conduction matrices using the connectivity between the element nodes and the global nodes, and **F** is the global heat load vector also assembled from the element heat load vectors.

## Example 11.1

Consider 1D steady-state heat conduction subjected to the following boundary conditions: $T = 0$ at $x = 0$ and $q = q_L$ at $x = L$. In addition, a constant heat source of $f_s(x) = BA$ is present over the body of uniform cross-sectional area. Use an FE model with three elements of equal length to do the following.

(a) Construct the global conduction matrix.
(b) Construct the element load vectors due to the heat source and assemble the global load vector.
(c) Determine the nodal temperatures. Plot the temperature distribution along the body and compare with the exact solution.
(d) Determine the heat flux in each element. Plot the heat flux distribution along the body and compare with the exact solution.

## Solution:

(a) For an element #e:

$$\begin{bmatrix} k_{11}^e & k_{12}^e \\ k_{21}^e & k_{22}^e \end{bmatrix} = \frac{kA}{l} \begin{bmatrix} 1 & -1 \\ -1 & 1 \end{bmatrix}, \quad l = \frac{L}{3}.$$

After assembly, the global conduction matrix is

$$\mathbf{K} = \frac{3kA}{L} \begin{bmatrix} 1 & -1 & 0 & 0 \\ -1 & 2 & -1 & 0 \\ 0 & -1 & 2 & -1 \\ 0 & 0 & -1 & 1 \end{bmatrix}.$$

(b) Element load vectors due to the constant heat source are determined using Eq. (11.1.25) as follows. For element #e:

$$Q_1^e = l \int_{s=0}^{s=1} (1-s)f_s ds = BAl \int_{s=0}^{s=1} (1-s)ds = \frac{BAl}{2} = \frac{BAL}{6},$$

$$Q_2^e = l \int_{s=0}^{s=1} sf_s ds = BAl \int_{s=0}^{s=1} sds = \frac{BAl}{2} = \frac{BAL}{6}.$$

For the three-element model, Eq. (11.1.11) can be expressed as

$$\delta W_B = -(q_L A)\delta q_4 + (R_1 A)\delta q_1,$$

where $\delta q_4 = (\delta T)_{x=L}$, $\delta q_1 = (\delta T)_{x=0}$, and $R_1 = q_{x=0}$ is the heat flux corresponding to the prescribed temperature. The global load vector is assembled as

$$\mathbf{F} = \frac{BAL}{6}\begin{Bmatrix} 1 \\ 2 \\ 2 \\ 1 \end{Bmatrix} + \begin{Bmatrix} R_1 A \\ 0 \\ 0 \\ -q_L A \end{Bmatrix} = \frac{BAL}{6}\begin{Bmatrix} 1 + 6R_1/(BL) \\ 2 \\ 2 \\ 1 - 6q_L/(BL) \end{Bmatrix}.$$

(c) With the global conduction matrix assembled in part (a) and the global load vector assembled in part (b), the FE equation is

$$\frac{3kA}{L}\begin{bmatrix} 1 & -1 & 0 & 0 \\ -1 & 2 & -1 & 0 \\ 0 & -1 & 2 & -1 \\ 0 & 0 & -1 & 1 \end{bmatrix}\begin{bmatrix} q_1 \\ q_2 \\ q_3 \\ q_4 \end{bmatrix} = \frac{BAL}{6}\begin{Bmatrix} 1 + 6R_1/(BL) \\ 2 \\ 2 \\ 1 - 6q_L/(BL) \end{Bmatrix}.$$

Setting $T_1 = 0$ to apply the temperature boundary condition, and deleting the first equation, the reduced equation is

$$\frac{3k}{L}\begin{bmatrix} 2 & -1 & 0 \\ -1 & 2 & -1 \\ 0 & -1 & 1 \end{bmatrix}\begin{Bmatrix} q_2 \\ q_3 \\ q_4 \end{Bmatrix} = \frac{BL}{6}\begin{Bmatrix} 2 \\ 2 \\ 1 - 6q_L/(BL) \end{Bmatrix}$$

$$\rightarrow \begin{bmatrix} 2 & -1 & 0 \\ -1 & 2 & -1 \\ 0 & -1 & 1 \end{bmatrix}\begin{Bmatrix} q_2 \\ q_3 \\ q_4 \end{Bmatrix} = \frac{BL^2}{18k}\begin{Bmatrix} 2 \\ 2 \\ 1 - 6\bar{q}_L \end{Bmatrix}, \bar{q}_L = \frac{q_L}{BL}.$$

Solving the equation above, we get

$$\begin{Bmatrix} q_2 \\ q_3 \\ q_4 \end{Bmatrix} = \frac{BL^2}{18k}\begin{Bmatrix} 5 - 6\bar{q}_L \\ 8 - 12\bar{q}_L \\ 9 - 18\bar{q}_L \end{Bmatrix}.$$

As an example, for $\bar{q}_L = 0$

$$\begin{Bmatrix} q_2 \\ q_3 \\ q_4 \end{Bmatrix} = \frac{BL^2}{18k}\begin{Bmatrix} 5 \\ 8 \\ 9 \end{Bmatrix}.$$

From Eq. (11.1.5), we can determine the exact temperature as

$$T = -\frac{BL^2}{k}\left[\frac{1}{2}\left(\frac{x}{L}\right)^2 + (\bar{q}_L - 1)\frac{x}{L}\right].$$

For $\bar{q}_L = 0$:

$$T = -\frac{BL^2}{k}\left[\frac{1}{2}\left(\frac{x}{L}\right)^2 - \left(\frac{x}{L}\right)\right] = \frac{BL^2}{2k}\left[2\left(\frac{x}{L}\right) - \left(\frac{x}{L}\right)^2\right].$$

We may plot non-dimensional temperature $\bar{T}$ vs. non-dimensional length $\bar{x}$ as shown in Figure 11.3, where

$$\bar{T} = \frac{T}{\left(\frac{BL^2}{2k}\right)}, \quad \bar{x} = \frac{x}{L}.$$

**Figure 11.3** Temperature distribution along the length

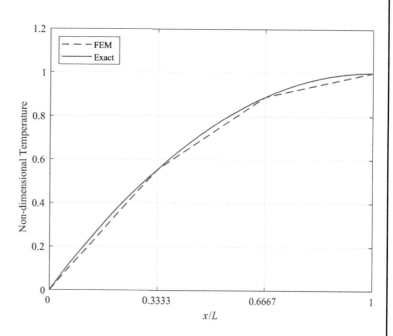

(d) Heat flux is related to temperature gradient such that

$$q = -k\frac{\partial T}{\partial x}.$$

For element #$e$:

$$q^e = -k\frac{\partial T}{\partial x} = -k\frac{1}{l}(T_2 - T_1) = -\frac{3k}{L}(T_2 - T_1).$$

Accordingly, using the connectivity and the global nodal temperatures already determined:

$$\begin{Bmatrix} q^1_{} \\ q^2_{} \\ q^3_{} \end{Bmatrix} = -\frac{3k}{L} \begin{Bmatrix} q_2 - q_1 \\ q_3 - q_2 \\ q_4 - q_3 \end{Bmatrix} = BL \begin{Bmatrix} \bar{q}_L - 5/6 \\ \bar{q}_L - 1/2 \\ \bar{q}_L - 1/6 \end{Bmatrix}.$$

For $\bar{q}_L = 0$, $\begin{Bmatrix} q^1_{} \\ q^2_{} \\ q^3_{} \end{Bmatrix} = -\frac{BL}{6} \begin{Bmatrix} 5 \\ 3 \\ 1 \end{Bmatrix}.$

The exact heat flux is

$$q = -k\frac{\partial T}{\partial x} = BL\left[\frac{x}{L} + (\bar{q} - 1)\right].$$

For $\bar{q}_L = 0$, $q = BL(x/L - 1) = -BL(1 - x/L)$.

We may plot non-dimensional flux $\bar{q}$ vs. $\bar{x}$ as shown in Figure 11.4, where

$$\bar{q} = \frac{q}{(-BL)}.$$

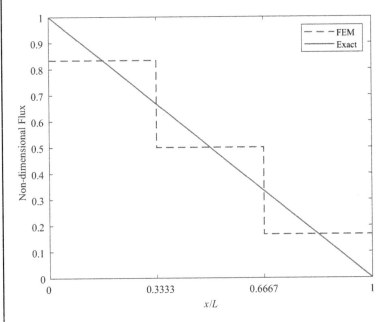

**Figure 11.4** Heat flux along the length

### Convection at the Boundary

For convection via moving fluid or gas at $x = L$, the heat flux can be expressed as

$$q = h(T - T_\infty), \tag{11.1.27}$$

where $h$ is the convection coefficient $(\text{W/m}^2 \cdot {}^\circ\text{C})$ and $T_\infty$ is the ambient temperature away from the solid boundary. Then, substituting the above equation into Eq. (11.1.11):

$$-\delta W_B = [Ah(T - T_\infty)\delta T]_{x=L} - (qA\delta T)_{x=0}$$
$$= (AhT\delta T)_{x=L} - (AhT_\infty \delta T)_{x=L} - (qA\delta T)_{x=0}. \tag{11.1.28}$$

We observe that the first term on the right-hand side of the above equation is dependent on the unknown temperature $T$. Accordingly, it contributes to the conduction matrix while the remaining two terms contribute to the load vector.

## Example 11.2

Consider 1D steady-state heat conduction subjected to the following boundary conditions: $T = 0$ at $x = 0$ and $q = h(T - T_\infty)$ at $x = L$. The body of uniform cross-sectional area is modeled with three elements of equal length.

(a) Construct the global heat transfer matrix.
(b) Construct the global load vector.

## Solution:

(a) For the three-element model, the first term on the right-hand side of Eq. (11.1.28) is

$$(AhT\delta T)_{x=L} = \delta q_4 (Ah) q_4.$$

Accordingly, the global heat transfer matrix is

$$\mathbf{K} = \frac{3kA}{L} \begin{bmatrix} 1 & -1 & 0 & 0 \\ -1 & 2 & -1 & 0 \\ 0 & -1 & 2 & -1 \\ 0 & 0 & -1 & 1 + h(L/3k) \end{bmatrix},$$

where the second term in $K_{44}$ is the contribution from the boundary convection.

(b) For the second term on the right-hand side of Eq. (11.1.28):

$$(AhT_\infty \delta T)_{x=L} = \delta q_4 (AhT_\infty).$$

Accordingly

$$\mathbf{F} = \frac{BAL}{6} \begin{Bmatrix} 1 \\ 2 \\ 2 \\ 1 \end{Bmatrix} + \begin{Bmatrix} R_1 A \\ 0 \\ 0 \\ AhT_\infty \end{Bmatrix} = \frac{BAL}{6} \begin{Bmatrix} 1 + R_1(6/BL) \\ 2 \\ 2 \\ 1 + T_\infty h(6/BL) \end{Bmatrix}.$$

**Convection on the Side Surface**

For thin fins, convection on the side surface along the body due to fluid or gas motion can be approximated as a negative heat source. This is distinct from the boundary convection treated previously because the side surface convection is felt along the entire 1D domain. The heat energy loss per unit length may then be expressed as

$$(f_s)_c = -Ph(T - T_\infty),$$
(11.1.29)

where $P$ is the perimeter of the cross-section. Equation (11.1.9) is then modified as

$$\int_{x=0}^{x=L} k\frac{\partial \delta T}{\partial x}\frac{\partial T}{\partial x}A dx - \int_{x=0}^{x=L} \delta T f_s dx - \int_{x=0}^{x=L} \delta T(f_s)_c dx - \delta W_B = 0.$$
(11.1.30)

Substituting Eq. (11.1.29) into the above equation:

$$\int_{x=0}^{x=L} k\frac{\partial \delta T}{\partial x}\frac{\partial T}{\partial x}A dx - \int_{x=0}^{x=L} \delta T f_s dx + \int_{x=0}^{x=L} \delta T Ph(T - T_\infty) dx - \delta W_B = 0.$$
(11.1.31)

Finally

$$\int_{x=0}^{x=L} k\frac{\partial \delta T}{\partial x}\frac{\partial T}{\partial x}A dx - \int_{x=0}^{x=L} \delta T f_s dx + \int_{x=0}^{x=L} \delta T h T Pdx - \int_{x=0}^{x=L} \delta T(hT_\infty)Pdx - \delta W_B = 0.$$
(11.1.32)

Observe that the third term depends on $T$ and thus adds to the conduction matrix, while the fourth term adds to the load vector. For the two-node element:

$$T = (1 - s)T_1 + sT_2 = N_1T_1 + N_2T_2,$$
(11.1.33)

where

$$N_1 = 1 - s, \quad N_2 = s$$
(11.1.34)

and

$$\delta T = N_1\delta T_1 + N_2\delta T_2 = \lfloor \delta T_1 \quad \delta T_2 \rfloor \begin{Bmatrix} N_1 \\ N_2 \end{Bmatrix}.$$
(11.1.35)

Then

$$\int_{x=x_1}^{x=x_2} \delta T h T Pdx = \int_{s=0}^{s=1} \lfloor \delta T_1 \quad \delta T_2 \rfloor \begin{Bmatrix} N_1 \\ N_2 \end{Bmatrix} \lfloor N_1 \quad N_2 \rfloor \begin{Bmatrix} T_1 \\ T_2 \end{Bmatrix} hPl ds$$

$$= \lfloor \delta T_1 \quad \delta T_2 \rfloor \left( l \int_{s=0}^{s=1} \begin{bmatrix} N_1^2 & N_1N_2 \\ N_2N_1 & N_2^2 \end{bmatrix} hP ds \right) \begin{Bmatrix} T_1 \\ T_2 \end{Bmatrix}.$$
(11.1.36)

The above equation can be expressed as

$$\int_{x=x_1}^{x=x_2} \delta Th TP dx = \delta \mathbf{d}^T \mathbf{k}_c^e \mathbf{d}, \tag{11.1.37}$$

where

$$\mathbf{k}_c^e = \left( l \int_{s=0}^{s=1} \begin{bmatrix} N_1^2 & N_1 N_2 \\ N_2 N_1 & N_2^2 \end{bmatrix} hP ds \right) : \text{element convection matrix} \tag{11.1.38}$$

$$\mathbf{d} = \left\{ \begin{array}{c} T_1 \\ T_2 \end{array} \right\} : \text{element DOF vector.} \tag{11.1.39}$$

For uniform $hP$, it can be shown that

$$\mathbf{k}_c^e = \frac{hPl}{6} \begin{bmatrix} 2 & 1 \\ 1 & 2 \end{bmatrix}. \tag{11.1.40}$$

For the fourth term in Eq. (11.1.32):

$$\int_{x=x_1}^{x=x_2} \delta T (hT_\infty) P dx = \int_{s=0}^{s=1} [(1-s)\delta T_1 + s\delta T_2](hT_\infty) Pl ds$$

$$= \delta T_1 \int_{s=0}^{s=1} (1-s)(hT_\infty) Pl ds + \delta T_2 \int_{s=0}^{s=1} s(hT_\infty) Pl ds. \tag{11.1.41}$$

The above equation can be expressed as

$$\int_{x=x_1}^{x=x_2} \delta T (hT_\infty) P dx = \lfloor \delta T_1 \quad \delta T_2 \rfloor \left\{ \begin{array}{c} Q_{1,c}^e \\ Q_{2,c}^e \end{array} \right\}, \tag{11.1.42}$$

where

$$Q_{1,c}^e = l \int_{s=0}^{s=1} (1-s)(hT_\infty) P ds, \quad Q_{2,c}^e = l \int_{s=0}^{s=1} s(hT_\infty) P ds. \tag{11.1.43}$$

For uniform $(hT_\infty)P$:

$$Q_{1,c}^e = Q_{2,c}^e = \frac{hT_\infty Pl}{2}. \tag{11.1.44}$$

## ▶ Example 11.3

Consider an aluminum fin, 0.35 cm thick and 8.75 cm long, protruding from a wall. The temperature at the end $(x = 0)$ attached to the wall is maintained at 260°C and the ambient temperature is 30°C. For the fin, $k = 200$ W/m · °C and the convection coefficient is $h = 10$ W/m² · °C. The fin is very large in the third dimension relative to its length. Use a model with four elements of equal length to do the following.

(a) Determine the temperature distribution along the fin.
(b) Determine the heat loss from the fin per unit depth.

## Solution:

(a) For a unit depth of 1 m, perimeter $P = 2 + 2 \times 0.35 \times 100^{-2} = 2.007$. The element matrix due to heat conduction and surface convection, and the element load vector is as follows.
  For elements 1 and 2:

$$\mathbf{k}^e = \frac{kA}{l}\begin{bmatrix} 1 & -1 \\ -1 & 1 \end{bmatrix} + \frac{hPl}{6}\begin{bmatrix} 2 & 1 \\ 1 & 2 \end{bmatrix}, \quad \mathbf{f}^e = \frac{hT_\infty Pl}{2}\begin{Bmatrix} 1 \\ 1 \end{Bmatrix}.$$

  For element 3:

$$\mathbf{k}^e = \frac{kA}{l}\begin{bmatrix} 1 & -1 \\ -1 & 1 \end{bmatrix} + \frac{hPl}{6}\begin{bmatrix} 2 & 1 \\ 1 & 2 \end{bmatrix} + hA\begin{bmatrix} 0 & 0 \\ 0 & 1 \end{bmatrix},$$

$$\mathbf{f}^e = \frac{hT_\infty Pl}{2}\begin{Bmatrix} 1 \\ 1 \end{Bmatrix} + AhT_\infty\begin{Bmatrix} 0 \\ 1 \end{Bmatrix}.$$

  The global matrix and the global load vector are assembled to construct the governing equation and then the boundary condition at $x = 0$ is prescribed. The nodal temperature is determined as

$$\mathbf{q} = \begin{Bmatrix} 260.00 \\ 249.56 \\ 242.14 \\ 237.64 \\ 236.23 \end{Bmatrix} °C,$$

  where the first entry is the prescribed temperature at $x = 0$.
(b) The heat loss per unit depth is determined as

$$\int_S qds = 2 \times 1 \cdot \int_{x=0}^{x=L} h(T - T_\infty)dx + 1 \cdot t_2 \times h(T_{x=L} - T_\infty) = 382.36 \text{ W/m}.$$

## 11.2 Unsteady Heat Transfer in the 1D Domain

Consider again the heat energy balance for the segment shown in Figure 11.2. For unsteady time-dependent heat conduction:

$$F(x) + f_s(x)dx - F(x+dx) = c\rho A(x)dx\frac{\partial T}{\partial t}, \qquad (11.2.1)$$

where

$\rho$: mass density per volume $(\mathrm{kg/m^3})$
$c$: specific heat $(\mathrm{J/kg \cdot {}^\circ C})$.

From Eq. (11.2.1):

$$F + \left(f_s - \rho cA\frac{\partial T}{\partial t}\right)dx - \left(F + \frac{\partial F}{\partial x}dx\right) = 0 \rightarrow \left(\frac{\partial F}{\partial x} - \hat{f_s}\right)dx = 0 \rightarrow \frac{\partial F}{\partial x} - \hat{f_s} = 0, \quad (11.2.2)$$

where

$$\hat{f_s} = f_s - c\rho A\frac{\partial T}{\partial t}. \qquad (11.2.3)$$

Equation (11.1.9) is then modified as

$$\int_{x=0}^{x=L} k\frac{\partial \delta T}{\partial x}\frac{\partial T}{\partial x}Adx - \int_{x=0}^{x=L} \delta T\hat{f_s}dx - \delta W_B = 0. \qquad (11.2.4)$$

Introducing Eq. (11.2.3), the above equation is expressed as

$$\int_{x=0}^{x=L} k\frac{\partial \delta T}{\partial x}\frac{\partial T}{\partial x}Adx - \int_{x=0}^{x=L} \delta Tf_sdx + \int_{x=0}^{x=L} \delta T\frac{\partial T}{\partial t}c\rho Adx - \delta W_B = 0. \qquad (11.2.5)$$

For the FE formulation, the third term in the above equation can be expressed as

$$\int_{x=0}^{x=L} \delta T\frac{\partial T}{\partial t}c\rho Adx = \sum_e \int_{x=x_1}^{x=x_2} \delta T\frac{\partial T}{\partial t}c\rho Adx. \qquad (11.2.6)$$

For the two-node element with the assumed temperature in Eq. (11.1.33):

$$\dot{T} = \frac{\partial T}{\partial t} = N_1\dot{T}_1 + N_2\dot{T}_2 = \lfloor N_1 \quad N_2 \rfloor \begin{Bmatrix} \dot{T}_1 \\ \dot{T}_2 \end{Bmatrix}. \qquad (11.2.7)$$

Substituting Eqs (11.1.35) and (11.2.7) into the element in Eq. (11.2.6):

$$\int_{x=x_1}^{x=x_2} \delta T \frac{\partial T}{\partial t} c\rho A \, dx = \int_{s=0}^{s=1} \lfloor \delta T_1 \quad \delta T_2 \rfloor \begin{Bmatrix} N_1 \\ N_2 \end{Bmatrix} \lfloor N_1 \quad N_2 \rfloor \begin{Bmatrix} \dot{T}_1 \\ \dot{T}_2 \end{Bmatrix} c\rho A l \, ds$$

$$= \lfloor \delta T_1 \quad \delta T_2 \rfloor \left( l \int_{s=0}^{s=1} \begin{bmatrix} N_1^2 & N_1 N_2 \\ N_2 N_1 & N_2^2 \end{bmatrix} c\rho A \, ds \right) \begin{Bmatrix} \dot{T}_1 \\ \dot{T}_2 \end{Bmatrix} \qquad (11.2.8)$$

$$= \lfloor \delta T_1 \quad \delta T_2 \rfloor \begin{bmatrix} c_{11}^e & c_{12}^e \\ c_{21}^e & c_{22}^e \end{bmatrix} \begin{Bmatrix} \dot{T}_1 \\ \dot{T}_2 \end{Bmatrix} = \delta \mathbf{d}^T \mathbf{c}^e \dot{\mathbf{d}},$$

where

$$\mathbf{c}^e = \begin{bmatrix} c_{11}^e & c_{12}^e \\ c_{21}^e & c_{22}^e \end{bmatrix} = l \int_{s=0}^{s=1} \begin{bmatrix} N_1^2 & N_1 N_2 \\ N_2 N_1 & N_2^2 \end{bmatrix} c\rho A \, ds : \text{element capacitance matrix.} \quad (11.2.9)$$

For $\rho c A$ uniform over the element, it can be shown that

$$\mathbf{c}^e = c(\rho A l) \frac{1}{6} \begin{bmatrix} 2 & 1 \\ 1 & 2 \end{bmatrix}. \qquad (11.2.10)$$

For the entire body:

$$\int_{x=0}^{x=L} \delta T \frac{\partial T}{\partial t} c\rho A \, dx = \sum_e \delta \mathbf{d}^T \mathbf{c}^e \dot{\mathbf{d}} = \delta \mathbf{q}^T \mathbf{C} \dot{\mathbf{q}}, \qquad (11.2.11)$$

where $\mathbf{C}$ is the global capacitance matrix assembled from the element capacitance matrices. The capacitance matrix is similar to the mass matrix seen in Chapter 4. Subsequently, the FE formulation based on Eq. (11.2.5) results in

$$\delta \mathbf{q}^T (\mathbf{Kq} + \mathbf{C}\dot{\mathbf{q}} - \mathbf{F}) = 0 \rightarrow \mathbf{Kq} + \mathbf{C}\dot{\mathbf{q}} - \mathbf{F} = \mathbf{0} \qquad (11.2.12)$$

or

$$\mathbf{C}\dot{\mathbf{q}} + \mathbf{Kq} = \mathbf{F}. \qquad (11.2.13)$$

The initial condition at time $t = 0$ can be expressed as

$$\mathbf{q}(0) = \mathbf{q}_0, \qquad (11.2.14)$$

where $\mathbf{q}_0$ is the vector of initial nodal temperatures. The above first-order (in time) equation can be solved using various numerical integration methods, such as the fourth-order Runge–Kutta (RK4) method or the trapezoidal rule.


**The Trapezoidal Rule**

In the trapezoidal rule, we assume that

$$\mathbf{q}_{n+1} = \mathbf{q}_n + \frac{1}{2} (\dot{\mathbf{q}}_n + \dot{\mathbf{q}}_{n+1}) \Delta t \qquad (11.2.15)$$

over a time increment $\Delta t$. From the above equation:

$$\dot{\mathbf{q}}_{n+1} = \frac{2}{\Delta t}\left(\mathbf{q}_{n+1} - \mathbf{q}_n\right) - \dot{\mathbf{q}}_n. \tag{11.2.16}$$

At time $t = t_{n+1}$, the governing equation is

$$\mathbf{C}\dot{\mathbf{q}}_{n+1} + \mathbf{K}\mathbf{q}_{n+1} = \mathbf{F}_{n+1}. \tag{11.2.17}$$

Substituting Eq. (11.2.16) into Eq. (11.2.17):

$$\mathbf{C}\left[\frac{2}{\Delta t}\left(\mathbf{q}_{n+1} - \mathbf{q}_n\right) - \dot{\mathbf{q}}_n\right] + \mathbf{K}\mathbf{q}_{n+1} = \mathbf{F}_{n+1}. \tag{11.2.18}$$

Rearranging:

$$\left[\mathbf{K} + \frac{2}{\Delta t}\mathbf{C}\right]\mathbf{q}_{n+1} = \mathbf{F}_{n+1} + \mathbf{C}\left[\frac{2}{\Delta t}\mathbf{q}_n + \dot{\mathbf{q}}_n\right]. \tag{11.2.19}$$

The above equation can be used to determine $\mathbf{q}_{n+1}$ for given values of $\mathbf{q}_n$ and $\dot{\mathbf{q}}_n$ in a recursive manner. We may then use Eq. (11.2.16) to solve for $\dot{\mathbf{q}}_{n+1}$. Note that the trapezoidal rule involves two steps in time. To start the solution process, we may choose a time segment $\Delta t$ and set $n = 0$ in Eq. (11.2.19). Then

$$\left[\mathbf{K} + \frac{2}{\Delta t}\mathbf{C}\right]\mathbf{q}_1 = \mathbf{F}_1 + \mathbf{C}\left[\frac{2}{\Delta t}\mathbf{q}_0 + \dot{\mathbf{q}}_0\right], \tag{11.2.20}$$

where $\mathbf{q}_0$ is prescribed at $t = 0$ and $\dot{\mathbf{q}}_0$ is obtained from Eq. (11.2.13) at $t = 0$, expressed as

$$\mathbf{C}\dot{\mathbf{q}}_0 = \mathbf{F}_0 - \mathbf{K}\mathbf{q}_0. \tag{11.2.21}$$

We can then solve Eq. (11.2.20) for $\mathbf{q}_1$ and determine $\dot{\mathbf{q}}_1$ using Eq. (11.2.16). To march further in time, we set $n = 1$ and use Eq. (11.2.19) to solve for $\mathbf{q}_2$. This process can be repeated until the time span of interest is covered.

### The RK4 Method

From Eq. (11.2.13):

$$\dot{\mathbf{q}} = \mathbf{C}^{-1}(\mathbf{F} - \mathbf{K}\mathbf{q}). \tag{11.2.22}$$

We can then apply the RK4 method to the equation above. The RK4 method is convenient for problems with temperature-dependent heat conductivity and specific heat.

 ## Example 11.4

Consider a 1D time-dependent heat conduction problem for a body of uniform cross-section, described as follows:

Boundary conditions: $T(0, t) = 0°C, \quad T(L, t) = 0°C.$
Initial conditions: $T(x, 0) = 0°C.$
Heat source: $f_s(x) = BA.$

Carry out FE analysis to determine the temperature distribution with time until steady state is reached. Use a model with 10 elements of equal length. For integration in time, use the trapezoidal rule or RK4 method. Plot the temperature distribution using non-dimensional time $\tau$ as a parameter. The non-dimensional time $\tau$ is defined as follows:

$$\tau = at \text{ where } a = \frac{k}{\rho c L^2}.$$

such that

$$\frac{dT}{dt} = \frac{dT}{d\tau}\frac{d\tau}{dt} = a\frac{dT}{d\tau}.$$

## Solution:

We may express the element capacitance matrix in Eq. (11.2.10) and the element conduction matrix in Eq. (11.1.21) as

$$\mathbf{c}^e = c(\rho Al)\frac{1}{6}\begin{bmatrix} 2 & 1 \\ 1 & 2 \end{bmatrix} = c(\rho AL)\left(\frac{l}{L}\right)\frac{1}{6}\begin{bmatrix} 2 & 1 \\ 1 & 2 \end{bmatrix} = c(\rho AL)\bar{\mathbf{c}}^e,$$

$$\mathbf{k}^e = \frac{kA}{l}\begin{bmatrix} 1 & -1 \\ -1 & 1 \end{bmatrix} = \frac{kA}{L}\left(\frac{L}{l}\right)\begin{bmatrix} 1 & -1 \\ -1 & 1 \end{bmatrix} = \frac{kA}{L}\bar{\mathbf{k}}^e,$$

where

$$\bar{\mathbf{c}}^e = \left(\frac{l}{L}\right)\frac{1}{6}\begin{bmatrix} 2 & 1 \\ 1 & 2 \end{bmatrix}, \quad \bar{\mathbf{k}}^e = \left(\frac{L}{l}\right)\begin{bmatrix} 1 & -1 \\ -1 & 1 \end{bmatrix}.$$

For the 10-element model, $L/l = 10$. As shown in Example 11.1, the element load vector due to a constant heat source of $BA$ is

$$\begin{Bmatrix} Q_1^e \\ Q_2^e \end{Bmatrix} = BAl\frac{1}{2}\begin{Bmatrix} 1 \\ 1 \end{Bmatrix} = BAL\left(\frac{l}{L}\right)\frac{1}{2}\begin{Bmatrix} 1 \\ 1 \end{Bmatrix} = BAL\begin{Bmatrix} \bar{Q}_1^e \\ \bar{Q}_2^e \end{Bmatrix},$$

where

$$\begin{Bmatrix} \bar{Q}_1^e \\ \bar{Q}_2^e \end{Bmatrix} = \left(\frac{l}{L}\right)\frac{1}{2}\begin{Bmatrix} 1 \\ 1 \end{Bmatrix}.$$

These element matrices and load vectors are assembled to construct the global heat conduction equation, expressed symbolically as

$$\rho ALc\bar{\mathbf{C}}\dot{\mathbf{q}} + \frac{kA}{L}\bar{\mathbf{K}}\mathbf{q} = BAL\bar{\mathbf{F}}.$$

For a unit cross-sectional area:

$$\rho Lc\bar{\mathbf{C}}\dot{\mathbf{q}} + \frac{k}{L}\bar{\mathbf{K}}\mathbf{q} = BL\bar{\mathbf{F}} \rightarrow \frac{\rho cL^2}{k}\bar{\mathbf{C}}\dot{\mathbf{q}} + \bar{\mathbf{K}}\mathbf{q} = \frac{BL^2}{k}\bar{\mathbf{F}}.$$

After applying the boundary conditions, we can use the above equation to solve for the response for given material properties and the initial condition. Alternatively, we may introduce non-dimensional time to express the governing equation as

$$\bar{\mathbf{C}}\mathbf{q}' + \bar{\mathbf{K}}\mathbf{q} = \frac{BL^2}{k}\bar{\mathbf{F}},$$

where

$$\mathbf{q}' = \frac{d\mathbf{q}}{d\tau}.$$

After applying the boundary conditions, the above equation can be solved in time for given $B, L, k$ or $BL^2/k = 1$ until a steady-state condition is reached. The exact solution for the steady-state solution is

$$T = \frac{B}{2K}x(L-x) = \frac{BL^2}{2K}\frac{x}{L}\left(1 - \frac{x}{L}\right),$$

which can be determined from Eq. (11.1.5). The temperature distribution can be non-dimensionalized such that

$$\bar{T} = \frac{T}{\left(\frac{BL^2}{8K}\right)}, \quad \bar{x} = \frac{x}{L}.$$

## 11.3  Heat Conduction Equation in the 2D and 3D Domains

We will first consider the heat conduction equation in the 3D domain and then reduce it for study in the 2D domain under simplifying assumptions. Based on the heat energy balance for the 1D case described in Section 11.1.1, we may consider a heat energy balance for an infinitesimal rectangular cuboid of side lengths $dx, dy$, and $dz$ in the 3D domain such that

$$-\left[\frac{\partial}{\partial x}(q_x dydz)dx + \frac{\partial}{\partial y}\left(q_y dxdz\right)dy + \frac{\partial}{\partial z}(q_z dxdy)dz\right] + Hdxdydz = c\rho dxdydz\frac{\partial T}{\partial t}$$
$$\rightarrow \left[\frac{\partial}{\partial x}(q_x) + \frac{\partial}{\partial y}\left(q_y\right) + \frac{\partial}{\partial z}(q_z) - H + c\rho\frac{\partial T}{\partial t}\right]dxdydz = 0, \tag{11.3.1}$$

where $H$ is the heat source per unit volume. From the above equation, it follows that

$$\frac{\partial}{\partial x}(q_x) + \frac{\partial}{\partial y}\left(q_y\right) + \frac{\partial}{\partial z}(q_z) - H + c\rho\frac{\partial T}{\partial t} = 0. \tag{11.3.2}$$

According to Fourier's law, heat fluxes are related to temperature gradients such that

$$q_x = -k_x \frac{\partial T}{\partial x}, \quad q_y = -k_y \frac{\partial T}{\partial y}, \quad q_z = -k_z \frac{\partial T}{\partial z}, \tag{11.3.3}$$

where $k_x, k_y, k_z$ are heat conductivities in the $x, y,$ and $z$ directions. Introducing the above equation into Eq. (11.3.2):

$$\frac{\partial}{\partial x}\left(k_x \frac{\partial T}{\partial x}\right) + \frac{\partial}{\partial y}\left(k_y \frac{\partial T}{\partial y}\right) + \frac{\partial}{\partial z}\left(k_z \frac{\partial T}{\partial z}\right) + H = c\rho \frac{\partial T}{\partial t}. \tag{11.3.4}$$

Introducing a heat flux vector $\mathbf{g}$ defined as

$$\mathbf{g} = k_x \frac{\partial T}{\partial x}\mathbf{i} + k_y \frac{\partial T}{\partial y}\mathbf{j} + k_z \frac{\partial T}{\partial z}\mathbf{k}, \tag{11.3.5}$$

Eq. (11.3.4) can be written as

$$\nabla \cdot \mathbf{g} + \hat{H} = 0, \tag{11.3.6}$$

where

$$\nabla = \mathbf{i}\frac{\partial}{\partial x} + \mathbf{j}\frac{\partial}{\partial y} + \mathbf{k}\frac{\partial}{\partial z} \tag{11.3.7}$$

and

$$\hat{H} = H - c\rho \frac{\partial T}{\partial t}. \tag{11.3.8}$$

Introducing a virtual temperature or weight function $\delta T$, which is arbitrary:

$$\int_V \delta T (\nabla \cdot \mathbf{g} + \hat{H}) dV = 0. \tag{11.3.9}$$

According to the divergence theorem:

$$\int_V \nabla \cdot \mathbf{h} dV = \int_S \mathbf{h} \cdot \mathbf{n} dS. \tag{11.3.10}$$

Setting $\mathbf{h} = \delta T \mathbf{g}$ in the above equation:

$$\int_V \nabla \cdot (\delta T \mathbf{g}) dV = \int_S \delta T \mathbf{g} \cdot \mathbf{n} dS \rightarrow \int_V \delta T \nabla \cdot \mathbf{g} dV + \int_V \nabla \delta T \cdot \mathbf{g} dV = \int_S \delta T \mathbf{g} \cdot \mathbf{n} dS$$

$$\rightarrow \int_V \delta T \nabla \cdot \mathbf{g} dV = \int_S \delta T \mathbf{g} \cdot \mathbf{n} dS - \int_V \nabla \delta T \cdot \mathbf{g} dV. \tag{11.3.11}$$

Introducing the result of Eq. (11.3.11) into Eq. (11.3.9) and changing sign:

$$\int_V \nabla \delta T \cdot \mathbf{g} dV - \int_V \delta T \hat{H} dV - \int_S \delta T \mathbf{g} \cdot \mathbf{n} dS = 0. \tag{11.3.12}$$

Using Eq. (11.3.5), the first integral in the above equation can be expressed as

$$\int_V \nabla \delta T \cdot \mathbf{g} dV = \int_V \delta \mathbf{T}_d^T \mathbf{D} \mathbf{T}_d dV, \tag{11.3.13}$$

where

$$\delta \mathbf{T}_d = \left\{ \begin{array}{c} \dfrac{\partial \delta T}{\partial x} \\[2mm] \dfrac{\partial \delta T}{\partial y} \\[2mm] \dfrac{\partial \delta T}{\partial z} \end{array} \right\}, \quad \mathbf{D} = \begin{bmatrix} k_x & 0 & 0 \\ 0 & k_y & 0 \\ 0 & 0 & k_z \end{bmatrix}, \quad \mathbf{T}_d = \left\{ \begin{array}{c} \dfrac{\partial T}{\partial x} \\[2mm] \dfrac{\partial T}{\partial y} \\[2mm] \dfrac{\partial T}{\partial z} \end{array} \right\}. \tag{11.3.14}$$

Substituting Eqs. (11.3.8) and (11.3.13) into Eq. (11.3.12):

$$\int_V \delta \mathbf{T}_d^T \mathbf{D} \mathbf{T}_d dV - \int_V \delta T H dV + \int_V \delta T \frac{\partial T}{\partial t} c \rho dV - \delta W_B = 0, \tag{11.3.15}$$

where

$$\delta W_B = \int_S \delta T \mathbf{g} \cdot \mathbf{n} dS = \int_{S_T} \delta T \mathbf{g} \cdot \mathbf{n} dS + \int_{S_F} \delta T \mathbf{g} \cdot \mathbf{n} dS. \tag{11.3.16}$$

Notice that in the above equation, the boundary surface is divided into $S_T$ where temperature is prescribed and $S_F$ where heat flux is prescribed.

Consider now 2D problems with no dependence on the $z$-coordinate, such that

$$T = T(x, y), H = H(x, y). \tag{11.3.17}$$

Equations (11.3.5), (11.3.15), and (11.3.16) then simplify to

$$\mathbf{g} = k_x \frac{\partial T}{\partial x} \mathbf{i} + k_y \frac{\partial T}{\partial y} \mathbf{j} \tag{11.3.18}$$

and

$$\int_A \delta \mathbf{T}_d^T \mathbf{D} \mathbf{T}_d t_z dA - \int_A \delta T H t_z dA + \int_A \delta T \frac{\partial T}{\partial t} c \rho t_z dA - \delta W_B = 0, \tag{11.3.19}$$

where

$$\delta \mathbf{T}_d = \left\{ \begin{array}{c} \dfrac{\partial \delta T}{\partial x} \\[2mm] \dfrac{\partial \delta T}{\partial y} \end{array} \right\}, \quad \mathbf{T}_d = \left\{ \begin{array}{c} \dfrac{\partial T}{\partial x} \\[2mm] \dfrac{\partial T}{\partial y} \end{array} \right\}, \quad \mathbf{D} = \begin{bmatrix} k_x & 0 \\ 0 & k_y \end{bmatrix}, \quad t_z = \int dz. \tag{11.3.20}$$

Also

$$\delta W_B = \int_S \delta T \mathbf{g} \cdot \mathbf{n} dS = \int_{l_B} \delta T \mathbf{g} \cdot \mathbf{n} t_z dl_B, \qquad (11.3.21)$$

where $l_B$ is a coordinate along the 2D domain boundary.

## 11.4 Finite Element Formulation for Heat Transfer in the 2D Domain

To illustrate how the FE equation is constructed for heat transfer problems, we will first consider 2D problems.

### Element Conduction Matrix

For the FE approximation, the first term in Eq. (11.3.19) can be expressed as

$$\int_A \delta \mathbf{T}_d^{\mathsf{T}} \mathbf{D} \mathbf{T}_d t_z dA = \sum_e \int_{A_e} \delta \mathbf{T}_d^{\mathsf{T}} \mathbf{D} \mathbf{T}_d t_z dA, \qquad (11.4.1)$$

where "$e$" stands for element. As shown in Chapter 7, the mapping and assumed temperature for an individual element can be expressed as follows.

**Mapping:**

$$x = \sum_{i=1}^n N_i(\xi, \eta) x_i = x(\xi, \eta), \quad y = \sum_{i=1}^n N_i(\xi, \eta) y_i = y(\xi, \eta). \qquad (11.4.2)$$

**Assumed temperature:**

$$T = \sum_{i=1}^n N_i(\xi, \eta) T_i = T(\xi, \eta), \qquad (11.4.3)$$

where $n$ is the total number of nodes in each element.

In Eq. (11.3.20), the derivatives are taken with respect to the physical $x, y$-coordinates while the assumed temperature is a function of the natural $\xi, \eta$-coordinates. Accordingly, we use the chain rule of differentiation to relate the derivatives in the two coordinate systems as

$$\frac{\partial T}{\partial \xi} = \frac{\partial T}{\partial x}\frac{\partial x}{\partial \xi} + \frac{\partial T}{\partial y}\frac{\partial y}{\partial \xi}, \quad \frac{\partial T}{\partial \eta} = \frac{\partial T}{\partial x}\frac{\partial x}{\partial \eta} + \frac{\partial T}{\partial y}\frac{\partial y}{\partial \eta}. \qquad (11.4.4)$$

In matrix form:

$$\begin{Bmatrix} \dfrac{\partial T}{\partial \xi} \\ \dfrac{\partial T}{\partial \eta} \end{Bmatrix} = \begin{bmatrix} \dfrac{\partial x}{\partial \xi} & \dfrac{\partial y}{\partial \xi} \\ \dfrac{\partial x}{\partial \eta} & \dfrac{\partial y}{\partial \eta} \end{bmatrix} \begin{Bmatrix} \dfrac{\partial T}{\partial x} \\ \dfrac{\partial T}{\partial y} \end{Bmatrix} = \mathbf{J}_2 \begin{Bmatrix} \dfrac{\partial T}{\partial x} \\ \dfrac{\partial T}{\partial y} \end{Bmatrix}, \qquad (11.4.5)$$

where

$$\mathbf{J}_2 = \begin{bmatrix} \dfrac{\partial x}{\partial \xi} & \dfrac{\partial y}{\partial \xi} \\[2mm] \dfrac{\partial x}{\partial \eta} & \dfrac{\partial y}{\partial \eta} \end{bmatrix} = \begin{bmatrix} \displaystyle\sum_{i=1}^{n} \dfrac{\partial N_i}{\partial \xi} x_i & \displaystyle\sum_{i=1}^{n} \dfrac{\partial N_i}{\partial \xi} y_i \\[4mm] \displaystyle\sum_{i=1}^{n} \dfrac{\partial N_i}{\partial \eta} x_i & \displaystyle\sum_{i=1}^{n} \dfrac{\partial N_i}{\partial \eta} y_i \end{bmatrix} \tag{11.4.6}$$

is the Jacobian matrix in the 2D domain. From Eq. (11.4.5):

$$\begin{Bmatrix} \dfrac{\partial T}{\partial x} \\[2mm] \dfrac{\partial T}{\partial y} \end{Bmatrix} = \mathbf{J}_2^{-1} \begin{Bmatrix} \dfrac{\partial T}{\partial \xi} \\[2mm] \dfrac{\partial T}{\partial \eta} \end{Bmatrix} = \begin{bmatrix} Y_{11} & Y_{12} \\ Y_{21} & Y_{22} \end{bmatrix} \begin{Bmatrix} \dfrac{\partial T}{\partial \xi} \\[2mm] \dfrac{\partial T}{\partial \eta} \end{Bmatrix} = \mathbf{Y} \begin{Bmatrix} \dfrac{\partial T}{\partial \xi} \\[2mm] \dfrac{\partial T}{\partial \eta} \end{Bmatrix}, \tag{11.4.7}$$

where

$$\mathbf{Y} = \mathbf{J}_2^{-1} = \begin{bmatrix} Y_{11} & Y_{12} \\ Y_{21} & Y_{22} \end{bmatrix}. \tag{11.4.8}$$

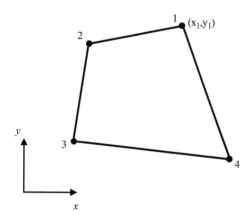

As an example, we now consider the four-node element. Extensions to higher-order elements are straightforward. According to Eq. (7.2.6) in Chapter 7, mapping/shape functions for the four-node element shown in Figure 11.5 are

$$N_1 = \frac{1}{4}(1 + \xi)(1 + \eta), \quad N_2 = \frac{1}{4}(1 - \xi)(1 + \eta),$$

$$N_3 = \frac{1}{4}(1 - \xi)(1 - \eta), \quad N_4 = \frac{1}{4}(1 + \xi)(1 - \eta). \tag{11.4.9}$$

**Figure 11.5** Four-node quadrilateral element

According to Eq. (11.4.6), the Jacobian matrix is then

$$\mathbf{J}_2 = \begin{bmatrix} \dfrac{\partial x}{\partial \xi} & \dfrac{\partial y}{\partial \xi} \\[2mm] \dfrac{\partial x}{\partial \eta} & \dfrac{\partial y}{\partial \eta} \end{bmatrix} = \begin{bmatrix} \displaystyle\sum_{i=1}^{4} \dfrac{\partial N_i}{\partial \xi} x_i & \displaystyle\sum_{i=1}^{4} \dfrac{\partial N_i}{\partial \xi} y_i \\[4mm] \displaystyle\sum_{i=1}^{4} \dfrac{\partial N_i}{\partial \eta} x_i & \displaystyle\sum_{i=1}^{4} \dfrac{\partial N_i}{\partial \eta} y_i \end{bmatrix}. \tag{11.4.10}$$

From Eq. (11.4.7):

$$
\begin{Bmatrix} \dfrac{\partial T}{\partial x} \\ \dfrac{\partial T}{\partial y} \end{Bmatrix} = \mathbf{Y} \begin{Bmatrix} \dfrac{\partial T}{\partial \xi} \\ \dfrac{\partial T}{\partial \eta} \end{Bmatrix} = \mathbf{Y} \begin{Bmatrix} \dfrac{\partial N_1}{\partial \xi} T_1 + \dfrac{\partial N_2}{\partial \xi} T_2 + \dfrac{\partial N_3}{\partial \xi} T_3 + \dfrac{\partial N_4}{\partial \xi} T_4 \\ \dfrac{\partial N_1}{\partial \eta} T_1 + \dfrac{\partial N_2}{\partial \eta} T_2 + \dfrac{\partial N_3}{\partial \eta} T_3 + \dfrac{\partial N_4}{\partial \eta} T_4 \end{Bmatrix},
\tag{11.4.11}
$$

where the $\mathbf{Y}$ matrix is obtained by inverting the $\mathbf{J}_2$ matrix as shown in Eq. (11.4.8). The above equation can be expressed as

$$
\begin{Bmatrix} \dfrac{\partial T}{\partial x} \\ \dfrac{\partial T}{\partial y} \end{Bmatrix} = \mathbf{Y} \begin{bmatrix} \dfrac{\partial N_1}{\partial \xi} & \dfrac{\partial N_2}{\partial \xi} & \dfrac{\partial N_3}{\partial \xi} & \dfrac{\partial N_4}{\partial \xi} \\ \dfrac{\partial N_1}{\partial \eta} & \dfrac{\partial N_2}{\partial \eta} & \dfrac{\partial N_3}{\partial \eta} & \dfrac{\partial N_4}{\partial \eta} \end{bmatrix} \begin{Bmatrix} T_1 \\ T_2 \\ T_3 \\ T_4 \end{Bmatrix} = \mathbf{Bd},
\tag{11.4.12}
$$

where

$$
\mathbf{B} = \mathbf{Y} \begin{bmatrix} \dfrac{\partial N_1}{\partial \xi} & \dfrac{\partial N_2}{\partial \xi} & \dfrac{\partial N_3}{\partial \xi} & \dfrac{\partial N_4}{\partial \xi} \\ \dfrac{\partial N_1}{\partial \eta} & \dfrac{\partial N_2}{\partial \eta} & \dfrac{\partial N_3}{\partial \eta} & \dfrac{\partial N_4}{\partial \eta} \end{bmatrix} = \dfrac{1}{4} \mathbf{Y} \begin{bmatrix} 1+\eta & -(1+\eta) & -(1-\eta) & (1-\eta) \\ 1+\xi & 1-\xi & -(1-\xi) & -(1+\xi) \end{bmatrix}
\tag{11.4.13}
$$

and

$$
\mathbf{d} = \begin{Bmatrix} T_1 \\ T_2 \\ T_3 \\ T_4 \end{Bmatrix} : 4 \times 1 \text{ element nodal temperature vector.}
\tag{11.4.14}
$$

Also, setting $\delta T = \sum N_i \delta T_i$:

$$
\begin{Bmatrix} \dfrac{\partial \delta T}{\partial x} \\ \dfrac{\partial \delta T}{\partial y} \end{Bmatrix} = \mathbf{B} \delta \mathbf{d},
\tag{11.4.15}
$$

where

$$
\delta \mathbf{d} = \begin{Bmatrix} \delta T_1 \\ \delta T_2 \\ \delta T_3 \\ \delta T_4 \end{Bmatrix}.
\tag{11.4.16}
$$

Accordingly

$$
\int_A \delta \mathbf{T}_d^{\mathrm{T}} \mathbf{D} \mathbf{T}_d t_z dA = \int_A \delta \mathbf{d}^{\mathrm{T}} \mathbf{B}^{\mathrm{T}} \mathbf{D} \mathbf{B} \mathbf{d} t_z dA = \delta \mathbf{d}^{\mathrm{T}} \left( \int_{A_e} \mathbf{B}^{\mathrm{T}} \mathbf{D} \mathbf{B} t_z dA \right) \mathbf{d} = \delta \mathbf{d}^{\mathrm{T}} \mathbf{k}^e \mathbf{d},
\tag{11.4.17}
$$

where

$$\mathbf{k}^e = \int_{A_e} \mathbf{B}^{\mathrm{T}} \mathbf{D} \mathbf{B} t_z \, dA \tag{11.4.18}$$

is the element conduction matrix which is symmetric. For an element of square and rectangular geometries, the integration in the above equation can be analytically carried out to generate the element conduction matrix. For an element of arbitrary geometry, the element conduction matrix can be constructed using the numerical integration schemes discussed in Chapter 7, such that

$$\mathbf{k}^e = \int_{-1}^{1} \int_{-1}^{1} \mathbf{B}^{\mathrm{T}} \mathbf{D} \mathbf{B} t_z (\det \mathbf{J}_2) \, d\xi d\eta = \sum_{i=1}^{} \sum_{j=1}^{} \left( \mathbf{B}^{\mathrm{T}} \mathbf{D} \mathbf{B} t_z \det \mathbf{J}_2 \right)_{\substack{\xi = a_i \\ \eta = a_j}} W_i W_j. \tag{11.4.19}$$

### Element Load Vector Due to Heat Source

Consider the second term in Eq. (11.3.19). Introducing the virtual temperature field into an element:

$$\int_{A_e} \delta T H t_z \, dA = \delta \mathbf{d}^{\mathrm{T}} \int_{A_e} \mathbf{N}^{\mathrm{T}} H t_z \, dA = \delta \mathbf{d}^{\mathrm{T}} \mathbf{f}_H^e, \tag{11.4.20}$$

where

$$\mathbf{f}_H^e = \int_{A_e} \mathbf{N}^{\mathrm{T}} H t_z \, dA. \tag{11.4.21}$$

For the four-node element, $\mathbf{f}_H^e$ is the $4 \times 1$ element load vector due to a heat source.

### Element Load Vector Due to Heat Flux Applied to the Boundary

Consider the last expression in Eq. (11.3.21). Introducing a virtual temperature field into an element with heat flux applied on the boundary:

$$\int_{l_{B_e}} \delta T \mathbf{g} \cdot \mathbf{n} t_z \, dl_B = \delta \mathbf{d}^{\mathrm{T}} \int_{l_{B_e}} \mathbf{N}^{\mathrm{T}} \mathbf{g} \cdot \mathbf{n} t_z \, dl_B = \delta \mathbf{d}^{\mathrm{T}} \mathbf{f}_s^e, \tag{11.4.22}$$

where

$$\mathbf{f}_s^e = \int_{l_{B_e}} \mathbf{N}^{\mathrm{T}} \mathbf{g} \cdot \mathbf{n} t_z \, dl_B \tag{11.4.23}$$

is the element load vector due to applied heat flux. The integration is over an element boundary.

**Convection on the Boundary**

For convection over a 2D domain boundary, heat flux $q_n$ normal to the surface can be expressed as

$$q_n = h(T - T_\infty). \tag{11.4.24}$$

It can then be shown that, as demonstrated for the 1D problem, the $hT$ term generates the element convection matrix that adds to the element conduction matrix, while $hT_\infty$ contributes to the element load vector.

**Construction of Element Capacitance Matrix**

For the FE formulation, the third term in Eq. (11.3.19) is expressed as

$$\int_A \delta T \frac{\partial T}{\partial t} c\rho t_z dA = \sum_e \int_{A_e} \delta T \frac{\partial T}{\partial t} c\rho t_z dA. \tag{11.4.25}$$

For FE modeling, the assumed temperature can be expressed as

$$T = \mathbf{Nd}. \tag{11.4.26}$$

Then

$$\delta T = \mathbf{N}\delta\mathbf{d}, \quad \dot{T} = \mathbf{N}\dot{\mathbf{d}}. \tag{11.4.27}$$

Introducing Eqs (11.4.26) and (11.4.27) into the expression for an element in Eq. (11.4.25):

$$\int_{A_e} \delta T \frac{\partial T}{\partial t} c\rho t_z dA = \int_{A_e} \delta\mathbf{d}^\mathsf{T}\mathbf{N}^\mathsf{T}\mathbf{N}\dot{\mathbf{d}} c\rho t_z dA = \delta\mathbf{d}^\mathsf{T}\left(\int_{A_e} \mathbf{N}^\mathsf{T}\mathbf{N} c\rho t_z dA\right)\dot{\mathbf{d}} = \delta\mathbf{d}^\mathsf{T}\mathbf{c}^e\dot{\mathbf{d}}, \tag{11.4.28}$$

where

$$\mathbf{c}^e = \int_{A_e} \mathbf{N}^\mathsf{T}\mathbf{N} c\rho t_z dA : \text{element capacitance matrix}. \tag{11.4.29}$$

For the entire body:

$$\sum_e \int_{A_e} \delta T \frac{\partial T}{\partial t} c\rho t_z dA = \delta\mathbf{q}^\mathsf{T}\mathbf{C}\dot{\mathbf{q}}, \tag{11.4.30}$$

where $\mathbf{C}$ is the global capacitance matrix assembled from the element capacitance matrices. Subsequently, the FE formulation based on Eq. (11.3.19) results in

$$\delta\mathbf{q}^{\mathrm{T}}(\mathbf{Kq} + \mathbf{C\dot{q}} - \mathbf{F}) = 0 \rightarrow \mathbf{Kq} + \mathbf{C\dot{q}} - \mathbf{F} = 0 \qquad (11.4.31)$$

or

$$\mathbf{C\dot{q}} + \mathbf{Kq} = \mathbf{F}. \qquad (11.4.32)$$

The initial condition at time $t = 0$ can be expressed as

$$\mathbf{q}(0) = \mathbf{q}_0. \qquad (11.4.33)$$

For the steady-state condition ($\dot{\mathbf{q}} = \mathbf{0}$), the equation reduces to

$$\mathbf{Kq} = \mathbf{F}. \qquad (11.4.34)$$

 **Example 11.5**

The nodal $(x, y)$ coordinates of a four-node element are given as node 1: $(2, 2)$, node 2: $(0, 2)$, node 3: $(0, 0)$, node 4: $(2, 0)$. For the material, $k_x = k_y = k$.

(a) Construct the element heat conduction matrix.
(b) Construct the element load vector due to a heat source which is constant over the element.
(c) Construct the element capacitance matrix for uniform $\rho c$.

**Solution:**

(a) For this element of square geometry, we can analytically carry out the integration in Eq. (11.4.19) to obtain

$$\mathbf{k}^e = \frac{kt_z}{6}\begin{bmatrix} 4 & -1 & -2 & -1 \\ -1 & 4 & -1 & -2 \\ -2 & -1 & 4 & -1 \\ -1 & -2 & -1 & 4 \end{bmatrix}.$$

(b) For a constant heat source, we can analytically integrate Eq. (11.4.21) such that

$$\mathbf{f}^e_H = \frac{HA_e t_z}{4}\begin{Bmatrix} 1 \\ 1 \\ 1 \\ 1 \end{Bmatrix},$$

where $A_e$ is the element area.

(c) For uniform $\rho c$, Eq. (11.4.29) can be analytically integrated such that

$$\mathbf{c}^e = \frac{c\rho A_e t_z}{36}\begin{bmatrix} 4 & 2 & 1 & 2 \\ 2 & 4 & 2 & 1 \\ 1 & 2 & 4 & 2 \\ 2 & 1 & 2 & 4 \end{bmatrix}.$$

## Example 11.6

For the square domain shown in Figure 11.6, the $x = 0$ side and the $y = 0$ side are insulated with no heat flux, and the temperature on the remaining sides is maintained at 75°C. A heat source of $H = 50\,\mathrm{MW/m^3}$ is present in the body. For the body, $k_x = k_y = 20\,\mathrm{W/m \cdot {}^\circ C}$ and $L = 2.5$ cm. Determine the temperature distribution in the body using a model with the $2 \times 2$ mesh of equal size elements.

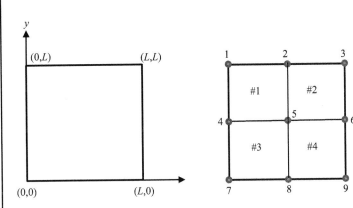

**Figure 11.6** A square 2D-domain model with a $2 \times 2$ mesh of four-node elements

### Solution:

From the FE model, we can obtain the global nodal temperature as

$$\begin{Bmatrix} q_4 \\ q_5 \\ q_7 \\ q_8 \end{Bmatrix} = \begin{Bmatrix} 451.67 \\ 376.34 \\ 560.49 \\ 451.67 \end{Bmatrix} {}^\circ C.$$

## 11.5 Finite Element Formulation for Heat Transfer in the 3D Domain

The procedure described for the 2D formulation can easily be extended to the 3D formulation.

**Construction of Element Conduction Matrix**

For FE approximation, the first term in Eq. (11.3.15) can be expressed as

$$\int_V \delta \mathbf{T}_d^\mathrm{T} \mathbf{D} \mathbf{T}_d dV = \sum_e \int_{V_e} \delta \mathbf{T}_d^\mathrm{T} \mathbf{D} \mathbf{T}_d dV, \qquad (11.5.1)$$

where "*e*" stands for element. According to the chain rule of differentiation in 3D space:

$$
\left\{\begin{array}{c} \dfrac{\partial T}{\partial \xi} \\[2ex] \dfrac{\partial T}{\partial \eta} \\[2ex] \dfrac{\partial T}{\partial \zeta} \end{array}\right\} = \begin{bmatrix} \dfrac{\partial x}{\partial \xi} & \dfrac{\partial y}{\partial \xi} & \dfrac{\partial z}{\partial \xi} \\[2ex] \dfrac{\partial x}{\partial \eta} & \dfrac{\partial y}{\partial \eta} & \dfrac{\partial z}{\partial \eta} \\[2ex] \dfrac{\partial x}{\partial \zeta} & \dfrac{\partial y}{\partial \zeta} & \dfrac{\partial z}{\partial \zeta} \end{bmatrix} \left\{\begin{array}{c} \dfrac{\partial T}{\partial x} \\[2ex] \dfrac{\partial T}{\partial y} \\[2ex] \dfrac{\partial T}{\partial z} \end{array}\right\} = \mathbf{J}_3 \left\{\begin{array}{c} \dfrac{\partial T}{\partial x} \\[2ex] \dfrac{\partial T}{\partial y} \\[2ex] \dfrac{\partial T}{\partial z} \end{array}\right\}, \tag{11.5.2}
$$

where

$$
\mathbf{J}_3 = \begin{bmatrix} \dfrac{\partial x}{\partial \xi} & \dfrac{\partial y}{\partial \xi} & \dfrac{\partial z}{\partial \xi} \\[2ex] \dfrac{\partial x}{\partial \eta} & \dfrac{\partial y}{\partial \eta} & \dfrac{\partial z}{\partial \eta} \\[2ex] \dfrac{\partial x}{\partial \zeta} & \dfrac{\partial y}{\partial \zeta} & \dfrac{\partial z}{\partial \zeta} \end{bmatrix} : \text{Jacobian matrix in the 3D space.} \tag{11.5.3}
$$

Multiplying Eq. (11.5.2) with the inverse of the Jacobian matrix:

$$
\left\{\begin{array}{c} \dfrac{\partial T}{\partial x} \\[2ex] \dfrac{\partial T}{\partial y} \\[2ex] \dfrac{\partial T}{\partial z} \end{array}\right\} = \mathbf{J}_3^{-1} \left\{\begin{array}{c} \dfrac{\partial T}{\partial \xi} \\[2ex] \dfrac{\partial T}{\partial \eta} \\[2ex] \dfrac{\partial T}{\partial \zeta} \end{array}\right\} = \mathbf{Y} \left\{\begin{array}{c} \dfrac{\partial T}{\partial \xi} \\[2ex] \dfrac{\partial T}{\partial \eta} \\[2ex] \dfrac{\partial T}{\partial \zeta} \end{array}\right\} = \begin{bmatrix} Y_{11} & Y_{12} & Y_{13} \\[1ex] Y_{21} & Y_{22} & Y_{23} \\[1ex] Y_{31} & Y_{32} & Y_{33} \end{bmatrix} \left\{\begin{array}{c} \dfrac{\partial T}{\partial \xi} \\[2ex] \dfrac{\partial T}{\partial \eta} \\[2ex] \dfrac{\partial T}{\partial \zeta} \end{array}\right\}, \tag{11.5.4}
$$

where $\mathbf{Y} = \mathbf{J}_3^{-1}$. As an illustration, we will now consider the eight-node element. Extensions to other types of elements are straightforward.

For the eight-node element:

$$
\left\{\begin{array}{c} \dfrac{\partial T}{\partial x} \\[2ex] \dfrac{\partial T}{\partial y} \\[2ex] \dfrac{\partial T}{\partial z} \end{array}\right\} = \mathbf{Y} \left\{\begin{array}{c} \dfrac{\partial T}{\partial \xi} \\[2ex] \dfrac{\partial T}{\partial \eta} \\[2ex] \dfrac{\partial T}{\partial \zeta} \end{array}\right\} = \mathbf{Y} \left\{\begin{array}{c} \displaystyle\sum_{i=1}^{8} \dfrac{\partial N_i}{\partial \xi} T_i \\[3ex] \displaystyle\sum_{i=1}^{8} \dfrac{\partial N_i}{\partial \eta} T_i \\[3ex] \displaystyle\sum_{i=1}^{8} \dfrac{\partial N_i}{\partial \zeta} T_i \end{array}\right\}, \tag{11.5.5}
$$

and

$$
\mathbf{T}_d = \left\{ \begin{array}{c} \dfrac{\partial T}{\partial x} \\[2mm] \dfrac{\partial T}{\partial y} \\[2mm] \dfrac{\partial T}{\partial z} \end{array} \right\} = \mathbf{Y} \begin{bmatrix} \dfrac{\partial N_1}{\partial \xi} & \dfrac{\partial N_2}{\partial \xi} & \dfrac{\partial N_3}{\partial \xi} & \dfrac{\partial N_4}{\partial \xi} & \dfrac{\partial N_5}{\partial \xi} & \dfrac{\partial N_6}{\partial \xi} & \dfrac{\partial N_7}{\partial \xi} & \dfrac{\partial N_8}{\partial \xi} \\[3mm] \dfrac{\partial N_1}{\partial \eta} & \dfrac{\partial N_2}{\partial \eta} & \dfrac{\partial N_3}{\partial \eta} & \dfrac{\partial N_4}{\partial \eta} & \dfrac{\partial N_5}{\partial \eta} & \dfrac{\partial N_6}{\partial \eta} & \dfrac{\partial N_7}{\partial \eta} & \dfrac{\partial N_8}{\partial \eta} \\[3mm] \dfrac{\partial N_1}{\partial \zeta} & \dfrac{\partial N_2}{\partial \zeta} & \dfrac{\partial N_3}{\partial \zeta} & \dfrac{\partial N_4}{\partial \zeta} & \dfrac{\partial N_5}{\partial \zeta} & \dfrac{\partial N_6}{\partial \zeta} & \dfrac{\partial N_7}{\partial \zeta} & \dfrac{\partial N_8}{\partial \zeta} \end{bmatrix} \left\{ \begin{array}{c} T_1 \\ T_2 \\ T_3 \\ T_4 \\ T_5 \\ T_6 \\ T_7 \\ T_8 \end{array} \right\} = \mathbf{Bd},
$$

$$(11.5.6)$$

where $\mathbf{d}$ is an $8 \times 1$ column vector of nodal temperatures. Similarly

$$
\delta \mathbf{T}_d = \left\{ \begin{array}{c} \dfrac{\partial \delta T}{\partial x} \\[2mm] \dfrac{\partial \delta T}{\partial y} \\[2mm] \dfrac{\partial \delta T}{\partial z} \end{array} \right\} = \mathbf{B} \delta \mathbf{d}.
$$

$$(11.5.7)$$

Introducing Eqs (11.5.6) and (11.5.7) into an element in Eq. (11.5.1):

$$
\int_{V_e} \delta \mathbf{T}_d^{\mathrm{T}} \mathbf{D} \mathbf{T}_d dV = \delta \mathbf{d}^{\mathrm{T}} \left( \int_{V_e} \mathbf{B}^{\mathrm{T}} \mathbf{D} \mathbf{B} dV \right) \mathbf{d} = \delta \mathbf{d}^{\mathrm{T}} \mathbf{k}_e \mathbf{d},
$$

$$(11.5.8)$$

where

$$
\mathbf{k}^e = \int_{V_e} \mathbf{B}^{\mathrm{T}} \mathbf{D} \mathbf{B} dV
$$

$$(11.5.9)$$

is the $8 \times 8$ element conduction matrix for the eight-node element. We can use Eq. (7.4.24) to transform the integration over the element volume in the physical 3D space into that in the mapped 3D space. We can then use Eq. (7.5.19) to apply the numerical integration rule as follows:

$$
\mathbf{k}^e = \int_{-1}^{1} \int_{-1}^{1} \int_{-1}^{1} \mathbf{B}^{\mathrm{T}} \mathbf{D} \mathbf{B} J_3 d\xi d\eta d\zeta = \sum_{i=1} \sum_{j=1} \sum_{k=1} \left[ \mathbf{B}^{\mathrm{T}} \mathbf{D} \mathbf{B} J_3 \right]_{\substack{\xi = a_i \\ \eta = a_j \\ \zeta = a_k}} W_i W_j W_k.
$$

$$(11.5.10)$$

### Element Load Vector Due to Heat Source

Consider the second term in Eq. (11.3.15). Introducing the virtual temperature field into an element:

$$\int_{V_e} \delta T H dV = \delta \mathbf{d}^{\mathrm{T}} \int_{V_e} \mathbf{N}^{\mathrm{T}} H dV = \delta \mathbf{d}^{\mathrm{T}} \mathbf{f}_H^e, \qquad (11.5.11)$$

where

$$\mathbf{f}_H^e = \int_{V_e} \mathbf{N}^{\mathrm{T}} H dV. \qquad (11.5.12)$$

For the eight-node element, $\mathbf{f}_H^e$ is the $8 \times 1$ element load vector due to heat source.

### Element Load Vector Due to Surface Heat Flux

Consider the second term in the last part of Eq. (11.3.16). Introducing a virtual temperature field into an element with heat flux applied on the surface:

$$\int_{S_{F_e}} \delta T \mathbf{g} \cdot \mathbf{n} dS = \delta \mathbf{d}^{\mathrm{T}} \int_{S_{F_e}} \mathbf{N}^{\mathrm{T}} \mathbf{g} \cdot \mathbf{n} dS = \delta \mathbf{d}^{\mathrm{T}} \mathbf{f}_s^e, \qquad (11.5.13)$$

where

$$\mathbf{f}_s^e = \int_{S_{F_e}} \mathbf{N}^{\mathrm{T}} \mathbf{g} \cdot \mathbf{n} dS \qquad (11.5.14)$$

is the element load vector due to applied surface flux. Integration over an element surface in the physical 3D space can be transformed into that in the mapped surface, as shown in Section 7.4.4 of Chapter 7.

### Convection on the Boundary

For convection over a 3D surface, heat flux $q_n$ normal to the surface can be expressed as

$$q_n = h(T - T_\infty). \qquad (11.5.15)$$

As in the 1D and 2D cases, the $hT$ term leads to the element convection matrix that adds to the element conduction matrix, while $hT_\infty$ contributes to the element load vector.

### Construction of Element Capacitance Matrix

For FE formulation, the third term in Eq. (11.3.15) is expressed as

$$\int_V \delta T \frac{\partial T}{\partial t} c\rho dV = \sum_e \int_{V_e} \delta T \frac{\partial T}{\partial t} c\rho dV. \qquad (11.5.16)$$

For FE modeling, the assumed temperature in the element can be expressed as

$$T = \mathbf{Nd}. \tag{11.5.17}$$

Then

$$\delta T = \mathbf{N}\delta\mathbf{d}, \quad \dot{T} = \mathbf{N}\dot{\mathbf{d}}. \tag{11.5.18}$$

Introducing Eqs (11.5.17) and (11.5.18) into the expression for an element in Eq. (11.5.16):

$$\int_{V_e} \delta T \frac{\partial T}{\partial t} c\rho dV = \int_{V} \delta\mathbf{d}^{T}\mathbf{N}^{T}\mathbf{N}\dot{\mathbf{d}} c\rho dV = \delta\mathbf{d}^{T}\left(\int_{V_e} \mathbf{N}^{T}\mathbf{N}c\rho dV\right)\dot{\mathbf{d}} = \delta\mathbf{d}^{T}\mathbf{c}^{e}\dot{\mathbf{d}}, \tag{11.5.19}$$

where

$$\mathbf{c}^{e} = \int_{V_e} \mathbf{N}^{T}\mathbf{N}c\rho dV : \text{element capacitance matrix}. \tag{11.5.20}$$

For the entire body:

$$\int_{V} \delta T \frac{\partial T}{\partial t} c\rho dV = \sum_{e} \int_{V_e} \delta T \frac{\partial T}{\partial t} c\rho dV = \delta\mathbf{q}^{T}\mathbf{C}\dot{\mathbf{q}}, \tag{11.5.21}$$

where $\mathbf{C}$ is the global capacitance matrix assembled from the element capacitance matrices.

### Finite Element Equation for Heat Transfer in the 3D Domain

After assembly over all elements, the FE formulation based on Eq. (11.3.15) results in

$$\delta\mathbf{q}^{T}(\mathbf{Kq} + \mathbf{C}\dot{\mathbf{q}} - \mathbf{F}) = 0 \rightarrow \mathbf{Kq} + \mathbf{C}\dot{\mathbf{q}} - \mathbf{F} = \mathbf{0} \tag{11.5.22}$$

or

$$\mathbf{C}\dot{\mathbf{q}} + \mathbf{Kq} = \mathbf{F}. \tag{11.5.23}$$

The initial condition at time $t = 0$ can be expressed as

$$\mathbf{q}(0) = \mathbf{q}_0. \tag{11.5.24}$$

With initial conditions prescribed, Eq. (11.5.23) can be solved for global nodal temperatures in time, using various numerical time integration methods.

## PROBLEMS

**11.1** Consider a 1D steady-state heat conduction problem for a body of uniform cross-section, described as follows:

Boundary conditions: $T = 0$ at $x = 0$ and $x = L$.
Heat source given as $f_s = BA$, where $B$ is a given constant.

(a) Use a model with four elements of equal length to determine the temperature distribution. Plot non-dimensional temperature vs. non-dimensional length.

(b) Determine the heat flux distribution and plot non-dimensional heat flux.

**11.2** Repeat Problem 11.1 using a model with eight elements of equal length.

**11.3** Consider 1D steady-state heat conduction subjected to the following boundary conditions: $T = 0$ at $x = 0$ and $q = q_L$ at $x = L$. In addition, a constant heat source of $f_s(x) = BA$ is present over the body of uniform cross-section. Use an FE model with six elements of equal length to do as follows.

(a) Construct the global conduction matrix.

(b) Construct the element load vectors due to the heat source and assemble the global load vector.

(c) Determine the nodal temperatures. Plot the temperature distribution along the body and compare with the exact solution.

(d) Determine the heat flux in each element. Plot heat flux along the body and compare with the exact solution.

**11.4** Consider a body of uniform cross-section subjected to the following boundary conditions: $T = 0$ at $x = 0$ and $q = h(T - T_\infty)$ at $x = L$. In addition, a constant heat source of $f_s(x) = BA$ is present over the body. Use an FE model with three elements of equal length to do as follows.

(a) Construct the global heat transfer matrix.

(b) Construct the element load vectors due to the heat source and assemble the global load vector.

(c) Determine the nodal temperatures.

(d) Determine the heat flux in each element.

(e) Plot the temperature distribution along the body and compare with the exact solution. Non-dimensionalize the temperature such that the maximum value of exact temperature is equal to unity.

(f) Plot the heat flux distribution along the body and compare with the exact solution. Non-dimensionalize the heat flux such that the maximum value of exact heat flux is equal to unity.

**11.5** Consider a 1D time-dependent problem described as follows:

Boundary conditions: $T(0, t) = T_0$, $\quad T(L, t) = 2T_0$
Initial conditions: $T(x, 0) = 0°C$.

Carry out FE analysis to determine the temperature distribution with time until a steady state is reached. For convenience, introduce non-dimensional time $\tau$ defined as follows:

$$\tau = at \quad \text{where} \quad a = \frac{k}{\rho c L^2}$$

such that

$$\frac{dT}{dt} = \frac{dT}{d\tau}\frac{d\tau}{dt} = a\frac{dT}{d\tau}.$$

Note that the steady-state solution is $T = T_0(1 + x/L)$.

(a) Try with a model with 10 elements of equal length. Use the RK4 method for integration in time.

(b) Plot $T/T_0$ vs. $x/L$ using non-dimensional time $\tau$ as a parameter.

**11.6** Repeat Problem 11.5 using a model with 20 elements of equal length.

**11.7** For the three-node element shown in Figure 11.7, the mapping and assumed temperature are expressed as

Mapping: $x = N_1 x_1 + N_2 x_2 + N_3 x_3$, $\quad y = N_1 y_1 + N_2 y_2 + N_3 y_3$

Assumed temperature: $T = N_1 T_1 + N_2 T_2 + N_3 T_3$ where $N_1 = 1 - \xi - \eta$, $N_2 = \xi$, $N_3 = \eta$.

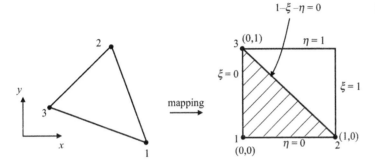

**Figure 11.7** For Problem 11.7

(a) Show that

$$\mathbf{J}_2 = \begin{bmatrix} \dfrac{\partial x}{\partial \xi} & \dfrac{\partial y}{\partial \xi} \\ \dfrac{\partial x}{\partial \eta} & \dfrac{\partial y}{\partial \eta} \end{bmatrix} = \begin{bmatrix} x_2 - x_1 & y_2 - y_1 \\ x_3 - x_1 & y_3 - y_1 \end{bmatrix},$$

which is constant for the element.

(b) Show that

$$\left\{ \begin{array}{c} \dfrac{\partial T}{\partial x} \\ \dfrac{\partial T}{\partial y} \end{array} \right\} = \mathbf{Bd},$$

where

$$\mathbf{B} = \mathbf{Y} \begin{bmatrix} -1 & 1 & 0 \\ -1 & 0 & 1 \end{bmatrix}, \quad \mathbf{Y} = \mathbf{J}_2^{-1}, \mathbf{d} = \left\{ \begin{array}{c} T_1 \\ T_2 \\ T_3 \end{array} \right\}.$$

(c) For $\mathbf{D}$ and $t_z$ constant in the element, show that the element conduction matrix is

$$\mathbf{k}^e = \mathbf{B}^{\mathrm{T}} \mathbf{D} \mathbf{B} t_z A_e,$$

where $A_e$ (element area) can be expressed as

$$A_e = \int_{A_e} dA = \int J_2 d\xi d\eta = \frac{1}{2}J_2 = \frac{1}{2}\det \mathbf{J}_2.$$

**11.8** Consider a three-node 2D element with nodal $(x,y)$ coordinates given as node 1: $(0,0)$, node 2: $(a,0)$, node 3: $(0,a)$. Show that

(a)
$$\mathbf{J}_2 = \begin{bmatrix} x_2 - x_1 & y_2 - y_1 \\ x_3 - x_1 & y_3 - y_1 \end{bmatrix} = a\begin{bmatrix} 1 & 0 \\ 0 & 1 \end{bmatrix}.$$

(b)
$$\mathbf{B} = \frac{1}{a}\begin{bmatrix} -1 & 1 & 0 \\ -1 & 0 & 1 \end{bmatrix}.$$

(c) For a thermally isotropic material with $k_x = k_y = k_z = k$, the element conduction matrix is

$$\mathbf{k}^e = \mathbf{B}^{\mathrm{T}}\mathbf{D}\mathbf{B}t_z A_e = \frac{kt_z}{2}\begin{bmatrix} 2 & -1 & -1 \\ -1 & 1 & 0 \\ -1 & 0 & 1 \end{bmatrix}.$$

**11.9** Solve the problem described in Example 1.6 using two different meshes of the four-node elements.
(a) Uniform $3 \times 3$ mesh.
(b) Uniform $4 \times 4$ mesh.

**11.10** Solve the problem described in Example 1.6 using two different meshes of the three-node triangular elements as follows.
(a) Uniform $3 \times 3$ mesh.
(b) Uniform $4 \times 4$ mesh.

# APPENDIX 1

# Fundamentals of Solid and Structural Mechanics

In this appendix we review some mathematical foundations for the study of deformable solids and structures in a 3D Cartesian coordinate system. For this, we use the original undeformed configuration as the reference to introduce 15 variables of three displacement components, six strain components and six stress components. The 15 variables are accompanied by a total of 15 equations. They are three equilibrium equations, six strain–displacement relations, and six equations called constitutive equations which relate the strain and stress components through material properties.

There are various measures of strain and stress. In this chapter, we introduce strain components under the assumption of small deformation and small rigid-body rotation. This results in strain–displacement relations which are linear. For definition of stress components, we neglect the effect of deformation, which allows us to assume that there is no distinction between the deformed configuration and the undeformed configuration when force equilibrium and moment equilibrium are considered.

For linear elastic constitutive equations which account for material properties, we consider three classes of materials, called orthotropic, isotropic, and transversely isotropic. Depending on the planes of symmetry that exist in these materials, we can determine the number of independent material constants that need to be obtained by experiments for each class of material.

Strain components, stress components, and thus constitutive equations are dependent on the coordinate system selected to describe a body. However, one can develop transformation rules which allow us to relate strain components, stress components, and constitutive equations in two different coordinate systems through the use of direction cosines.

## A1.1  Strain

Strain is a non-dimensional measure of relative deformation under applied loads. There are six components in the strain. Consider a structural body in the original undeformed configuration and in the deformed configuration. In the undeformed body, one may choose a generic point labeled P and then draw three infinitesimal lengths $dx$, $dy$, and $dz$ in the $x$, $y$, and $z$ direction of the Cartesian coordinate system as shown in Figure A1.1. As a result of the deformation, point P moves to P′ while points A, B, and C move to points A′, B′, and C′.

In the figure, we observe that line vector $\overrightarrow{PA}$ changes to $\overrightarrow{P'A'}$, line vector $\overrightarrow{PB}$ changes to $\overrightarrow{P'B'}$, and line vector $\overrightarrow{PC}$ changes to $\overrightarrow{P'C'}$. We also note that the angle between $\overrightarrow{PA}$ and $\overrightarrow{PB}$ changes to the angle between $\overrightarrow{P'A'}$ and $\overrightarrow{P'B'}$, the angle between $\overrightarrow{PB}$ and $\overrightarrow{PC}$ changes to the

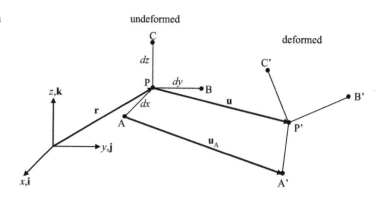

**Figure A1.1** Infinitesimal line vectors in the original undeformed configuration and the deformed configuration

angle between $\overrightarrow{P'B'}$ and $\overrightarrow{P'C'}$, and the angle between $\overrightarrow{PA}$ and $\overrightarrow{PC}$ changes to the angle between $\overrightarrow{P'A'}$ and $\overrightarrow{P'C'}$.

Strain is defined by considering the length changes of the three line vectors and the changes in the angles between the line vectors. For this, consider the following position and displacement vectors in the Cartesian coordinate system:

$\mathbf{r} = x\mathbf{i} + y\mathbf{j} + z\mathbf{k}$: position vector of point P
$\mathbf{u} = u\mathbf{i} + v\mathbf{j} + w\mathbf{k}$: displacement vector of point P
$\mathbf{u}_A = u_A\mathbf{i} + v_A\mathbf{j} + w_A\mathbf{k}$: displacement vector of point A
$\mathbf{u}_B = u_B\mathbf{i} + v_B\mathbf{j} + w_B\mathbf{k}$: displacement vector of point B
$\mathbf{u}_C = u_C\mathbf{i} + v_C\mathbf{j} + w_C\mathbf{k}$: displacement vector of point C

where $\mathbf{i}, \mathbf{j}, \mathbf{k}$ are the unit vectors. Then, we note that

$$\overrightarrow{PA} = dx\mathbf{i}, \mathrm{PB} = dy\mathbf{j}, \ \overrightarrow{PC} = dz\mathbf{k}, \tag{A1.1.1}$$

and

$$\mathbf{u} + \overrightarrow{P'A'} = \overrightarrow{PA} + \mathbf{u}_A. \tag{A1.1.2}$$

Then

$$\overrightarrow{P'A'} = \overrightarrow{PA} + \mathbf{u}_A - \mathbf{u} = dx\mathbf{i} + \left(\mathbf{u} + \frac{\partial \mathbf{u}}{\partial x}dx\right) - \mathbf{u} = dx\mathbf{i} + \frac{\partial}{\partial x}(u\mathbf{i} + v\mathbf{j} + w\mathbf{k})dx, \tag{A1.1.3}$$

$$\overrightarrow{P'A'} = \left[\left(1 + \frac{\partial u}{\partial x}\right)\mathbf{i} + \frac{\partial v}{\partial x}\mathbf{j} + \frac{\partial w}{\partial x}\mathbf{k}\right]dx. \tag{A1.1.4}$$

Also

$$\mathbf{u} + \overrightarrow{P'B'} = \overrightarrow{PB} + \mathbf{u}_B, \tag{A1.1.5}$$

$$\overrightarrow{P'B'} = \overrightarrow{PB} + \mathbf{u}_B - \mathbf{u} = dy\mathbf{j} + \left(\mathbf{u} + \frac{\partial \mathbf{u}}{\partial y}dy\right) - \mathbf{u} = dy\mathbf{j} + \frac{\partial}{\partial y}(u\mathbf{i} + v\mathbf{j} + w\mathbf{k})dy, \tag{A1.1.6}$$

$$\overrightarrow{P'B'} = \left[ \frac{\partial u}{\partial y}\mathbf{i} + \left(1 + \frac{\partial v}{\partial y}\right)\mathbf{j} + \frac{\partial w}{\partial y}\mathbf{k} \right] dy. \tag{A1.1.7}$$

Similarly

$$\overrightarrow{P'C'} = \left[ \frac{\partial u}{\partial z}\mathbf{i} + \frac{\partial v}{\partial z}\mathbf{j} + \left(1 + \frac{\partial w}{\partial z}\right)\mathbf{k} \right] dz. \tag{A1.1.8}$$

### Extensional Strains

Extensional strains are introduced as a measure of non-dimensional length changes. Consider the non-dimensional length change of line vector $\overrightarrow{PA} = dx\mathbf{i}$ defined as

$$\varepsilon_{xx} = \frac{\left|\overrightarrow{P'A'}\right| - \left|\overrightarrow{PA}\right|}{\left|\overrightarrow{PA}\right|}, \tag{A1.1.9}$$

where

$$\left|\overrightarrow{P'A'}\right| = \sqrt{\left(\overrightarrow{P'A'} \cdot \overrightarrow{P'A'}\right)} = dx\sqrt{\left(1 + \frac{\partial u}{\partial x}\right)^2 + \left(\frac{\partial v}{\partial x}\right)^2 + \left(\frac{\partial w}{\partial x}\right)^2} \tag{A1.1.10}$$

is the length in the deformed configuration. Then

$$\varepsilon_{xx} = \frac{dx\sqrt{\left(1 + \frac{\partial u}{\partial x}\right)^2 + \left(\frac{\partial v}{\partial x}\right)^2 + \left(\frac{\partial w}{\partial x}\right)^2} - dx}{dx} = \sqrt{\left(1 + \frac{\partial u}{\partial x}\right)^2 + \left(\frac{\partial v}{\partial x}\right)^2 + \left(\frac{\partial w}{\partial x}\right)^2} - 1$$

$$= \sqrt{1 + 2\frac{\partial u}{\partial x} + \left(\frac{\partial u}{\partial x}\right)^2 + \left(\frac{\partial v}{\partial x}\right)^2 + \left(\frac{\partial w}{\partial x}\right)^2} - 1. \tag{A1.1.11}$$

Assuming small strain and small rigid-body rotation:

$$\left|\frac{\partial u}{\partial x}\right|, \left|\frac{\partial v}{\partial x}\right|, \left|\frac{\partial w}{\partial x}\right| << 1.$$

Then

$$\varepsilon_{xx} \simeq \sqrt{1 + 2\frac{\partial u}{\partial x}} - 1 \simeq \left(1 + \frac{\partial u}{\partial x}\right) - 1 = \frac{\partial u}{\partial x}. \tag{A1.1.12}$$

Similarly, considering non-dimensional length change of $\overrightarrow{PB} = dy\,\mathbf{j}$, one can show that

$$\varepsilon_{yy} = \frac{\left|\overrightarrow{P'B'}\right| - \left|\overrightarrow{PB}\right|}{\left|\overrightarrow{PB}\right|} \simeq \left(1 + \frac{\partial v}{\partial y}\right) - 1 = \frac{\partial v}{\partial y}. \tag{A1.1.13}$$

320    Appendices

Also, considering non-dimensional length change of $\overrightarrow{PC}=d z\mathbf{k}$:

$$\varepsilon_{zz}=\frac{\left|\overrightarrow{P'C'}\right|-\left|\overrightarrow{PC}\right|}{\left|\overrightarrow{PC}\right|}\cong\left(1+\frac{\partial w}{\partial z}\right)-1=\frac{\partial w}{\partial z}. \qquad (A1.1.14)$$

## Shear Strains

Shear strains $\varepsilon_{xy}, \varepsilon_{yz}$, and $\varepsilon_{zx}$ are introduced as a measure of relative angle changes. Consider the angle change between $\overrightarrow{PA}$ and $\overrightarrow{PB}$ due to deformation, as shown in Figure A1.2. For positive $\varepsilon_{xy}$, the angle between $\overrightarrow{P'A'}$ and $\overrightarrow{P'B'}$ is smaller than 90°. Using the dot product formula:

$$\left(\overrightarrow{P'A'}\right)\cdot\left(\overrightarrow{P'B'}\right)=\left|\overrightarrow{P'A'}\right|\left|\overrightarrow{P'B'}\right|\cos\left(\frac{\pi}{2}-\varepsilon_{xy}\right). \qquad (A1.1.15)$$

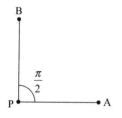

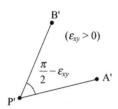

Before deformation          After deformation

**Figure A1.2** Changes in the angle between the two infinitesimal line vectors

Then

$$\cos\left(\frac{\pi}{2}-\varepsilon_{xy}\right)=\frac{\left(\overrightarrow{P'A'}\right)\cdot\left(\overrightarrow{P'B'}\right)}{\left|\overrightarrow{P'A'}\right|\left|\overrightarrow{P'B'}\right|}$$

$$=\frac{\left(1+\frac{\partial u}{\partial x}\right)\frac{\partial u}{\partial y}+\frac{\partial v}{\partial x}\left(1+\frac{\partial v}{\partial y}\right)+\frac{\partial w}{\partial x}\frac{\partial w}{\partial y}}{\sqrt{\left(1+\frac{\partial u}{\partial x}\right)^2+\left(\frac{\partial v}{\partial x}\right)^2+\left(\frac{\partial w}{\partial x}\right)^2}\sqrt{\left(\frac{\partial u}{\partial y}\right)^2+\left(1+\frac{\partial v}{\partial y}\right)^2+\left(\frac{\partial w}{\partial y}\right)^2}} \qquad (A1.1.16)$$

and

$$\sin\varepsilon_{xy}=\frac{\frac{\partial u}{\partial y}+\frac{\partial v}{\partial x}+\text{higher-order terms (HOT)}}{\sqrt{1+\text{HOT}}\sqrt{1+\text{HOT}}}. \qquad (A1.1.17)$$

For small angle change, $\sin\varepsilon_{xy}=\varepsilon_{xy}$ and neglecting the higher-order terms:

$$\varepsilon_{xy} = \frac{\partial u}{\partial y} + \frac{\partial v}{\partial x}. \qquad (A1.1.18)$$

Similarly, for the angle change between $\overrightarrow{PB}$ and $\overrightarrow{PC}$:

$$\varepsilon_{yz} = \frac{\partial v}{\partial z} + \frac{\partial w}{\partial y} \qquad (A1.1.19)$$

and for the angle change between $\overrightarrow{PC}$ and $\overrightarrow{PA}$:

$$\varepsilon_{zx} = \frac{\partial w}{\partial x} + \frac{\partial u}{\partial z}. \qquad (A1.1.20)$$

### Relative Volume Change

The original volume of a rectangular cuboid with side lengths $dx$, $dy$, and $dz$ is

$$dV_0 = dxdydz. \qquad (A1.1.21)$$

According to the triple scalar product formula, the volume in the deformed configuration is

$$dV = \overrightarrow{P'A'} \cdot \left( \overrightarrow{P'B'} \times \overrightarrow{P'C'} \right). \qquad (A1.1.22)$$

As a non-dimensional measure of volume change, RVC is defined as

$$\text{RVC} = \frac{dV - dV_0}{dV_0}. \qquad (A1.1.23)$$

Placing Eqs (A1.1.4), (A1.1.7), and (A1.1.8) into Eq. (A1.1.23) and dropping higher-order terms under the assumption of small strain and small rigid-body rotation:

$$\text{RVC} = \frac{\partial u}{\partial x} + \frac{\partial v}{\partial y} + \frac{\partial w}{\partial z} = \varepsilon_{xx} + \varepsilon_{yy} + \varepsilon_{zz}. \qquad (A1.1.24)$$

## A1.2  Stress

A solid body deforms when an external load is applied. Accordingly, equilibrium must be considered over the body in the deformed configuration. However, in many practical situations the difference between the original undeformed configuration and the deformed configuration is very small and can be neglected when equilibrium is considered. We will adopt this simplifying assumption in this section and the following section.

To define a stress vector at a point in the solid body, consider imaginary cutting planes that contain the point. There are infinite choices of such cutting planes, and each cutting plane produces a pair of matching surfaces. For example, we may select the cutting plane which is normal to the $x$-axis. The surface with $\mathbf{n} = \mathbf{i}$ is then called the positive $x$ surface, while the opposite surface with $\mathbf{n} = -\mathbf{i}$ is called the negative $x$ surface.

The stress vector $\boldsymbol{\sigma}_x$ acting on the positive $x$ surface at a generic point P in the solid body is defined as

$$\boldsymbol{\sigma}_x = \frac{d\mathbf{F}_x}{dA} = \frac{d\mathbf{F}_x}{dydz},$$

(A1.2.1)

where $dA = dydz$ is the infinitesimally small area as shown in Figure A1.3 and $d\mathbf{F}_x$ is the force vector acting on the area. In component form, the stress vector can be expressed as

$$\boldsymbol{\sigma}_x = \sigma_{xx}\mathbf{i} + \sigma_{xy}\mathbf{j} + \sigma_{xz}\mathbf{k}.$$

(A1.2.2)

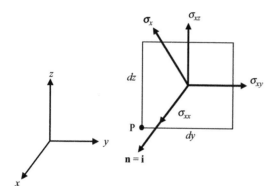

**Figure A1.3** Positive stress components on the positive $x$ surface

The first subscript in the stress components represents the surface on which it is acting, while the second subscript indicates the direction. Note that $\sigma_{xx}$ is normal to the surface, while $\sigma_{xy}$ and $\sigma_{xz}$ are tangent to the surface. The opposite surface with $\mathbf{n} = -\mathbf{i}$ is called the negative $x$ surface. As shown in Figure A1.4, the stress vector acting on the negative $x$ surface is

$$-\boldsymbol{\sigma}_x = \sigma_{xx}(-\mathbf{i}) + \sigma_{xy}(-\mathbf{j}) + \sigma_{xz}(-\mathbf{k}).$$

(A1.2.3)

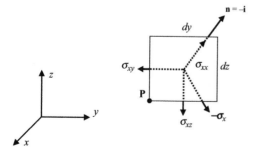

**Figure A1.4** Positive stress components on the negative $x$ surface

Similarly, the stress vector $\boldsymbol{\sigma}_y$ acting on the positive $y$ surface with $\mathbf{n} = \mathbf{j}$ is

$$\boldsymbol{\sigma}_y = \sigma_{yx}\mathbf{i} + \sigma_{yy}\mathbf{j} + \sigma_{yz}\mathbf{k}$$

(A1.2.4)

while the stress vector acting on the negative $y$ surface with $\mathbf{n} = -\mathbf{j}$ is

$$-\boldsymbol{\sigma}_y = \sigma_{yx}(-\mathbf{i}) + \sigma_{yy}(-\mathbf{j}) + \sigma_{yz}(-\mathbf{k}).$$

(A1.2.5)

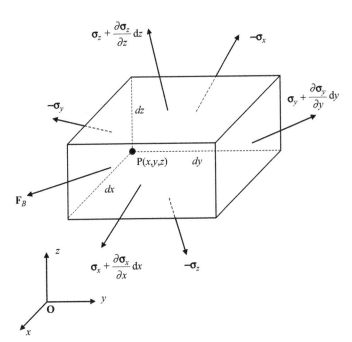

**Figure A1.5** Stress vectors and body force vector acting on the free body

Similarly, the stress vector acting over the positive $z$ surface with $\mathbf{n} = \mathbf{k}$ is

$$\boldsymbol{\sigma}_z = \sigma_{zx}\mathbf{i} + \sigma_{zy}\mathbf{j} + \sigma_{zz}\mathbf{k} \qquad (A1.2.6)$$

while the stress vector acting over the negative $z$ surface with $\mathbf{n} = -\mathbf{k}$ is

$$-\boldsymbol{\sigma}_z = \sigma_{zx}(-\mathbf{i}) + \sigma_{zy}(-\mathbf{j}) + \sigma_{zz}(-\mathbf{k}). \qquad (A1.2.7)$$

Note that $\sigma_{xx}, \sigma_{yy}, \sigma_{zz}$ are called normal stresses and $\sigma_{xy}, \sigma_{yz}$, etc. are called shear stresses.

## A1.3  Equilibrium

To consider equilibrium, we draw at point P three lines with length $dx, dy, dz$ to form a rectangular cuboid. The cuboid is then isolated from the rest of the body to create a free body as shown in Figure A1.5. The body is subjected to the surface forces and the body force vector $\mathbf{F}_B$ per unit volume.

### Force Equilibrium

For force equilibrium, the summation of all force vectors is equal to zero. Accordingly

$$\left(\boldsymbol{\sigma}_x + \frac{\partial \boldsymbol{\sigma}_x}{\partial x}dx\right)dydz - \boldsymbol{\sigma}_x dydz + \left(\boldsymbol{\sigma}_y + \frac{\partial \boldsymbol{\sigma}_y}{\partial y}dy\right)dxdz - \boldsymbol{\sigma}_y dxdz$$
$$+ \left(\boldsymbol{\sigma}_z + \frac{\partial \boldsymbol{\sigma}_z}{\partial z}dz\right)dxdy - \boldsymbol{\sigma}_z dxdy + \mathbf{F}_B dxdydz = \mathbf{0}. \qquad (A1.3.1)$$

Canceling and collecting terms:

$$\left(\frac{\partial\boldsymbol{\sigma}_x}{\partial x}+\frac{\partial\boldsymbol{\sigma}_y}{\partial y}+\frac{\partial\boldsymbol{\sigma}_z}{\partial z}+\mathbf{F}_B\right)dxdydz = \mathbf{0}. \tag{A1.3.2}$$

Accordingly

$$\frac{\partial\boldsymbol{\sigma}_x}{\partial x}+\frac{\partial\boldsymbol{\sigma}_y}{\partial y}+\frac{\partial\boldsymbol{\sigma}_z}{\partial z}+\mathbf{F}_B = \mathbf{0}. \tag{A1.3.3}$$

## Moment Equilibrium

For moment equilibrium, the sum of all moment vectors acting on the free body is equal to zero. It can then be shown that shear stresses are symmetric in the sense that

$$\sigma_{xy} = \sigma_{yx}, \quad \sigma_{yz} = \sigma_{zy}, \quad \sigma_{xz} = \sigma_{zx}. \tag{A1.3.4}$$

Accordingly, there are six independent stress components to match six strain components. To avoid clutter, let us consider first the moments due to the forces acting over the $y$ surfaces and the body force as shown in Figure A1.6, where points H and K are at the surface centroids and point O is located at the volume centroid. In the figure, we consider three length vectors as follows:

$$\overrightarrow{PH}=\frac{1}{2}dx\mathbf{i}+\frac{1}{2}dz\mathbf{k}, \quad \overrightarrow{PK}=\overrightarrow{PH}+dy\mathbf{j}, \quad \overrightarrow{PO}=\frac{1}{2}dx\mathbf{i}+\frac{1}{2}dy\mathbf{j}+\frac{1}{2}dz\mathbf{k}. \tag{A1.3.5}$$

For convenience, we may assume without loss of generality that surface force vectors are acting on the surface centroids and the body force vector is acting at the body centroid. Then, moment vector $\mathbf{M}_p$ about point P due to the forces on the $\pm y$ surfaces and the body force is

$$\mathbf{M}_p = \overrightarrow{PH} \times \left(-\boldsymbol{\sigma}_y\right)dxdz + \left(\overrightarrow{PH} + dy\mathbf{j}\right) \times \left(\boldsymbol{\sigma}_y+\frac{\partial\boldsymbol{\sigma}_y}{\partial y}dy\right)dxdz + \overrightarrow{PO} \times \mathbf{F}_Bdxdydz$$

$$= dy\mathbf{j} \times \boldsymbol{\sigma}_y dxdz + dy\mathbf{j} \times \frac{\partial\boldsymbol{\sigma}_y}{\partial y}dxdydz + \overrightarrow{PH} \times \frac{\partial\boldsymbol{\sigma}_y}{\partial y}dydxdz + \overrightarrow{PO} \times \mathbf{F}_Bdxdydz. \tag{A1.3.6}$$

Dropping the last three terms which are of higher order as $dx, dy, dz$ approach zero in the limit:

**Figure A1.6** Surface force vectors located at the surface centroids and body force vector at the volume centroid

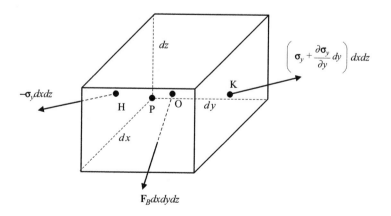

$$\mathbf{M}_p = dy\mathbf{j} \times \boldsymbol{\sigma}_y dxdz. \tag{A1.3.7}$$

Including the moments due to the forces acting over the $\pm x$ surface and $\pm z$ surface, and dropping higher-order terms:

$$
\begin{aligned}
(dy\mathbf{j}) \times (\boldsymbol{\sigma}_y dxdz) + (dx\mathbf{i}) \times (\boldsymbol{\sigma}_x dydz) + (dz\mathbf{k}) \times (\boldsymbol{\sigma}_z dxdy) &= \mathbf{0} \\
\rightarrow (\mathbf{j} \times \boldsymbol{\sigma}_y + \mathbf{i} \times \boldsymbol{\sigma}_x + \mathbf{k} \times \boldsymbol{\sigma}_z) dxdydz &= \mathbf{0} \\
\rightarrow \mathbf{j} \times \boldsymbol{\sigma}_y + \mathbf{i} \times \boldsymbol{\sigma}_x + \mathbf{k} \times \boldsymbol{\sigma}_z &= \mathbf{0}.
\end{aligned}
\tag{A1.3.8}
$$

One can then show the symmetry of shear stress components via expanding the last of the above equation as follows:

$$
\begin{aligned}
\mathbf{j} \times (\sigma_{yx}\mathbf{i} + \sigma_{yy}\mathbf{j} + \sigma_{yz}\mathbf{k}) + \mathbf{i} \times (\sigma_{xx}\mathbf{i} + \sigma_{xy}\mathbf{j} + \sigma_{xz}\mathbf{k}) + \mathbf{k} \times (\sigma_{zx}\mathbf{i} + \sigma_{zy}\mathbf{j} + \sigma_{zz}\mathbf{k}) &= \mathbf{0} \\
\rightarrow \mathbf{i}(\sigma_{yz} - \sigma_{zy}) + \mathbf{j}(\sigma_{zx} - \sigma_{xz}) + \mathbf{k}(\sigma_{xy} - \sigma_{yx}) &= \mathbf{0}.
\end{aligned}
\tag{A1.3.9}
$$

### Traction Vector

As shown in Figure A1.7, the traction vector $\mathbf{T}$ is a stress vector acting over an arbitrarily oriented surface with unit vector $\mathbf{n}$ outward normal to the surface. Expressed in component form:

$$\mathbf{T} = T_x\mathbf{i} + T_y\mathbf{j} + T_z\mathbf{k}, \mathbf{n} = l\mathbf{i} + m\mathbf{j} + n\mathbf{k}. \tag{A1.3.10}$$

It can then be shown that

$$\mathbf{T} = \boldsymbol{\sigma}_x l + \boldsymbol{\sigma}_y m + \boldsymbol{\sigma}_z n. \tag{A1.3.11}$$

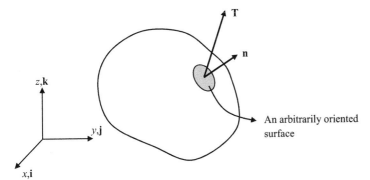

**Figure A1.7** Traction vector acting over an arbitrarily oriented surface

To derive the above equation, consider a tetrahedron as shown in Figure A1.8. Note that as $dx, dy, dz \rightarrow 0$ with the direction of the $\mathbf{n}$ vector fixed, surface ABC approaches point P. In the figure:

$\frac{1}{2}dA_n$: area of surface ABC, $\frac{1}{2}dA_x$: area of surface PBC,
$\frac{1}{2}dA_y$: area of surface PAC, $\frac{1}{2}dA_z$: area of surface PAB.

Summing all force vectors acting on the tetrahedron:

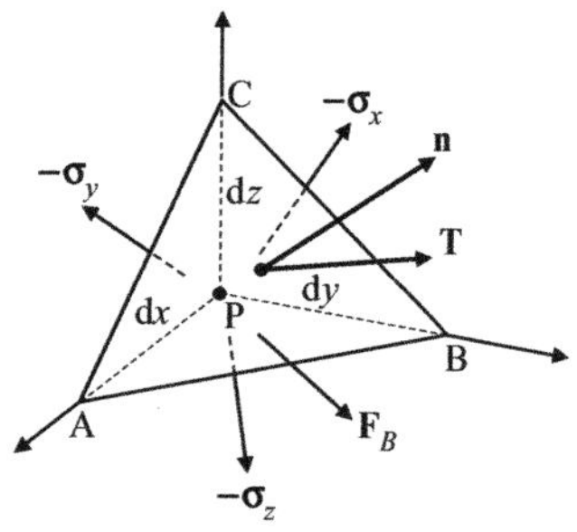

**Figure A1.8** Stress vectors and body force vector acting on a tetrahedron

$$\mathbf{T}\left(\frac{1}{2}dA_n\right) + \left(-\boldsymbol{\sigma}_x\frac{1}{2}dA_x\right) + \left(-\boldsymbol{\sigma}_y\frac{1}{2}dA_y\right) + \left(-\boldsymbol{\sigma}_z\frac{1}{2}dA_z\right) + \mathbf{F}_B\left(\frac{1}{6}dxdydz\right) = \mathbf{0}.$$

(A1.3.12)

Dropping the last term which is of high order and manipulating further:

$$\mathbf{T} = \boldsymbol{\sigma}_x\frac{dA_x}{dA_n} + \boldsymbol{\sigma}_y\frac{dA_y}{dA_n} + \boldsymbol{\sigma}_z\frac{dA_z}{dA_n}.$$

(A1.3.13)

Equation (A1.3.11) follows if it can be shown that

$$\frac{dA_x}{dA_n} = \mathbf{i}\cdot\mathbf{n} = l, \quad \frac{dA_y}{dA_n} = \mathbf{j}\cdot\mathbf{n} = m, \quad \frac{dA_z}{dA_n} = \mathbf{k}\cdot\mathbf{n} = n.$$

(A1.3.14)

To verify Eq. (A1.3.14), we look at the tetrahedron shown in Figure A1.8 and note that

$$\overrightarrow{AB} \times \overrightarrow{AC} = dA_n\mathbf{n} \rightarrow \left(\overrightarrow{AP} + \overrightarrow{PB}\right) \times \left(\overrightarrow{AP} + \overrightarrow{PC}\right) = dA_n\mathbf{n}$$
$$\rightarrow \left(\overrightarrow{AP} \times \overrightarrow{PC}\right) + \left(\overrightarrow{PB} \times \overrightarrow{AP}\right) + \left(\overrightarrow{PB} \times \overrightarrow{PC}\right) = dA_n\mathbf{n}$$
$$\rightarrow dA_y\mathbf{j} + dA_z\mathbf{k} + dA_x\mathbf{i} = dA_n\mathbf{n}.$$

(A1.3.15)

From the last of the above equation:

$$\mathbf{n} = \frac{dA_x}{dA_n}\mathbf{i} + \frac{dA_y}{dA_n}\mathbf{j} + \frac{dA_z}{dA_n}\mathbf{k},$$

(A1.3.16)

from which Eq. (A1.3.14) follows.

## A1.4 Linear Elastic Constitutive Equations

In general, the linear elastic relationship between six strain components and six stress components can be expressed in the following matrix form when there is no temperature change:

$$\begin{Bmatrix} \varepsilon_{xx} \\ \varepsilon_{yy} \\ \varepsilon_{zz} \\ \varepsilon_{yz} \\ \varepsilon_{zx} \\ \varepsilon_{xy} \end{Bmatrix} = \begin{bmatrix} \hat{C}_{11} & \hat{C}_{12} & \hat{C}_{13} & \hat{C}_{14} & \hat{C}_{15} & \hat{C}_{16} \\ \hat{C}_{21} & \hat{C}_{22} & \hat{C}_{23} & \hat{C}_{24} & \hat{C}_{25} & \hat{C}_{26} \\ \hat{C}_{31} & \hat{C}_{32} & \hat{C}_{33} & \hat{C}_{34} & \hat{C}_{35} & \hat{C}_{36} \\ \hat{C}_{41} & \hat{C}_{42} & \hat{C}_{43} & \hat{C}_{44} & \hat{C}_{45} & \hat{C}_{46} \\ \hat{C}_{51} & \hat{C}_{52} & \hat{C}_{53} & \hat{C}_{54} & \hat{C}_{55} & \hat{C}_{56} \\ \hat{C}_{61} & \hat{C}_{62} & \hat{C}_{63} & \hat{C}_{64} & \hat{C}_{65} & \hat{C}_{66} \end{bmatrix} \begin{Bmatrix} \sigma_{xx} \\ \sigma_{yy} \\ \sigma_{zz} \\ \sigma_{yz} \\ \sigma_{zx} \\ \sigma_{xy} \end{Bmatrix}$$

(A1.4.1)

or

$$\varepsilon = \hat{\mathbf{C}}\sigma, \tag{A1.4.2}$$

where $\varepsilon$ is a column vector of six strain components, $\sigma$ is a column vector of six stress components, and $\hat{\mathbf{C}}$ is a $6 \times 6$ matrix of elastic compliance coefficients as shown in Eq. (A1.4.1). Inverting the above equation:

$$\sigma = \mathbf{C}\varepsilon, \tag{A1.4.3}$$

where $\mathbf{C} = \hat{\mathbf{C}}^{-1}$ is a matrix of elastic stiffness coefficients.

The entries in the $\hat{\mathbf{C}}$ matrix can be determined by experiment. However, through a theoretical reasoning process, many of the entries in the $\hat{\mathbf{C}}$ matrix can be determined before conducting actual experiments. First of all, it can be shown that the $\hat{\mathbf{C}}$ matrix is symmetric. In addition, for materials with certain types of symmetry, many entries in the $\hat{\mathbf{C}}$ matrix are equal to zero. At the microscopic level, a material is an aggregate of small constituent particles, and the $\hat{\mathbf{C}}$ matrix represents a smeared property of the material. Depending on the manner in which these particles are distributed, it is possible to identify various planes of symmetry within a given material. Materials can then be classified according to types of symmetry. Examples are orthotropic material, isotropic material, and transversely isotropic material. In the following discussion we will assume without proof the symmetry of the $\hat{\mathbf{C}}$ matrix.

## Orthotropic Material

A material is called orthotropic if it has three planes of symmetry. For orthotropic material it can be shown that there exist nine nonzero independent entries in the $\hat{\mathbf{C}}$ matrix. Suppose the planes of symmetry are identified as the $xy$-plane, $yz$-plane, and $zx$-plane. To determine the number of independent material constants, we may consider a stress state in which only one stress component is nonzero.

For example, consider a state in which $\sigma_{xx} \neq 0$ and all other stresses are equal to zero. From Eq. (A1.4.1):

$$\varepsilon_{xx} = \hat{C}_{11}\sigma_{xx}, \quad \varepsilon_{yy} = \hat{C}_{21}\sigma_{xx}, \quad \varepsilon_{zz} = \hat{C}_{31}\sigma_{xx}, \tag{A1.4.4}$$
$$\varepsilon_{yz} = \hat{C}_{41}\sigma_{xx}, \quad \varepsilon_{zx} = \hat{C}_{51}\sigma_{xx}, \quad \varepsilon_{xy} = \hat{C}_{61}\sigma_{xx}.$$

In addition, due to material symmetry, we can conclude that

$$\varepsilon_{yz} = 0 \rightarrow \hat{C}_{41} = 0, \quad \varepsilon_{zx} = 0 \rightarrow \hat{C}_{51} = 0, \quad \varepsilon_{xy} = 0 \rightarrow \hat{C}_{61} = 0. \tag{A1.4.5}$$

From experiments using a specimen of finite size in which stress $\sigma_{xx}$ is uniform:

$$\varepsilon_{xx} = \frac{1}{E_x}\sigma_{xx} = \hat{C}_{11}\sigma_{xx} \rightarrow \hat{C}_{11} = \frac{1}{E_x}, \tag{A1.4.6}$$

where $E_x$ is Young's modulus as shown in Figure A1.9.

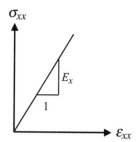

$\sigma_{xx}$

$E_x$

1

$\varepsilon_{xx}$

Poisson's ratio $v_{xy}$ is defined as $v_{xy} = -\varepsilon_{yy}/\varepsilon_{xx}$. Then

$$\varepsilon_{yy} = -v_{xy}\varepsilon_{xx} = -\frac{v_{xy}}{E_x}\sigma_{xx} = \hat{C}_{21}\sigma_{xx} \rightarrow \hat{C}_{21} = -\frac{v_{xy}}{E_x}. \tag{A1.4.7}$$

Similarly, Poisson's ratio $v_{xz}$ is defined as $v_{xz} = -\varepsilon_{zz}/\varepsilon_{xx}$. Then

$$\varepsilon_{zz} = -v_{xz}\varepsilon_{xx} = -\frac{v_{xz}}{E_x}\sigma_{xx} = \hat{C}_{31}\sigma_{xx} \rightarrow \hat{C}_{31} = -\frac{v_{xz}}{E_x}. \tag{A1.4.8}$$

Repeating similar procedures for all six stress components applied individually, one can show that nine independent material constants, which are determined experimentally, are three Young's moduli $(E_x, E_y, E_z)$, three Poisson ratios $(v_{yz}, v_{zx}, v_{xy})$, and three shear moduli $(G_{yz}, G_{zx}, G_{xy})$. Accordingly, for an orthotropic material:

$$\begin{Bmatrix} \varepsilon_{xx} \\ \varepsilon_{yy} \\ \varepsilon_{zz} \\ \varepsilon_{yz} \\ \varepsilon_{zx} \\ \varepsilon_{xy} \end{Bmatrix} = \begin{bmatrix} \frac{1}{E_x} & \frac{-v_{yx}}{E_y} & \frac{-v_{zx}}{E_z} & 0 & 0 & 0 \\ \frac{-v_{xy}}{E_x} & \frac{1}{E_y} & \frac{-v_{zy}}{E_z} & 0 & 0 & 0 \\ \frac{-v_{xz}}{E_x} & \frac{-v_{yz}}{E_y} & \frac{1}{E_z} & 0 & 0 & 0 \\ 0 & 0 & 0 & \frac{1}{G_{yz}} & 0 & 0 \\ 0 & 0 & 0 & 0 & \frac{1}{G_{zx}} & 0 \\ 0 & 0 & 0 & 0 & 0 & \frac{1}{G_{xy}} \end{bmatrix} \begin{Bmatrix} \sigma_{xx} \\ \sigma_{yy} \\ \sigma_{zz} \\ \sigma_{yz} \\ \sigma_{zx} \\ \sigma_{xy} \end{Bmatrix}. \tag{A1.4.9}$$

Note that

$$\frac{v_{xy}}{E_x} = \frac{v_{yx}}{E_y}, \quad \frac{v_{xz}}{E_x} = \frac{v_{zx}}{E_z}, \quad \frac{v_{yz}}{E_y} = \frac{v_{zy}}{E_z}. \tag{A1.4.10}$$

## Isotropic Material

A material is called isotropic if it has no preferred directional property. Accordingly, for an isotropic material, any arbitrarily chosen plane is a plane of symmetry. Then

$$E_x = E_y = E_z = E, \quad v_{yz} = v_{zx} = v_{xy} = v, \quad G_{yz} = G_{zx} = G_{xy} = G \tag{A1.4.11}$$

and

$$
\begin{Bmatrix} \varepsilon_{xx} \\ \varepsilon_{yy} \\ \varepsilon_{zz} \\ \varepsilon_{yz} \\ \varepsilon_{zx} \\ \varepsilon_{xy} \end{Bmatrix} =
\begin{bmatrix}
\dfrac{1}{E} & \dfrac{-v}{E} & \dfrac{-v}{E} & 0 & 0 & 0 \\[6pt]
\dfrac{-v}{E} & \dfrac{1}{E} & \dfrac{-v}{E} & 0 & 0 & 0 \\[6pt]
\dfrac{-v}{E} & \dfrac{-v}{E} & \dfrac{1}{E} & 0 & 0 & 0 \\[6pt]
0 & 0 & 0 & \dfrac{1}{G} & 0 & 0 \\[6pt]
0 & 0 & 0 & 0 & \dfrac{1}{G} & 0 \\[6pt]
0 & 0 & 0 & 0 & 0 & \dfrac{1}{G}
\end{bmatrix}
\begin{Bmatrix} \sigma_{xx} \\ \sigma_{yy} \\ \sigma_{zz} \\ \sigma_{yz} \\ \sigma_{zx} \\ \sigma_{xy} \end{Bmatrix}.
\tag{A1.4.12}
$$

In addition, it can be shown that

$$
G = \frac{E}{2(1+v)}.
\tag{A1.4.13}
$$

Accordingly, in the linear elastic range, $E$ and $v$ determine the property of an isotropic material. Derivation of Eq. (A1.4.13) is left as a problem at the end of this chapter.

## Transversely Isotropic Material

Suppose a material is isotropic in one plane, for example in the $yz$-plane. Then starting from the compliance matrix of an orthotropic material:

$$
E_y = E_z, \quad v_{xy} = v_{xz}, \quad G_{zx} = G_{xy}, \quad G_{yz} = \frac{E_y}{2(1+v_{yz})}.
\tag{A1.4.14}
$$

Accordingly, a transversely isotropic material can be characterized by five nonzero independent constants and

$$
\begin{Bmatrix} \varepsilon_{xx} \\ \varepsilon_{yy} \\ \varepsilon_{zz} \\ \varepsilon_{yz} \\ \varepsilon_{zx} \\ \varepsilon_{xy} \end{Bmatrix} =
\begin{bmatrix}
\dfrac{1}{E_x} & \dfrac{-v_{xy}}{E_x} & \dfrac{-v_{xy}}{E_x} & 0 & 0 & 0 \\[6pt]
\dfrac{-v_{xy}}{E_x} & \dfrac{1}{E_y} & \dfrac{-v_{yz}}{E_y} & 0 & 0 & 0 \\[6pt]
\dfrac{-v_{xy}}{E_x} & \dfrac{-v_{yz}}{E_y} & \dfrac{1}{E_y} & 0 & 0 & 0 \\[6pt]
0 & 0 & 0 & \dfrac{2(1+v_{yz})}{E_y} & 0 & 0 \\[6pt]
0 & 0 & 0 & 0 & \dfrac{1}{G_{xy}} & 0 \\[6pt]
0 & 0 & 0 & 0 & 0 & \dfrac{1}{G_{xy}}
\end{bmatrix}
\begin{Bmatrix} \sigma_{xx} \\ \sigma_{yy} \\ \sigma_{zz} \\ \sigma_{yz} \\ \sigma_{zx} \\ \sigma_{xy} \end{Bmatrix}.
\tag{A1.4.15}
$$

For a ply of glass/epoxy or graphite/epoxy composite materials, unidirectional fibers are embedded in a matrix of epoxy material. Usually, the cross-sections of fibers are randomly distributed over a plane (say the $yz$-plane). Composite materials can then be considered isotropic in the $yz$-plane when smeared properties are of interest, and any plane perpendicular to the $yz$-plane is a plane of symmetry.

## Thermal Strain

Consider a thermally orthotropic material in which the planes of symmetry are the $xy$-plane, the $yz$-plane, and the $zx$-plane. For an unconstrained rectangular cuboid subjected to uniform temperature change, thermally induced strains are

$$
\begin{aligned}
&\varepsilon^o_{xx} = \alpha_x \Delta T, \quad \varepsilon^o_{yy} = \alpha_y \Delta T, \quad \varepsilon^o_{zz} = \alpha_z \Delta T, \\
&\varepsilon^o_{xy} = \varepsilon^o_{yz} = \varepsilon^o_{zx} = 0,
\end{aligned}
\tag{A1.4.16}
$$

where $\Delta T$ is temperature rise ($\Delta T > 0$) or drop ($\Delta T < 0$) and $\alpha_x, \alpha_y, \alpha_z$ are material properties called coefficients of thermal expansion (CTE).

For a thermally isotropic material:

$$
\alpha_x = \alpha_y = \alpha_z = \alpha
\tag{A1.4.17}
$$

while for a material which is transversely isotropic in the $yz$-plane:

$$
\alpha_y = \alpha_z.
\tag{A1.4.18}
$$

Total strain is the sum of stress-induced strain and thermally induced strain. Accordingly

$$
\boldsymbol{\varepsilon} = \hat{\mathbf{C}}\boldsymbol{\sigma} + \boldsymbol{\varepsilon}^o.
\tag{A1.4.19}
$$

For an orthotropic material:

$$
\boldsymbol{\varepsilon}^o = \left\{
\begin{array}{c}
\alpha_x \Delta T \\
\alpha_y \Delta T \\
\alpha_z \Delta T \\
0 \\
0 \\
0
\end{array}
\right\}.
\tag{A1.4.20}
$$

From Eq. (A1.4.19):

$$
\boldsymbol{\sigma} = \mathbf{C}(\boldsymbol{\varepsilon} - \boldsymbol{\varepsilon}^o),
\tag{A1.4.21}
$$

where

$$
\mathbf{C} = \hat{\mathbf{C}}^{-1}.
\tag{A1.4.22}
$$

For isotropic material, one can show that

$$\mathbf{C} = \frac{E}{(1+v)(1-2v)} \begin{bmatrix} 1-v & v & v & 0 & 0 & 0 \\ v & 1-v & v & 0 & 0 & 0 \\ v & v & 1-v & 0 & 0 & 0 \\ 0 & 0 & 0 & \dfrac{1-2v}{2} & 0 & 0 \\ 0 & 0 & 0 & 0 & \dfrac{1-2v}{2} & 0 \\ 0 & 0 & 0 & 0 & 0 & \dfrac{1-2v}{2} \end{bmatrix}. \qquad (A1.4.23)$$

## A1.5 Transformation Rules

Strain components and stress components of a deformed body are dependent on the orientation of the chosen coordinate system. One can show that if strain or stress components are known for one coordinate system, strain and stress components in another coordinate system can be determined through transformation rules which depend on the direction cosines between the two coordinate systems.

### Strain Transformation

Let us consider strain components in the two different Cartesian coordinate systems as shown in Figure A1.10.

For the $x, y, z$ system:

$$\varepsilon_{xx} = \frac{\partial u}{\partial x}, \varepsilon_{yy} = \frac{\partial v}{\partial y}, \varepsilon_{xy} = \frac{\partial u}{\partial y} + \frac{\partial v}{\partial x}, \text{etc.} \qquad (A1.5.1)$$

For the $\tilde{x}, \tilde{y}, \tilde{z}$ system:

$$\varepsilon_{\tilde{x}\tilde{x}} = \frac{\partial \tilde{u}}{\partial \tilde{x}}, \varepsilon_{\tilde{y}\tilde{y}} = \frac{\partial \tilde{v}}{\partial \tilde{y}}, \varepsilon_{\tilde{x}\tilde{y}} = \frac{\partial \tilde{u}}{\partial \tilde{y}} + \frac{\partial \tilde{v}}{\partial \tilde{x}}, \text{etc.} \qquad (A1.5.2)$$

**Figure A1.10** Two different Cartesian coordinate systems

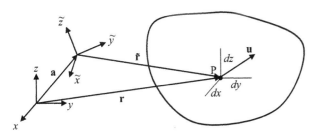

In matrix form, the transformation rule between the strain components in the $\tilde{x}, \tilde{y}, \tilde{z}$ system and those in the $x, y, z$ system can be expressed as

$$
\begin{bmatrix}
\varepsilon_{\tilde{x}\tilde{x}} & \frac{1}{2}\varepsilon_{\tilde{x}\tilde{y}} & \frac{1}{2}\varepsilon_{\tilde{x}\tilde{z}} \\[2mm]
\frac{1}{2}\varepsilon_{\tilde{y}\tilde{x}} & \varepsilon_{\tilde{y}\tilde{y}} & \frac{1}{2}\varepsilon_{\tilde{y}\tilde{z}} \\[2mm]
\frac{1}{2}\varepsilon_{\tilde{z}\tilde{x}} & \frac{1}{2}\varepsilon_{\tilde{z}\tilde{y}} & \varepsilon_{\tilde{z}\tilde{z}}
\end{bmatrix}
= \mathbf{T}^{\mathsf{T}}
\begin{bmatrix}
\varepsilon_{xx} & \frac{1}{2}\varepsilon_{xy} & \frac{1}{2}\varepsilon_{xz} \\[2mm]
\frac{1}{2}\varepsilon_{yx} & \varepsilon_{yy} & \frac{1}{2}\varepsilon_{yz} \\[2mm]
\frac{1}{2}\varepsilon_{zx} & \frac{1}{2}\varepsilon_{zy} & \varepsilon_{zz}
\end{bmatrix}
\mathbf{T},
\tag{A1.5.3}
$$

where

$$
\mathbf{T} =
\begin{bmatrix}
l_{x\tilde{x}} & l_{x\tilde{y}} & l_{x\tilde{z}} \\
l_{y\tilde{x}} & l_{y\tilde{y}} & l_{y\tilde{z}} \\
l_{z\tilde{x}} & l_{z\tilde{y}} & l_{z\tilde{z}}
\end{bmatrix}
\tag{A1.5.4}
$$

is the transformation matrix of direction cosines. For example:

$$
l_{x\tilde{y}} = \mathbf{i} \cdot \tilde{\mathbf{j}}
\tag{A1.5.5}
$$

is the cosine of the angle between the $x$-axis and the $\tilde{y}$-axis. The above transformation rule can be derived as follows. For the position vector of point P:

$$
\mathbf{r} = x\mathbf{i} + y\mathbf{j} + z\mathbf{k} = \tilde{x}\tilde{\mathbf{i}} + \tilde{y}\tilde{\mathbf{j}} + \tilde{z}\tilde{\mathbf{k}} + \mathbf{a}.
\tag{A1.5.6}
$$

For the relationship between the two coordinates:

$$
\begin{aligned}
x = \mathbf{r} \cdot \mathbf{i} &= \left(\tilde{x}\tilde{\mathbf{i}} + \tilde{y}\tilde{\mathbf{j}} + \tilde{z}\tilde{\mathbf{k}} + \mathbf{a}\right) \cdot \mathbf{i} = \tilde{x}\left(\tilde{\mathbf{i}} \cdot \mathbf{i}\right) + \tilde{y}\left(\tilde{\mathbf{j}} \cdot \mathbf{i}\right) + \tilde{z}\left(\tilde{\mathbf{k}} \cdot \mathbf{i}\right) + \mathbf{a} \cdot \mathbf{i} \\
&\to x = \tilde{x}l_{x\tilde{x}} + \tilde{y}l_{x\tilde{y}} + \tilde{z}l_{x\tilde{z}} + \mathbf{a} \cdot \mathbf{i},
\end{aligned}
\tag{A1.5.7}
$$

$$
\begin{aligned}
y = \mathbf{r} \cdot \mathbf{j} &= \left(\tilde{x}\tilde{\mathbf{i}} + \tilde{y}\tilde{\mathbf{j}} + \tilde{z}\tilde{\mathbf{k}} + \mathbf{a}\right) \cdot \mathbf{j} = \tilde{x}\left(\tilde{\mathbf{i}} \cdot \mathbf{j}\right) + \tilde{y}\left(\tilde{\mathbf{j}} \cdot \mathbf{j}\right) + \tilde{z}\left(\tilde{\mathbf{k}} \cdot \mathbf{j}\right) + \mathbf{a} \cdot \mathbf{j} \\
&\to y = \tilde{x}l_{y\tilde{x}} + \tilde{y}l_{y\tilde{y}} + \tilde{z}l_{y\tilde{z}} + \mathbf{a} \cdot \mathbf{j},
\end{aligned}
\tag{A1.5.8}
$$

$$
\begin{aligned}
z = \mathbf{r} \cdot \mathbf{k} &= \left(\tilde{x}\tilde{\mathbf{i}} + \tilde{y}\tilde{\mathbf{j}} + \tilde{z}\tilde{\mathbf{k}} + \mathbf{a}\right) \cdot \mathbf{k} = \tilde{x}\left(\tilde{\mathbf{i}} \cdot \mathbf{k}\right) + \tilde{y}\left(\tilde{\mathbf{j}} \cdot \mathbf{k}\right) + \tilde{z}\left(\tilde{\mathbf{k}} \cdot \mathbf{k}\right) + \mathbf{a} \cdot \mathbf{k} \\
&\to z = \tilde{x}l_{z\tilde{x}} + \tilde{y}l_{z\tilde{y}} + \tilde{z}l_{z\tilde{z}} + \mathbf{a} \cdot \mathbf{k},
\end{aligned}
\tag{A1.5.9}
$$

For the relationship between the displacements in the two coordinate systems:

$$
\begin{aligned}
\tilde{u} = \mathbf{u} \cdot \tilde{\mathbf{i}} &= \left(u\mathbf{i} + v\mathbf{j} + w\mathbf{k}\right) \cdot \tilde{\mathbf{i}} = u\left(\mathbf{i} \cdot \tilde{\mathbf{i}}\right) + v\left(\mathbf{j} \cdot \tilde{\mathbf{i}}\right) + w\left(\mathbf{k} \cdot \tilde{\mathbf{i}}\right) \\
&\to \tilde{u} = ul_{x\tilde{x}} + vl_{y\tilde{x}} + wl_{z\tilde{x}},
\end{aligned}
\tag{A1.5.10}
$$

$$
\begin{aligned}
\tilde{v} = \mathbf{u} \cdot \tilde{\mathbf{j}} &= \left(u\mathbf{i} + v\mathbf{j} + w\mathbf{k}\right) \cdot \tilde{\mathbf{j}} = u\left(\mathbf{i} \cdot \tilde{\mathbf{j}}\right) + v\left(\mathbf{j} \cdot \tilde{\mathbf{j}}\right) + w\left(\mathbf{k} \cdot \tilde{\mathbf{j}}\right) \\
&\to \tilde{v} = ul_{x\tilde{y}} + vl_{y\tilde{y}} + wl_{z\tilde{y}},
\end{aligned}
\tag{A1.5.11}
$$

$$
\begin{aligned}
\tilde{w} = \mathbf{u} \cdot \tilde{\mathbf{k}} &= \left(u\mathbf{i} + v\mathbf{j} + w\mathbf{k}\right) \cdot \tilde{\mathbf{k}} = u\left(\mathbf{i} \cdot \tilde{\mathbf{k}}\right) + v\left(\mathbf{j} \cdot \tilde{\mathbf{k}}\right) + w\left(\mathbf{k} \cdot \tilde{\mathbf{k}}\right) \\
&\to \tilde{w} = ul_{x\tilde{z}} + vl_{y\tilde{z}} + wl_{z\tilde{z}}.
\end{aligned}
\tag{A1.5.12}
$$

For strain $\varepsilon_{\tilde{x}\tilde{x}}$ in the $\tilde{x}, \tilde{y}, \tilde{z}$ system:

$$\varepsilon_{\tilde{x}\tilde{x}} = \frac{\partial \tilde{u}}{\partial \tilde{x}} = \frac{\partial \tilde{u}}{\partial x}\frac{\partial x}{\partial \tilde{x}} + \frac{\partial \tilde{u}}{\partial y}\frac{\partial y}{\partial \tilde{x}} + \frac{\partial \tilde{u}}{\partial z}\frac{\partial z}{\partial \tilde{x}}. \tag{A1.5.13}$$

Introducing Eqs (A1.5.7)–(A1.5.10) into Eq. (A1.5.13):

$$\varepsilon_{\tilde{x}\tilde{x}} = \frac{\partial}{\partial x}\left(ul_{x\tilde{x}} + vl_{y\tilde{x}} + wl_{z\tilde{x}}\right)l_{x\tilde{x}} + \frac{\partial}{\partial y}\left(ul_{x\tilde{x}} + vl_{y\tilde{x}} + wl_{z\tilde{x}}\right)l_{y\tilde{x}} + \frac{\partial}{\partial z}\left(ul_{x\tilde{x}} + vl_{y\tilde{x}} + wl_{z\tilde{x}}\right)l_{z\tilde{x}}$$

$$\rightarrow \varepsilon_{\tilde{x}\tilde{x}} = \frac{\partial u}{\partial x}l_{x\tilde{x}}l_{x\tilde{x}} + \frac{\partial v}{\partial y}l_{y\tilde{x}}l_{y\tilde{x}} + \frac{\partial w}{\partial z}l_{z\tilde{x}}l_{z\tilde{x}} + \left(\frac{\partial v}{\partial x} + \frac{\partial u}{\partial y}\right)l_{y\tilde{x}}l_{x\tilde{x}} + \cdots$$

$$\rightarrow \varepsilon_{\tilde{x}\tilde{x}} = \varepsilon_{xx}l_{x\tilde{x}}l_{x\tilde{x}} + \varepsilon_{yy}l_{y\tilde{x}}l_{y\tilde{x}} + \varepsilon_{zz}l_{z\tilde{x}}l_{z\tilde{x}} + \varepsilon_{xy}l_{y\tilde{x}}l_{x\tilde{x}} + \cdots . \tag{A1.5.14}$$

Written in matrix form:

$$\varepsilon_{\tilde{x}\tilde{x}} = \lfloor l_{x\tilde{x}} \quad l_{y\tilde{x}} \quad l_{z\tilde{x}} \rfloor \begin{bmatrix} \varepsilon_{xx} & \frac{1}{2}\varepsilon_{xy} & \frac{1}{2}\varepsilon_{xz} \\ \frac{1}{2}\varepsilon_{yx} & \varepsilon_{yy} & \frac{1}{2}\varepsilon_{yz} \\ \frac{1}{2}\varepsilon_{zx} & \frac{1}{2}\varepsilon_{zy} & \varepsilon_{zz} \end{bmatrix} \begin{Bmatrix} l_{x\tilde{x}} \\ l_{y\tilde{x}} \\ l_{z\tilde{x}} \end{Bmatrix}. \tag{A1.5.15}$$

Similar relations hold for other components to result in the strain transformation rule shown in Eq. (A1.5.3).

## Stress Transformation

If stress components at a point are given in the $x, y, z$ system, stress components in the $\tilde{x}, \tilde{y}, \tilde{z}$ system can be found by the stress transformation rule expressed as

$$\begin{bmatrix} \sigma_{\tilde{x}\tilde{x}} & \sigma_{\tilde{x}\tilde{y}} & \sigma_{\tilde{x}\tilde{z}} \\ \sigma_{\tilde{y}\tilde{x}} & \sigma_{\tilde{y}\tilde{y}} & \sigma_{\tilde{y}\tilde{z}} \\ \sigma_{\tilde{z}\tilde{x}} & \sigma_{\tilde{z}\tilde{y}} & \sigma_{\tilde{z}\tilde{z}} \end{bmatrix} = \mathbf{T}^{\mathrm{T}} \begin{bmatrix} \sigma_{xx} & \sigma_{xy} & \sigma_{xz} \\ \sigma_{yx} & \sigma_{yy} & \sigma_{yz} \\ \sigma_{zx} & \sigma_{zy} & \sigma_{zz} \end{bmatrix} \mathbf{T}, \tag{A1.5.16}$$

where the transformation matrix $\mathbf{T}$ is identical to that for the strain transformation. In order to derive the above stress transformation rule, let us look at the equilibrium of the forces acting on the tetrahedron cut out of the body, as shown in Figure A1.11. Note that this tetrahedron is similar to the one in Figure A1.8 with $\mathbf{n} = \tilde{\mathbf{i}}$.

Let us introduce the four triangular areas as follows:

$\frac{1}{2}dA_x$: area of surface PBC, $\frac{1}{2}dA_y$: area of surface PAC,

$\frac{1}{2}dA_z$: area of surface PAB, $\frac{1}{2}dA_{\tilde{x}}$: area of surface ABC.

From the force equilibrium:

$$\boldsymbol{\sigma}_{\tilde{x}}\left(\frac{1}{2}dA_{\tilde{x}}\right) + \left(-\boldsymbol{\sigma}_x\frac{1}{2}dA_x\right) + \left(-\boldsymbol{\sigma}_y\frac{1}{2}dA_y\right) + \left(-\boldsymbol{\sigma}_z\frac{1}{2}dA_z\right) + \mathbf{F}_B\left(\frac{1}{6}dxdydz\right) = \mathbf{0}. \tag{A1.5.17}$$

**Figure A1.11** Stress vectors and body force vector per volume acting on a tetrahedron of infinitesimal volume

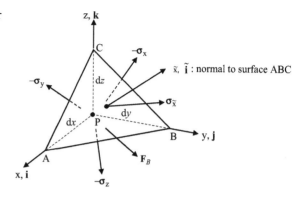

Dropping the last term, which is of higher order:

$$\boldsymbol{\sigma}_{\tilde{x}}dA_{\tilde{x}} = \boldsymbol{\sigma}_x dA_x + \boldsymbol{\sigma}_y dA_y + \boldsymbol{\sigma}_z dA_z$$

$$\rightarrow \boldsymbol{\sigma}_{\tilde{x}} = \boldsymbol{\sigma}_x \frac{dA_x}{dA_{\tilde{x}}} + \boldsymbol{\sigma}_y \frac{dA_y}{dA_{\tilde{x}}} + \boldsymbol{\sigma}_z \frac{dA_z}{dA_{\tilde{x}}} = \boldsymbol{\sigma}_x\left(\mathbf{i}\cdot\tilde{\mathbf{i}}\right) + \boldsymbol{\sigma}_y\left(\mathbf{j}\cdot\tilde{\mathbf{i}}\right) + \boldsymbol{\sigma}_z\left(\mathbf{k}\cdot\tilde{\mathbf{i}}\right) \qquad \text{(A1.5.18)}$$

$$\rightarrow \boldsymbol{\sigma}_{\tilde{x}} = \boldsymbol{\sigma}_x l_{x\tilde{x}} + \boldsymbol{\sigma}_y l_{y\tilde{x}} + \boldsymbol{\sigma}_z l_{z\tilde{x}}.$$

Then

$$\sigma_{\tilde{x}\tilde{x}} = \tilde{\mathbf{i}}\cdot\boldsymbol{\sigma}_{\tilde{x}} = \tilde{\mathbf{i}}\cdot\left(\sigma_{\tilde{x}\tilde{x}}\tilde{\mathbf{i}} + \sigma_{\tilde{x}\tilde{y}}\tilde{\mathbf{j}} + \sigma_{\tilde{x}\tilde{z}}\tilde{\mathbf{k}}\right) = \tilde{\mathbf{i}}\cdot\left(\sigma_{xx}\mathbf{i} + \sigma_{xy}\mathbf{j} + \sigma_{xz}\mathbf{k}\right)l_{x\tilde{x}}$$

$$+\tilde{\mathbf{i}}\cdot\left(\sigma_{yx}\mathbf{i} + \sigma_{yy}\mathbf{j} + \sigma_{yz}\mathbf{k}\right)l_{y\tilde{x}} + \tilde{\mathbf{i}}\cdot\left(\sigma_{zx}\mathbf{i} + \sigma_{zy}\mathbf{j} + \sigma_{zz}\mathbf{k}\right)l_{z\tilde{x}}$$

$$= \sigma_{xx}l_{x\tilde{x}}l_{x\tilde{x}} + \sigma_{xy}l_{x\tilde{x}}l_{y\tilde{x}} + \sigma_{xz}l_{x\tilde{x}}l_{z\tilde{x}} + \sigma_{yx}l_{y\tilde{x}}l_{x\tilde{x}} + \sigma_{yy}l_{y\tilde{x}}l_{y\tilde{x}}$$

$$+\sigma_{yz}l_{y\tilde{x}}l_{z\tilde{x}} + \sigma_{zx}l_{z\tilde{x}}l_{x\tilde{x}} + \sigma_{zy}l_{z\tilde{x}}l_{y\tilde{x}} + \sigma_{zz}l_{z\tilde{x}}l_{z\tilde{x}}. \qquad \text{(A1.5.19)}$$

Written in matrix from:

$$\sigma_{\tilde{x}\tilde{x}} = \lfloor l_{x\tilde{x}} \quad l_{y\tilde{x}} \quad l_{z\tilde{x}} \rfloor \begin{bmatrix} \sigma_{xx} & \sigma_{xy} & \sigma_{xz} \\ \sigma_{yx} & \sigma_{yy} & \sigma_{yz} \\ \sigma_{zx} & \sigma_{zy} & \sigma_{zz} \end{bmatrix} \begin{Bmatrix} l_{x\tilde{x}} \\ l_{y\tilde{x}} \\ l_{z\tilde{x}} \end{Bmatrix}. \qquad \text{(A1.5.20)}$$

Similarly

$$\sigma_{\tilde{x}\tilde{y}} = \tilde{\mathbf{j}}\cdot\boldsymbol{\sigma}_{\tilde{x}} = \cdots,$$

$$\sigma_{\tilde{x}\tilde{z}} = \tilde{\mathbf{k}}\cdot\boldsymbol{\sigma}_{\tilde{x}} = \cdots, \qquad \text{(A1.5.21)}$$

Following similar procedures for $\boldsymbol{\sigma}_{\tilde{y}}$ and $\boldsymbol{\sigma}_{\tilde{z}}$ leads to the stress transformation rule.

## Transformation of the Constitutive Equation

In the $x, y, z$ system (with $\Delta T = 0$):

$$\boldsymbol{\sigma} = \mathbf{C}\boldsymbol{\varepsilon}. \qquad \text{(A1.5.22)}$$

In the $\tilde{x}, \tilde{y}, \tilde{z}$ system:

$$\tilde{\boldsymbol{\sigma}} = \tilde{\mathbf{C}}\tilde{\boldsymbol{\varepsilon}}. \qquad \text{(A1.5.23)}$$

The two matrices, $\tilde{\mathbf{C}}$ and $\mathbf{C}$, are related via a transformation rule. To show this, consider the stress transformation rule expressed as

$$\tilde{\boldsymbol{\sigma}} = \mathbf{T}_\sigma \boldsymbol{\sigma}, \tag{A1.5.24}$$

where $\mathbf{T}_\sigma$ is a $6 \times 6$ matrix which is dependent on the direction cosines. Similarly, the transformation rule for strain can be expressed as

$$\boldsymbol{\varepsilon} = \mathbf{T}_\varepsilon \tilde{\boldsymbol{\varepsilon}}, \tag{A1.5.25}$$

where $\mathbf{T}_\varepsilon$ is also a $6 \times 6$ matrix which is dependent on the direction cosines. It turns out that

$$\mathbf{T}_\varepsilon = \mathbf{T}_\sigma^{\mathrm{T}}. \tag{A1.5.26}$$

Substituting Eq. (A1.5.22) into Eq. (A1.5.24):

$$\tilde{\boldsymbol{\sigma}} = \mathbf{T}_\sigma \boldsymbol{\sigma} = \mathbf{T}_\sigma \mathbf{C} \boldsymbol{\varepsilon}. \tag{A1.5.27}$$

Using Eqs (A1.5.25) and (A1.5.26):

$$\tilde{\boldsymbol{\sigma}} = \mathbf{T}_\sigma \mathbf{C} \mathbf{T}_\sigma^{\mathrm{T}} \tilde{\boldsymbol{\varepsilon}}. \tag{A1.5.28}$$

Comparing Eq. (A1.5.28) with Eq. (A1.5.23):

$$\tilde{\mathbf{C}} = \mathbf{T}_\sigma \mathbf{C} \mathbf{T}_\sigma^{\mathrm{T}}. \tag{A1.5.29}$$

## PROBLEMS

**A1.1** Proceed as follows to show that for isotropic material

$$G = \frac{E}{2(1 + v)}.$$

(a) Consider a stress state in which $\sigma_{xx} = k$, $\sigma_{yy} = -k$ and all other stresses are equal to zero. Then, from the strain–stress relation, show that

$$\varepsilon_{xx} = \frac{1}{E}(1 + v)k, \quad \varepsilon_{yy} = -\frac{1}{E}(1 + v)k, \quad \varepsilon_{zz} = 0,$$
$$\varepsilon_{yz} = \varepsilon_{zx} = \varepsilon_{xy} = 0.$$

(b) Consider the $\tilde{x}, \tilde{y}, \tilde{z}$ system with $\tilde{z} = z$, as shown in Figure A1.12.
From the stress transformation rule, show that $\sigma_{\tilde{x}\tilde{y}} = -k$ and all other stresses are equal to zero. Show also that

$$\varepsilon_{\tilde{x}\tilde{y}} = \frac{1}{G}\sigma_{\tilde{x}\tilde{y}} = \frac{1}{G}(-k).$$

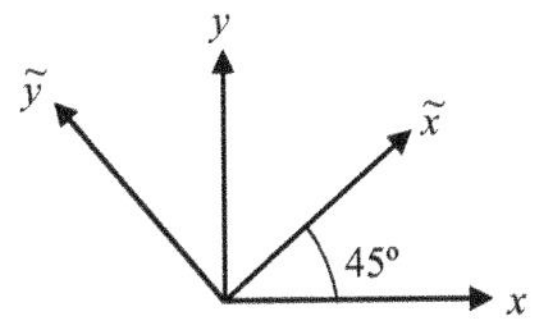

**Figure A1.12** For Problem A1.1

(c) From the strain transformation rule, show that

$$\varepsilon_{\tilde{x}\tilde{y}} = -\varepsilon_{xx} + \varepsilon_{yy}.$$

(d) Using the results in part (a), show that

$$\varepsilon_{\tilde{x}\tilde{y}} = \frac{2(1+v)}{E}(-k).$$

(e) Comparing the equation in part (b) with the one in part (d):

$$\frac{1}{G} = \frac{2(1+v)}{E} \rightarrow G = \frac{E}{2(1+v)}.$$

One may now proceed to show that the above relationship holds for any angle other than 45°.

**A1.2** Following the procedure described in Section A1.5, express strain $\varepsilon_{\tilde{x}\tilde{y}}$ in terms of strain components in the $x, y, z$ coordinate system and direction cosines.

**A1.3** A strain gage is a device utilizing the fact that electrical resistance changes when the strain gage material undergoes length changes. The strain gages are in the form of thin sheet or foil and are mounted on the surface of a structure. The arrangement of strain gages is called a strain rosette. For example, the configuration shown in Figure A1.13 is called a 45° rosette.

(a) With $\tilde{z} = z$, express $\varepsilon_{\tilde{x}\tilde{x}}$ in terms of the strain components in the $x, y, z$ system.

(b) Show that the shear strain $\varepsilon_{xy}$ can be determined from strains $\varepsilon_{xx}$, $\varepsilon_{yy}$, and $\varepsilon_{\tilde{x}\tilde{x}}$ measured by the strain gages as shown in the figure.

**Figure A1.13** For Problem A1.3

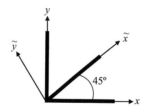

**A1.4** Following the procedure described in Section A1.5, express the stress $\sigma_{\tilde{x}\tilde{y}}$ in terms of the stress components in the $x, y, z$ coordinate system and direction cosines.

**A1.5** The stress transformation rule can be expressed as $\tilde{\sigma} = \mathbf{T}_\sigma \sigma$. Construct $\mathbf{T}_\sigma$, which is a $6 \times 6$ matrix of direction cosines multiplied.

**A1.6** The strain transformation rule can be expressed as $\tilde{\varepsilon} = \mathbf{T}_\varepsilon \varepsilon$. Construct $\mathbf{T}_\varepsilon$, which is a $6 \times 6$ matrix of direction cosines multiplied. Show that $\mathbf{T}_\varepsilon = \mathbf{T}_\sigma^\mathrm{T}$.

# APPENDIX 2

# Solution Methods

The solution of a system of linear algebraic equations is one of the most important steps in FE analysis. For a model with a large number of unknown DOF, the time spent solving the linear equation may account for a substantial part of the entire computing time. It is therefore important for students of FE methods to appreciate how an FE formulation can produce a system of equations that are more readily solvable. In this appendix, we discuss approaches that can be used for further study to exploit sparsity, bandedness, and bandwidth stemming from FE formulations.

## A2.1 Gaussian Elimination with Triple Factorization

Consider a system of simultaneous linear algebraic equations expressed in matrix form as

$$\mathbf{Kq} = \mathbf{F}, \tag{A2.1.1}$$

where $\mathbf{K}$ is an $n \times n$ symmetric and positive definite matrix, $\mathbf{q}$ is an $n \times 1$ column vector of unknowns, and $\mathbf{F}$ is an $n \times 1$ column vector which is given. The equation above can be solved for $\mathbf{q}$ via the Gaussian elimination method, which involves the triangularization process. One of the variants of the Gaussian elimination method is the triple factorization method, in which the $\mathbf{K}$ matrix is decomposed as

$$\mathbf{K} = \mathbf{LDL}^{\mathrm{T}}, \tag{A2.1.2}$$

where

$$\mathbf{L} = \begin{bmatrix} L_{11} & 0 & \cdots & 0 \\ L_{21} & L_{22} & \cdots & 0 \\ \vdots & \vdots & \vdots & \vdots \\ L_{n1} & L_{n2} & \cdots & L_{nn} \end{bmatrix} \tag{A2.1.3}$$

is a lower triangular matrix and $\mathbf{D}$ is a diagonal matrix. Equation (A2.1.1) can be expressed as

$$\mathbf{LDL}^{\mathrm{T}}\mathbf{q} = \mathbf{F}. \tag{A2.1.4}$$

Setting

$$\mathbf{DL}^{\mathrm{T}}\mathbf{q} = \mathbf{Y}, \tag{A2.1.5}$$

Eq. (A2.1.4) can be expressed as

$$\mathbf{LY} = \mathbf{F} \tag{A2.1.6}$$

or

$$
\begin{bmatrix}
L_{11} & 0 & \cdots & 0 \\
L_{21} & L_{22} & \cdots & 0 \\
\vdots & \vdots & \vdots & \vdots \\
L_{n1} & L_{n2} & \cdots & L_{nn}
\end{bmatrix}
\begin{Bmatrix}
y_1 \\ y_2 \\ \vdots \\ y_n
\end{Bmatrix}
=
\begin{Bmatrix}
F_1 \\ F_2 \\ \vdots \\ F_n
\end{Bmatrix}. \tag{A2.1.7}
$$

The above equation can be solved for $y_1, y_2, \ldots, y_n$. This process is called forward substitution. With $\mathbf{Y}$ determined and using Eq. (A2.1.5):

$$\mathbf{L}^T\mathbf{q} = \mathbf{D}^{-1}\mathbf{Y} = \mathbf{Y}^*, \tag{A2.1.8}$$

where $\mathbf{Y}^*$ is known. The above equation can be written in expanded form as

$$
\begin{bmatrix}
L_{11} & L_{21} & \cdots & L_{n1} \\
0 & L_{22} & \cdots & L_{n2} \\
\vdots & \vdots & \vdots & \vdots \\
0 & 0 & \cdots & L_{nn}
\end{bmatrix}
\begin{Bmatrix}
q_1 \\ q_2 \\ \vdots \\ q_n
\end{Bmatrix}
=
\begin{Bmatrix}
y_1^* \\ y_2^* \\ \vdots \\ y_n^*
\end{Bmatrix}. \tag{A2.1.9}
$$

The above equation can be solved for $q_n, q_{n-1}, \ldots, q_1$. This process is called backward substitution.

The solution process mentioned above assumes that the $\mathbf{L}$ and $\mathbf{D}$ matrices are known. To show how they are determined from the $\mathbf{K}$ matrix, consider a simple case with three unknowns. Then, from Eq. (A2.1.2):

$$
\begin{bmatrix}
L_{11} & 0 & 0 \\
L_{21} & L_{22} & 0 \\
L_{31} & L_{32} & L_{33}
\end{bmatrix}
\begin{bmatrix}
D_1 & 0 & 0 \\
0 & D_2 & 0 \\
0 & 0 & D_3
\end{bmatrix}
\begin{bmatrix}
L_{11} & L_{21} & L_{31} \\
0 & L_{22} & L_{32} \\
0 & 0 & L_{33}
\end{bmatrix}
=
\begin{bmatrix}
K_{11} & K_{12} & K_{13} \\
K_{21} & K_{22} & K_{23} \\
K_{31} & K_{32} & K_{33}
\end{bmatrix}. \tag{A2.1.10}
$$

Carrying out the multiplication on the left-hand side of the above equation:

$$
\begin{bmatrix}
L_{11}^2 D_1 & L_{21}L_{11}D_1 & L_{31}L_{11}D_1 \\
 & L_{21}^2 D_1 + L_{22}^2 D_2 & L_{31}L_{21}D_1 + L_{32}L_{22}D_2 \\
\text{symmetric} & & L_{31}^2 D_1 + L_{32}^2 D_2 + L_{33}^2 D_3
\end{bmatrix}
=
\begin{bmatrix}
K_{11} & K_{12} & K_{13} \\
K_{21} & K_{22} & K_{23} \\
K_{31} & K_{32} & K_{33}
\end{bmatrix}. 
$$
$$\tag{A2.1.11}$$

Matching the left side with the right side, one can evaluate the **L** matrix as follows:

$$L_{11}^2 D_1 = K_{11} \rightarrow \text{set } L_{11} = 1, D_1 = K_{11},$$

$$L_{21}L_{11}D_1 = K_{12} \rightarrow L_{21} = \frac{1}{D_1}K_{12},$$

$$L_{21}^2 D_1 + L_{22}^2 D_2 = K_{22} \rightarrow \text{set } L_{22} = 1, D_2 = K_{22} - L_{21}^2 D_1,$$

$$L_{31}L_{11}D_1 = K_{13} \rightarrow L_{31} = \frac{1}{D_1}K_{13},$$

$$L_{31}L_{21}D_1 + L_{32}L_{22}D_2 = K_{23} \rightarrow L_{32} = \frac{1}{D_2}(K_{23} - L_{31}L_{21}D_1),$$

$$L_{31}^2 D_1 + L_{32}^2 D_2 + L_{33}^2 D_3 = K_{33} \rightarrow \text{set } L_{33} = 1, D_3 = K_{33} - L_{31}^2 D_1 - L_{32}^2 D_2.$$

In general, for a global stiffness matrix of arbitrary size:

$$L_{ii} = 1,$$

$$D_i = K_{ii} - \sum_{l=1}^{i-1}(L_{il})^2 D_l,$$

$$L_{ij} = \frac{1}{D_j}\left(K_{ji} - \sum_{l=1}^{i-1}L_{il}L_{jl}D_l\right) \quad (i > j). \tag{A2.1.12}$$

## A2.2 Skyline Method

Sparse matrices are commonly encountered in FE formulations. To appreciate the sparseness of the global stiffness matrix, we consider a 1D problem modeled with four two-node elements as shown in Figure A2.1.

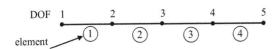

DOF 1 2 3 4 5
element no.

**Figure A2.1** 1D problem modeled with four two-node elements

Note that $K_{ij} = 0$ if DOF $i$ and DOF $j$ do not belong to the same element (i.e., if they are decoupled). Accordingly, we can identify zero entries in the global stiffness matrix, as shown below:

$$\mathbf{K} = \begin{bmatrix} K_{11} & K_{12} & 0 & 0 & 0 \\ & K_{22} & K_{23} & 0 & 0 \\ & & K_{33} & K_{34} & 0 \\ & \text{sym} & & K_{44} & K_{45} \\ & & & & K_{55} \end{bmatrix}. \tag{A2.2.1}$$

In general, the global **K** matrix is a sparse matrix with many zero entries, and we can define the population density (PD) as

$$PD = \frac{\text{nonzero entries in the upper triangular part}}{\text{total entries in the upper triangular part}}.$$

For the example case, PD $= 9/15 = 0.6 = 60\%$. For 3D problems, the PD can be as low as 5–10%.

## Column Heights

We may now define the half bandwidth or column height of each column in **K**, which is the number of entries in each column counted from the first nonzero entry to the diagonal entry. For the example in Figure A2.1, the column heights are as follows:

| Column or global DOF | Column height (or half bandwidth) |
|---|---|
| 1 | 1 |
| 2 | 2 |
| 3 | 2 |
| 4 | 2 |
| 5 | 2 |

We observe that the total sum of the column heights is equal to 9.

Consider now the same 1D problem modeled with four two-node elements, but with a different DOF numbering as shown in Figure A2.2.

**Figure A2.2** 1D problem modeled with a different DOF numbering system

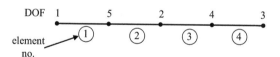

The zero entries in the global stiffness matrix are shown below:

$$\mathbf{K} = \begin{bmatrix} K_{11} & 0 & 0 & 0 & K_{15} \\ & K_{22} & 0 & K_{24} & K_{25} \\ & & K_{33} & K_{34} & 0 \\ & \text{sym} & & K_{44} & 0 \\ & & & & K_{55} \end{bmatrix}. \tag{A2.2.2}$$

For the DOF numbering in Figure A2.2, PD $= 9/15 = 60\%$. For column heights:

| Column or global DOF no. | Column height (or half bandwidth) |
|---|---|
| 1 | 1 |
| 2 | 1 |
| 3 | 1 |
| 4 | 3 |
| 5 | 5 |

The total sum of column heights is 11, which is larger than 9 in the previous case. We observe that population density is independent of DOF numbering. However, column heights are dependent on the DOF numbering. To save storage and computing time, a good solution scheme must utilize the sparsity of the global stiffness matrix and also use the "optimum" node numbering to reduce half bandwidths or column heights.

## Skyline Method

The skyline method is a scheme for efficient storage of sparse matrices in which the entries in the half bandwidth or height of each column are stored in an array, starting from the first column. As an example, consider the 1D problem modeled with four two-node elements with the DOF numbering shown in Figure A2.2. For each column in the global stiffness matrix, we can identify the row number corresponding to the first nonzero entry to define the "skyline" profile as shown in Figure A2.3.

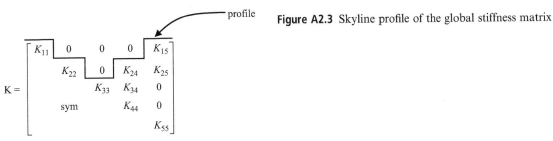

**Figure A2.3** Skyline profile of the global stiffness matrix

In the skyline method, the entries in the profile are stored as an array as follows:

$$K_{11}\, K_{22}\, K_{33}\, K_{24}\, K_{34}\, K_{44}\, K_{15}\, K_{25}\, K_{35}\, K_{45}\, K_{55},$$

where $K_{35} = K_{45} = 0$. Note that there are still zeros or "windows" within the skyline profile. The profile of $\mathbf{L}^T$ is equal to the profile of the upper triangular part of $\mathbf{K}$. That is, $L_{ji} = 0$ if $K_{ij} = 0$. Accordingly, $\mathbf{L}^T$ can overwrite $\mathbf{K}$ during decomposition.

The profile of $\mathbf{K}$ can be determined before the element matrices are computed and assembled. Accordingly, a method is needed to determine the location of the first nonzero entry in each column of $\mathbf{K}$. The row number of the first nonzero entry in each column can be determined using the connectivity between the element DOF and the global DOF. This is done by sweeping element by element, starting from the first element. The task is to determine the smallest row number $i$ for which $K_{ij}$ is nonzero for given column number $j$. As an example, consider the same example problem described in Figure A2.2.

We may define an array, called NCH here, which contains the row number of the first nonzero entries for all columns from column 1 to column 5. First, NCH is initialized as follows:

| 1 | 2 | 3 | 4 | 5 |
|---|---|---|---|---|

(1) Element 1. Connectivity between the element DOF and the global DOF for element 1 is as follows:

| element DOF | global DOF |
| --- | --- |
| 1 | 1 |
| 2 | 5 |

The nonzero entries are $K_{11}, K_{15}, K_{55}$, and NCH(5) = 1. This is smaller than the existing number, and thus we update the NCH array such that

| 1 | 2 | 3 | 4 | 1 |
| --- | --- | --- | --- | --- |

(2) Element 2. Connectivity between the element DOF and the global DOF for element 2 is as follows:

| element DOF | global DOF |
| --- | --- |
| 1 | 5 |
| 2 | 2 |

The nonzero entries are $K_{22}, K_{25}, K_{55}$, and NCH(5) = 2. This is larger than the existing number, and thus we ignore it and move to element 3.

(3) Element 3. Connectivity between the element DOF and the global DOF for element 3 is as follows:

| element DOF | global DOF |
| --- | --- |
| 1 | 2 |
| 2 | 4 |

The nonzero entries are $K_{22}, K_{24}, K_{44}$, and NCH(4) = 2. This is smaller than the existing number and we use it to update the NCH array such that

| 1 | 2 | 3 | 2 | 1 |
| --- | --- | --- | --- | --- |

(4) Element 4. Connectivity between the element DOF and the global DOF for element 4 is as follows:

| element DOF | global DOF |
|:---:|:---:|
| 1 | 4 |
| 2 | 3 |

The nonzero entries are $K_{33}, K_{34}, K_{44}$, and $NCH(4) = 3$. This is larger than the existing number and we ignore it.

The final entry in the NCH array is as follows:

| 1 | 2 | 3 | 2 | 1 |
|:---:|:---:|:---:|:---:|:---:|

Note that:

(1) For column number $j$, (column height) $= j - NCH(j) + 1$.
(2) To minimize the bandwidth or column height, the global DOF must be numbered such that (Max global DOF number − Min global DOF number) within an element is as small as possible.

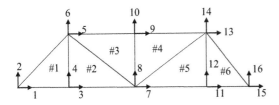

Figure A2.4 A plane stress problem modeled with six triangular elements

**Problem**: Consider a plane stress problem modeled with six three-node triangular elements as shown in Figure A2.4. The skyline method is used for this model.

(a) Following the procedure described in this appendix, create and update the NCH array to identify the row number of the first nonzero entry in each column of the **K** matrix.
(b) Determine the bandwidth (or height) of each column before reduction.
(c) Determine the minimum size of an array necessary to store the entries in the profile of the upper triangular part of the **K** matrix.

# Bibliography

## Chapter 1

Barlow, J. (1976). Optimal stress locations in finite element models. *International Journal for Numerical Methods in Engineering*, Vol. 10, pp. 243–251.

Cook, R.D., Malkus, D.S., Plesha, M.E., and Witt, R.J. (2002). *Concepts and Applications of Finite Element Analysis*, 4th edn. Wiley, New York.

Fish, J. and Belytschko, T. (2007). *A First Course in Finite Elements*. Wiley, Chichester.

Tauchert, T.R. (1974). *Energy Principles in Structural Mechanics*. McGraw-Hill, New York.

## Chapter 2

Hibbeler, R.C. (2016). *Engineering Mechanics: Statics*, 14th edn. Pearson Education, New York.

Megson, T.H.G. (2017). *Aircraft Structures for Engineering Students*, 6th edn. Butterworth-Heinemann, Oxford.

## Chapter 3

Beer, F., Johnston, E., DeWolf, J.T., and Mazurek, D.F. (2014). *Mechanics of Materials*, 7th edn. McGraw-Hill, New York.

Megson, T.H.G. (2017). *Aircraft Structures for Engineering Students*, 6th edn. Butterworth-Heinemann, Oxford.

Ugural, A.C. and Fenster, S.K. (2012). *Advanced Mechanics of Materials and Applied Elasticity*, 5th edn. Prentice-Hall, Englewood Cliffs, NJ.

## Chapter 4

Bathe, K.J. (1996). *Finite Element Procedures*. Prentice-Hall, Englewood Cliffs, NJ.

Hughes, T.J.R. (1987). *The Finite Element Method*. Prentice-Hall, Englewood Cliffs, NJ.

Tamma, K.K., Zhou, X., and Sha, D. (2001). A theory of development and design of generalized integration operators for computational structural dynamics. *International Journal for Numerical Methods in Engineering*, Vol. 50, No. 7, pp. 1619–1664.

Thorby, D. (2008). *Structural Dynamics and Vibration in Practice: An Engineering Handbook*. Butterworth-Heinemann, Oxford.

## Chapter 5

Bolotin, V.V. (1963). *Nonconservative Problems of the Theory of Elastic Stability*. Pergamon Press, New York.

Simites, G. and Hodges, D. (2006). *Fundamentals of Structural Stability*. Butterworth-Heinemann, Oxford.

Sugiyama, Y., Langthjem, M., and Katayama, K. (2019). *Dynamic Stability of Columns under Nonconservative Forces*. Springer Nature, New York.

Timoshenko, S.P. and Gere, J.M. (1961). *Theory of Elastic Stability*, 2nd edn. McGraw-Hill, New York.

Yoo, C.H. and Lee, S.C. (2011). *Stability of Structures*. Butterworth-Heinemann, Oxford.

## Chapter 6

Dym, C.L. and Shames, I.H. (1973). *Solid Mechanics: A Variational Approach*. McGraw-Hill, New York.

Tauchert, T.R. (1974). *Energy Principles in Structural Mechanics*. McGraw-Hill, New York.

Washizu, K. (1975). *Variational Methods in Elasticity and Plasticity*. Pergamon Press, Elmsford, NY.

## Chapter 7

Cowper, G.R. (1973). Gaussian quadrature formulas for triangles. *International Journal for Numerical Methods in Engineering*, Vol. 7, pp. 405–408.

Dunavant, D.A. (1985). High-degree efficient symmetrical Gaussian quadrature rules for the triangle. *International Journal for Numerical Methods in Engineering*, Vol. 21, No. 6, pp. 1129–1148.

Ergatoudis, I., Irons, B.M. and Zienkiewicz, O.C. (1968). Curved, isoparametric, "quadrilateral" elements for finite element analysis. *International Journal of Solids and Structures*, Vol. 4, pp. 21–42.

Hammer, P.C., Marlowe, O.J., and Stroud, A.H. (1956). Numerical integration over simplex and cones. *Mathematical Tables and Other Aids to Computation*, Vol. 10, pp. 130–137.

Keast, P. (1986). Moderate-degree tetrahedral quadrature formulas. *Computer Methods in Applied Mechanics and Engineering*, Vol. 55, No. 6, pp. 339–348.

Mamatha, T.M. and Venkatesh, B. (2015). Gauss quadrature rules for numerical integration over a standard tetrahedral element by decomposing into hexahedral elements.

*Applied Mathematics and Computation*, Vol. 271, pp. 1062–1070.

## Chapter 8

Akin, J.E. (2005). *Finite Element Analysis with Error Estimates*. Butterworth-Heinemann, Burlington, MA.

Bathe, K.J. (1996). *Finite Element Procedures*. Prentice-Hall, Englewood Cliffs, NJ.

Boroomand, B. and Zienkiewicz, O.C. (1997). Recovery by equilibrium in patches (REP). *International Journal for Numerical Methods in Engineering*, Vol. 40, No. 1, pp. 137–164.

Cuthill, E. and McKee, J. (1969). Reducing the bandwidth of sparse symmetric matrices. In *Proceedings of the 24th National Conference, Association for Computing Machinery*, pp. 157–172.

Hestenes, M.R. and Stiefel, E. (1952). Methods of conjugate gradients for solving linear systems. *Journal of Research of the National Bureau of Standards*, Vol. 40, No. 6, pp. 409–436.

Heubner, K.H., Thornton, E.A., and Byrom, T.G. (1994). *The Finite Element Method for Engineers*, 3rd edn. Wiley-Interscience, Hoboken, NJ.

Hughes, T.J.R. (1987). *The Finite Element Method*. Prentice-Hall, Englewood Cliffs, NJ.

Park, H.C., Shin, S.H., and Lee, S.W. (1999). A superconvergent stress recovery technique for accurate boundary stress extraction. *International Journal for Numerical Methods in Engineering*, Vol. 45, No. 9, pp. 1227–1242.

Sharma, R., Zhang, J., Langehaar, M., van Kuelen, F., and Aragon, A.M. (2017). An improved stress recovery technique for low-order 3D finite elements. *International Journal for Numerical Methods in Engineering*, Vol. 114, pp. 88–103.

Shewchuck, J.R. (1994). *An Introduction to the Conjugate Gradient Method without the*

*Agonizing Pain.* School of Computer Science, Carnegie Mellon University, Pittsburgh, PA.

Zienkiewicz, O.C. and Zhu, J.Z. (1992). The superconvergent patch recovery and a posteriori error estimates in the finite element method, Part I: The recovery technique. *International Journal for Numerical Methods in Engineering*, Vol. 33, No. 7, pp. 1331–1364.

Zienkiewicz, O.C., Taylor, R.L., and Zhu, J.Z. (2000). *The Finite Element Method: Its Basis and Fundamentals*, 6th edn. Butterworth-Heinemann, Oxford.

# Chapter 9

Ahmad, S., Irons, B.M., and Zienckiewicz, O.C. (1970). Analysis of thick and thin shell structures by curved elements. *International Journal for Numerical Methods in Engineering*, Vol. 2, pp. 419–451.

Ausserer, M.F. and Lee, S.W. (1988). An eighteen node solid element for thin shell analysis. *International Journal for Numerical Methods in Engineering*, Vol. 26, No. 6, pp. 1345–1364.

Carrera, E. (2003). Theories and finite elements for multilayered plates and shells: A unified compact formulation with numerical assessment and benchmarking. *Archives of Computational Methods in Engineering*, Vol. 10, pp. 215–296.

Radwanska, M., Stankiewicz, A., Wasatko, A., and Pamin, J. (2017). *Plate and Shell Structures: Selected Analytical and Finite Element Solutions*. Wiley, New York.

Reddy, J.N. (2007). *Theory and Analysis of Elastic Plates and Shells*, 2nd edn. Taylor & Francis, London.

Rhim, J. and Lee, S.W. (1998). A vectorial approach to computational modeling of beams undergoing finite rotations. *International Journal for Numerical Methods in Engineering*, Vol. 41, pp. 527–540.

Stemple, A.D. and Lee, S.W. (1988). A finite element model for composite beams with arbitrary cross-sectional warping. *AIAA Journal*, Vol. 26, No. 12, pp. 1512–1520.

Timoshenko, S.P. and Woinowsky-Krieger, S. (1959). *Theory of Plates and Shells*. McGraw-Hill, New York.

Yang, H.T.Y., Saigal, S., Masud, A., and Kapania, R.K. (2000). A survey of recent shell finite elements. *International Journal for Numerical Methods in Engineering*, Vol. 47, pp. 101–127.

# Chapter 10

Atluri, S.N., Gallagher, R.H., and Zienkiewicz, O.C. (1983). *Hybrid and Mixed Finite Element Methods*. Wiley, New York.

Hong, W., Kim, Y.H., and Lee, S.W. (2001). Assumed strain triangular solid shell elements with bubble function displacements for analysis of plates and shells. *International Journal for Numerical Methods in Engineering*, Vol. 52, No. 4, pp. 455–469.

Hughes, T.J.R. (1987). *The Finite Element Method.* Prentice-Hall, Englewood Cliffs, NJ.

Lee, K., Cho, C., and Lee, S.W. (2002). An assumed strain nine-node solid shell element with improved performance. *Computer Modeling in Engineering and Science*, Vol. 2, No. 3, pp. 339–349.

Lee, S.W. and Pian, T.H.H. (1978). Improvements of plate and shell finite elements by mixed formulations. *AIAA Journal*, Vol. 16, No. 1, pp. 29–34.

Lee, S.W. and Rhiu, J.J. (1986). A new efficient approach to the formulation of mixed finite element models for structural analysis. *International Journal for Numerical Methods in Engineering*, Vol. 22, No. 9, pp. 1629–1644.

Park, H.C. and Lee, S.W. (1995). A local coordinate system for assumed strain shell element formulation. *Computational Mechanics*, Vol. 15, No. 5, pp. 473–484.

Park, H.C., Cho, C.M., and Lee, S.W. (1995). An efficient assumed strain element model with six DOF per node for analysis of geometrically nonlinear shells. *International Journal for Numerical Methods in Engineering*, Vol. 38, No. 24, pp. 4101–4121.

Radwanska, M., Stankiewicz, A., Wasatko, A., and Pamin, J. (2017). *Plate and Shell Structures: Selected Analytical and Finite Element Solutions*. Wiley, New York.

Stolarski, H. and Belytschko, T. (1983). Shear and membrane locking in curved elements. *Computer Methods in Applied Mechanics and Engineering*, Vol. 41, No. 3, pp. 279–296.

Trinh, V., Abed-Meraim, F., and Combescure, A. (2011). A new assumed strain solid-shell formulation "SHB6" for prismatic finite element. *Journal of Mechanical Science and Technology*, Vol. 25, No. 9, pp. 2345–2364.

Yeom, C.H. and Lee, S.W. (1989). An assumed strain finite element model for large deflection composite shells. *International Journal for Numerical Methods in Engineering*, Vol. 28, No. 8, pp. 1749–1768.

# Chapter 11

Holman, J.P. (2009). *Heat Transfer*, 10th edn. McGraw-Hill, New York.

Lewis, R.W., Morgan, K., Thomas, H.R., and Seetharamu, K. (1996). *The Finite Element Methods in Heat Transfer Analysis*. Wiley, New York.

Reddy, J.N. and Gartling, D.K. (2010). *The Finite Element Method in Heat Transfer and Fluid Dynamics*, 3rd edn. CRC Press, Boca Raton, FL.

Tamma, K.K. and Namburu, R.R. (1997). Computational approaches with applications to non-classical and classical thermomechanical problems. *Applied Mechanics Reviews*, Vol. 50, No. 9, pp. 514–551.

# Index